## A Guided Tour of Mathematical Methods

Mathematical methods are essential tools for all physical scientists. This novel textbook provides a comprehensive guided tour of the mathematical knowledge and techniques needed by students in this area. In contrast to more traditional textbooks, all the material is presented in the form of problems. Within these problems the basic mathematical theory and its physical applications are well integrated. In this way the mathematical insights that the students acquire are driven by their physical insight. Topics that are covered include vector calculus (div, grad, curl, Laplacian), linear algebra, Fourier analysis, scale analysis, Green's functions, normal modes, tensor calculus and perturbation theory. This book can be used by undergraduates or by lower-level graduate students in the physical sciences. It can serve as a stand-alone text, or as a source of problems and examples to complement other textbooks. This guided tour of mathematical techniques is instructive, applied, and fun.

ROEL SNIEDER currently holds the Keck Foundation Endowed Chair of Basic Exploration Science at the Colorado School of Mines (USA). After finishing his undergraduate training in theoretical physics in 1982, he received a Masters degree in geophysical fluid dynamics from Princeton University in 1984, and a PhD in seismology from Utrecht University in 1987. Following this he became a postdoctoral fellow at the Institute de Physique du Globe in Paris. In 1993 he was appointed Professor of Seismology at Utrecht University, where from 1997–2000 he was chairman of the Faculty of Earth Sciences. In 1997 he was a visiting professor at the Colorado School of Mines (Golden CO, USA). Roel Snieder has served on the editorial boards of *Geophysical Journal International*, *Inverse Problems*, and *Reviews of Geophysics*. In 2000 he was elected Fellow of the American Geophysical Union for his important contributions to geophysical inverse theory, seismic tomography, and the theory of surface waves. Roel Snieder has published more than 80 papers and chapters in the internationally refereed literature, His research has focussed on inverse problems and wave propagation with applications in seismology.

# A Guided Tour of Mathematical Methods for the Physical Sciences

## Roel Snieder

Colorado School of Mines

**CAMBRIDGE** UNIVERSITY PRESS

PUBLISHED BY THE PRESS SYNDICATE OF THE UNIVERSITY OF CAMBRIDGE
The Pitt Building, Trumpington Street, Cambridge, United Kingdom

CAMBRIDGE UNIVERSITY PRESS
The Edinburgh Building, Cambridge CB2 2RU, UK
40 West 20th Street, New York, NY 10011–4211, USA
10 Stamford Road, Oakleigh, VIC 3166, Australia
Ruiz de Alarcón 13, 28014 Madrid, Spain
Dock House, The Waterfront, Cape Town 8001, South Africa

http://www.cambridge.org

First published 2001

Printed in the United Kingdom at the University Press, Cambridge

*Typeface* Computer Modern 11/13pt.    *System* LATEX 2$_\varepsilon$   [UPH]

*A catalogue record for this book is available from the British Library*

*Library of Congress Cataloguing in Publication data*

Snieder, Roel, 1958–
A guided tour of mathematical methods for the physical sciences / Roelof Kees Snieder.
     p.   cm.
Includes bibliographical references and index.
ISBN 0-521-78241-4 (hb) ISBN 0-521-78751-3 (pb)
1. Mathematical physics. I. Title.
QC20.S585 2001
530.15–dc21   00-045536   CIP

ISBN 0 521 78241 4 hardback
ISBN 0 521 78751 3 paperback

To Idske, Hylke, Hidde, and Julia

# Contents

# 1

---

# Introduction

The topic of this book is the application of mathematics to physical problems. Mathematics and physics are often taught separately. Despite the fact that education in physics relies on mathematics, it turns out that students consider mathematics to be disjoint from physics. Although this point of view may strictly be correct, it reflects an erroneous opinion when it concerns an education in physics or geophysics. The reason for this is that mathematics is the *only* language at our disposal for quantifying physical processes. One cannot learn a language by just studying a textbook. In order to truly learn how to use a language one has to go abroad and start using that language. By the same token one cannot learn how to use mathematics in the physical sciences by just studying textbooks or attending lectures, the only way to achieve this is to venture into the unknown and apply mathematics to physical problems.

It is the goal of this book to do exactly that; problems are presented in order to apply mathematical techniques and knowledge to physical concepts. These examples are not presented as well-developed theory. Instead, they are presented as a number of problems that elucidate the issues that are at stake. In this sense this book offers a guided tour: material for learning is presented but true learning will only take place by active exploration. In this process, the interplay of mathematics and physics is essential; mathematics is the natural language for physics while physical insight allows for a better understanding of the mathematics that is presented.

## How can you use this book most efficiently?

Since this book is written as a set of problems you may frequently want to consult other material as well to refresh or deepen your understanding of material. In many places we will refer to the book of Boas [15]. In addition, the books of Butkov [18], Riley *et al.* [72] and Arfken [5] on

1

mathematical physics are excellent. If you are a physics or geophysics student you should seriously consider buying a comprehensive textbook on mathematical physics; it will be of great benefit to you.

In addition to books, colleagues in either the same field or other fields can be a great source of knowledge and understanding. Therefore, don't hesitate to work together with others on these problems if you are in the fortunate position to do so. This may not only make the work more enjoyable, it may also help you in getting 'unstuck' at difficult moments and the different viewpoints of others may help to deepen yours.

### For who is this book written?

This book is set up with the goal of obtaining a good working knowledge of mathematical physics that is needed for students in physics or geophysics. A certain basic knowledge of calculus and linear algebra is required to digest the material presented here. For this reason, this book is meant for upper-level undergraduate students or lower-level graduate students, depending on the background and skill of the student. In addition, teachers can use this book as a source of examples and illustrations that can be used to enrich their courses.

### This book is evolving

This book will be improved regularly by adding new material, correcting errors and making the text clearer. The feedback of both teachers and students who use this material is vital in improving this text, please send your remarks to:

Roel Snieder

Dept of Geophysics
Colorado School of Mines
Golden CO 80401
USA

telephone: +1-303-273.3456
fax: +1-303-273.3478
email: rsnieder@mines.edu

Errata can be found at the following website:
www.mines.edu/~rsnieder/Errata.html

## Acknowledgements

This book resulted from two courses on mathematical physics that were taught at Utrecht University. The remarks, corrections and the encouragement of a large number of students have been very important in its development. It is impossible to thank all the students, but I especially want to thank the feedback from Jojanneke van den Berg, Jehudi Blom, Sterre Dortland, Thomas Geenen, Wiebe van Driel, Luuk van Gerven, Noor Hogeweg and Frederiek Siegenbeek. In their role as teaching assistants, Dirk Kraaipoel and Jesper Spetzler have helped greatly in improving this book. Huub Douma has spent numerous hours at sea correcting earlier drafts. A number of colleagues have helped me very much with their comments; I especially want to mention Freeman Gilbert, Alexander Kaufman, Antoine Khater and Jeannot Trampert. Many of the figures were drafted by Barbara McLenon, who I thank for her support. The help of Joop Hoofd, Everhard Muyzert and John Stockwell who patiently coped with my computer illiteracy allowed me to prepare this book electronically. The support and advice of Adam Black, Eoin O'Sullivan, Jayne Aldhouse, and Maureen Storey, of Cambridge University Press has been very helpful and stimulating during the preparation of this work. Lastly, I want to thank everybody who helped me in numerous ways to make writing this book a joy.

# 2

---

# Power series

## 2.1 The Taylor series

In many applications in mathematical physics it is extremely useful to write the quantity of interest as a sum of a large number of terms. To fix our mind, let us consider the motion of a particle that moves along a line as time progresses. The motion is completely described by giving the position $x(t)$ of the particle as a function of time. Consider the four different types of motion that are shown in figure 2.1.

The simplest motion is a particle that does not move, as shown in panel (a). In this case the position of the particle is constant:

$$x(t) = x_0 .\tag{2.1}$$

The value of the parameter $x_0$ follows by setting $t = 0$ in this expression; this immediately gives

$$x_0 = x(0) .\tag{2.2}$$

In panel (b) the situation for a particle that moves with a constant velocity is shown, thus the position is a linear function of time:

$$x(t) = x_0 + v_0 t .\tag{2.3}$$

Again, setting $t = 0$ gives the parameter $x_0$, which is given again by (2.2). The value of the parameter $v_0$ follows by differentiating (2.3) with respect to time and by setting $t = 0$.

**Problem a:** Do this and show that

$$v_0 = \frac{dx}{dt}(t = 0) .\tag{2.4}$$

This expression reflects that the velocity $v_0$ is given by the time-derivative of the position. Next, consider a particle moving with a constant acceleration $a_0$ as shown in panel (c). As you probably know from classical

4

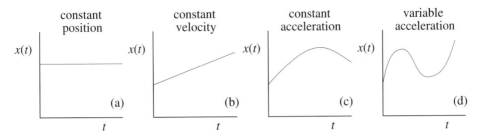

Fig. 2.1. Four different kinds of motion of a particle along a line as a function of time.

mechanics the motion in that case is a quadratic function of time:

$$x(t) = x_0 + v_0 t + \frac{1}{2} a_0 t^2 \; . \tag{2.5}$$

**Problem b:** Evaluate this expression at $t = 0$ to show that $x_0$ is given by (2.2). Differentiate (2.5) once with respect to time and evaluate the result at $t = 0$ to show that $v_0$ is again given by (2.4). Differentiate (2.5) twice with respect to time, set $t = 0$ to show that $a_0$ is given by

$$a_0 = \frac{d^2 x}{dt^2}(t = 0) \; . \tag{2.6}$$

This result reflects the fact that the acceleration is the second derivative of the position with respect to time.

Let us now consider the motion shown in panel (d) where the acceleration changes with time. In that case the displacement as a function of time is not a linear function of time (as in (2.3) for the case of a constant velocity) nor is it a quadratic function of time (as in (2.5) for the case of a constant acceleration). Instead, the displacement is in general a function of all possible powers in $t$:

$$x(t) = c_0 + c_1 t + c_2 t^2 + \cdots = \sum_{n=0}^{\infty} c_n t^n \; . \tag{2.7}$$

This series, in which a function is expressed as a sum of terms with increasing powers of the independent variable, is called a *Taylor series*. At this point we do not know what the constants $c_n$ are. These coefficients can be found in exactly the same way as in problem b in which you determined the coefficients $a_0$ and $v_0$ in the expansion (2.5).

**Problem c:** Determine the coefficient $c_m$ by differentiating expression (2.7) $m$ times with respect to $t$ and by evaluating the result at $t = 0$ to show that

$$c_m = \frac{1}{m!} \frac{d^m x}{dt^m}(x = 0) \; . \tag{2.8}$$

Of course there is no reason why the Taylor series can only be used to describe the displacement $x(t)$ as a function of time $t$. In the literature, the Taylor series is frequently used to describe a function $f(x)$ that depends on $x$. Of course it is immaterial what we call a function. By making the replacements $x \to f$ and $t \to x$ expressions (2.7) and (2.8) can also be written as:

$$f(x) = \sum_{n=0}^{\infty} c_n x^n , \tag{2.9}$$

with

$$c_n = \frac{1}{n!} \frac{d^n f}{dx^n}(x=0) . \tag{2.10}$$

You may also find this result in the literature written as

$$f(x) = \sum_{n=0}^{\infty} \frac{x^n}{n!} \frac{d^n f}{dx^n}(x=0) = f(0) + x \frac{df}{dx}(x=0) + \frac{x^2}{2} \frac{d^2 f}{dx^2}(x=0) + \cdots . \tag{2.11}$$

**Problem d:** By evaluating the derivatives of $f(x)$ at $x=0$ show that the Taylor series of the following functions are:

$$\sin(x) = x - \frac{1}{3!} x^3 + \frac{1}{5!} x^5 - \cdots ; \tag{2.12}$$

$$\cos(x) = 1 - \frac{1}{2} x^2 + \frac{1}{4!} x^4 - \cdots ; \tag{2.13}$$

$$e^x = 1 + x + \frac{1}{2!} x^2 + \frac{1}{3!} x^3 + \cdots = \sum_{n=0}^{\infty} \frac{1}{n!} x^n ; \tag{2.14}$$

$$\frac{1}{1-x} = 1 + x + x^2 + \cdots = \sum_{n=0}^{\infty} x^n ; \tag{2.15}$$

$$(1-x)^\alpha = 1 - \alpha x + \frac{1}{2!} \alpha (\alpha - 1) x^2 - \frac{1}{3!} \alpha (\alpha - 1) (\alpha - 2) x^3 + \cdots . \tag{2.16}$$

Up to this point the Taylor expansion has been made around the point $x = 0$. However, one can make a Taylor expansion of $f(x+h)$ around any arbitrary point $x$. The associated Taylor series can be obtained by replacing the distance $x$ that we move from the expansion point by a distance $h$ and by replacing the expansion point 0 by $x$. Making the replacements $x \to h$ and $0 \to x$ expansion (2.11) is given by

$$f(x+h) = \sum_{n=0}^{\infty} \frac{h^n}{n!} \frac{d^n f}{dx^n}(x) . \tag{2.17}$$

**Problem e:** Truncate this series after the second term and show that this leads to the following approximations:

$$f(x + h) - f(x) \approx h\frac{df}{dx}(x) , \tag{2.18}$$

$$\frac{df}{dx} \approx \frac{f(x + h) - f(x)}{h} . \tag{2.19}$$

These expressions may appear to you to be equivalent in a trivial way. However, we will make extensive use of them in different ways. Equation (2.18) makes it possible to estimate the change in a function when the independent variable is changed slightly, whereas (2.19) is very useful for estimating the derivative of a function given its values at neighboring points. The issue of estimating the derivative of a function is treated in much more detail in section 11.2. Figure 11.2 makes it possible to also derive the estimate (2.19) geometrically by using that the derivative of a function is just the slope of that function.

The Taylor series can also be used for functions of more than one variable. As an example consider a function $f(x, y)$ that depends on the variables $x$ and $y$. The generalization of the Taylor series (2.9) to functions of two variables is given by

$$f(x, y) = \sum_{n,m=0}^{\infty} c_{nm} x^n y^m . \tag{2.20}$$

At this point the coefficients $c_{nm}$ are not yet known. They follow in the same way as the coefficients of the Taylor series of a function that depends on a single variable by taking the partial derivatives of the Taylor series and evaluating the result at the point where the expansion is made.

**Problem f:** Take suitable partial derivatives of (2.20) with respect to $x$ and $y$ and evaluate the result at the expansion point $x = y = 0$ to show that up to second order the Taylor expansion (2.20) is given by

$$\begin{aligned}
f(x, y) &= f(0, 0) + \frac{\partial f}{\partial x}(0, 0) \, x + \frac{\partial f}{\partial y}(0, 0) \, y \\
&+ \frac{1}{2}\frac{\partial^2 f}{\partial x^2}(0, 0) \, x^2 + \frac{\partial^2 f}{\partial x \partial y}(0, 0) \, xy \\
&+ \frac{1}{2}\frac{\partial^2 f}{\partial y^2}(0, 0) \, y^2 + \cdots .
\end{aligned} \tag{2.21}$$

**Problem g:** This is the Taylor expansion of $f(x, y)$ around the point $x = y = 0$. Make suitable substitutions in this result to show that the Taylor expansion around an arbitrary point $(x, y)$ is given by

$$
\begin{aligned}
f(x + h_x, y + h_y) \;=\;& f(x, y) + \frac{\partial f}{\partial x}(x, y)\, h_x + \frac{\partial f}{\partial y}(x, y)\, h_y \\
+\;& \frac{1}{2}\frac{\partial^2 f}{\partial x^2}(x, y)\, h_x^2 + \frac{\partial^2 f}{\partial x \partial y}(x, y)\, h_x h_y \\
+\;& \frac{1}{2}\frac{\partial^2 f}{\partial y^2}(x, y)\, h_y^2 + \cdots .
\end{aligned}
\tag{2.22}
$$

Let us now return to the Taylor series (2.9) with the coefficients $c_m$ given by (2.10). This series hides a very intriguing result. Equations (2.9) and (2.10) suggest that a function $f(x)$ is specified for all values of its argument $x$ when all the derivatives are known at a single point $x = 0$. This means that the global behavior of a function is completely contained in the properties of the function at a single point. In fact, this is not always true.

First, the series (2.9) is an infinite series, and the sum of infinitely many terms does not necessarily lead to a finite answer. As an example look at the series (2.15). A series can only converge when the terms go to zero as $n \to \infty$, because otherwise every additional term changes the sum. The terms in the series (2.15) are given by $x^n$; these terms only go to zero as $n \to \infty$ when $|x| < 1$. In general, the Taylor series (2.9) only *converges* when $x$ is smaller than a certain critical value called the *radius of convergence*. Details on the criteria for the convergence of series can be found in for example Boas [15] or Butkov [18].

The second reason why the derivatives at one point do not necessarily constrain the function everywhere is that a function may change its character over the range of parameter values that is of interest. As an example let us return to a moving particle and consider a particle at position $x(t)$ that is at rest until a certain time $t_0$ and that then starts moving with a uniform velocity $v \neq 0$:

$$
x(t) = \begin{cases} x_0 & \text{for } t \le t_0 \\ x_0 + v(t - t_0) & \text{for } t > t_0 \end{cases}.
\tag{2.23}
$$

The motion of the particle is sketched in figure 2.2. A straightforward application of (2.8) shows that all the coefficients $c_n$ of this function vanish except $c_0$ which is given by $x_0$. The Taylor series (2.7) is therefore given by $x(t) = x_0$ which clearly differs from (2.23). The reason for this is that the function (2.23) changes its character at $t = t_0$ in such a way that nothing in the behavior for times $t < t_0$ predicts the sudden change in the motion at time $t = t_0$. Mathematically things go wrong because the first and higher derivatives of the function are not defined at time $t = t_0$.

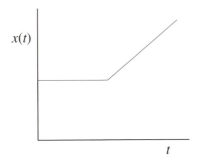

Fig. 2.2.   The motion of a particle that suddenly changes character at time $t_0$.

**Problem h:** What is the second derivative of $x(t)$ at $t = t_0$ ?

The function (2.23) is said to be not analytic at the point $t = t_0$. The issue of analytic functions is treated in more detail in the sections 15.1 and 16.1.

**Problem i:** Try to compute the Taylor series of the function $x(t) = 1/t$ using (2.7) and (2.8). Draw this function and explain why the Taylor series cannot be used for this function.

**Problem j:** Do the same for the function $x(t) = \sqrt{t}$.

The examples in the last two problems show that when a function is not analytic at a certain point, the coefficients of the Taylor series are not defined. This signals that such a function cannot be represented by a Taylor series around that point.

Frequently the result of a calculation can be obtained by summing a series. In section 2.3 this is used to study the behavior of a bouncing ball. The bounces are 'natural' units for analyzing the problem at hand. In section 2.4 the reverse is done when studying the total reflection of a stack of reflective layers. In this case a series expansion actually gives physical insight into a complex expression.

## 2.2 Growth of the Earth by cosmic dust

In this section we use the growth of the Earth by the accretion of cosmic dust as an example to illustrate the usefulness of the (first order) Taylor series. The Earth is continuously bombarded from space by meteorites. Some of these meteorites can be large and lead to massive impact craters. As an example the gravity anomaly over the Chicxulub impact crater in Mexico is shown figure 20.1. The diameter of this impact crater is about 100 km. However, the bulk of the cosmic dust that falls from space onto the Earth is in the form of many small particles. The total mass of all the cosmic dust that falls on the Earth is estimated by Love and Brownlee [51] to be approximately $5 \times 10^7$ kg/a. (The unit a stands for annum (or year); this means that the unit used here is kilogram per year.) However, this estimate is not very accurate and we can probably only trust the first decimal of this number. This means that in subsequent calculations it is pointless to aim for an accuracy of more than one significant figure.

Since the cosmic dust increases the mass of the Earth, the size of the Earth will increase. In this section we determine the growth of the Earth's radius per year due to the bombardment of our planet by cosmic dust.

**Problem a:** Assuming a density of meteorites given by $\rho = 2.5 \times 10^3$ kg/m$^3$ [51] show that the annual growth of the volume of the Earth is given by

$$\delta V = 2 \times 10^4 \ \text{m}^3 \ . \tag{2.24}$$

Also show that this corresponds to a block of $27 \times 27 \times 27$ m$^3$.

We assume that the Earth is a perfect sphere so that the volume and the radius $r$ of the Earth are related by the relation

$$V = \frac{4\pi}{3}r^3 \ . \tag{2.25}$$

From this relation we can deduce that the annual change $\delta r$ of the radius of the Earth can be computed from the expression

$$\delta r = \left[\frac{3\left(V + \delta V\right)}{4\pi}\right]^{1/3} - \left(\frac{3V}{4\pi}\right)^{1/3} \ . \tag{2.26}$$

**Problem b:** Assume the Earth's radius is given by $r = 7000$ km. Insert this number and the value of $\delta V$ from expression (2.24) into (2.26) and use a calculator to compute the increase $\delta r$ of the radius of the Earth.

You have probably found that the annual the increase Earth's radius is equal to zero. This cannot of course be true because we know that $\delta V$ is not equal to zero. The reason that your calculator has given you a wrong answer is that in (2.26) we are subtracting two very large numbers that differ by a small amount. The volume $V$ of the Earth is of the order $10^{21}$ m$^3$ while according to (2.24) the annual increase of the volume is of the order $10^4$ m$^3$. You calculator carries out all the calculations with a relatively small number of digits; it will use probably between 6 and 10 decimals. When you subtract two numbers that are very large and that have a very small difference, this difference will be truncated after say 6 or 10 decimals. In our problem, the first ten decimals of both terms in

(2.26) are identical, hence your calculator tells you that the radius of the Earth is not growing because of the accretion of cosmic dust.

In general, subtracting two large numbers that have a difference that is much smaller leads to numerical inaccuracies. Clearly a trick is needed to obtain the desired growth of the Earth's radius. The cause of this problem is that the annual change in the volume is so small. We can turn this problem to our advantage by using that the Taylor series that we introduced in the previous section is extremely accurate when the independent variable is changed by a very small amount. Here we will compute the increase of the Earth's radius using expression (2.18).

**Problem c:** Show that this expression can also be written as

$$\delta f \approx \frac{\partial f}{\partial x} \delta x \ , \tag{2.27}$$

where $\delta f$ is the change in the function $f(x)$ due to a change $\delta x$ in the independent variable $x$.

**Problem d:** Apply this result to the function $r(V) = (3V/4\pi)^{1/3}$ that gives the radius as function of the volume to derive that

$$\delta r = \frac{1}{3} r \frac{\delta V}{V} \ . \tag{2.28}$$

**Problem e:** Use this result to compute the annual increase of the radius of the Earth due to the accretion of cosmic dust and show that the result is of the order of 1 ångström per year (1 ångström is $10^{-10}$ m).

**Problem f:** Can you think of an object that is the size of 1 ångström?

**Problem g:** How much has the Earth's radius increased over the age of the Earth? In this calculation you may assume that the age of the Earth is 4.5 billion years.

The upshot of this calculation is that the growth of the Earth due to the present-day accretion of cosmic dust is negligible. However, the technique of using the first order Taylor series to determine the small change in a quantity is extremely powerful. In fact, you have encountered in this section an example that demonstrates that an approximation can provide a more meaningful answer than a calculation carried out using a calculator or computer.

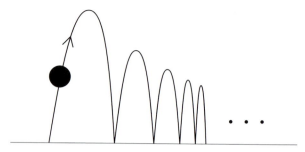

Fig. 2.3.   The motion of a bouncing ball that looses energy with every bounce.
To visualize the motion of the ball better, the ball is a given a constant horizontal
velocity that is conserved during the bouncing.

## 2.3 The bouncing ball

In this section we study a rubber ball that bounces on a flat surface
and slowly comes to rest as sketched in figure 2.3. You will know from
experience that the ball bounces more and more rapidly with time. The
question we address here is whether the ball can actually bounce infinitely
many times in a finite amount of time. This problem is not an easy one.
In general with large difficult problems it is a useful strategy to divide the
large and difficult problem that you cannot solve in smaller and simpler
problems that you can solve. By assembling these smaller sub-problems
one can then often solve the large problem. This is exactly what we will
do here. First we will find how much time it takes for the ball to bounce
once given its velocity. Given a prescription of the energy loss in one
bounce we will determine a relation between the velocity of subsequent
bounces. From these ingredients we can determine the relation between
the times needed for subsequent bounces. By summing this series over
an infinite number of bounces we can determine the total time that the
ball has bounced. *Keep this general strategy in mind when solving complex
problems. Almost all of us are better at solving a number of small problems
rather than a single large problem!*

**Problem a:** A ball moves upward from the level $z = 0$ with velocity $v$
and is subject to a constant gravitational acceleration $g$. Determine
the height the ball reaches and the time it takes for the ball to return
to its starting point.

At this point we have determined the relevant properties for a single
bounce. During each bounce the ball loses energy due to the fact that the
ball is deformed inelastically during the bounce. We assume that during
each bounce the ball loses a fraction $\gamma$ of its energy.

**Problem b:** Let the velocity at the beginning of the $n$th bounce be $v_n$.

Show that with the assumed rule for energy loss this velocity is related to the velocity $v_{n-1}$ of the previous bounce by

$$v_n = \sqrt{1-\gamma}\, v_{n-1}. \qquad (2.29)$$

Hint: when the ball bounces upward from $z = 0$ all its energy is kinetic energy $\frac{1}{2}mv^2$.

In problem a you determined the time it took the ball to bounce once, given the initial velocity, while expression (2.29) gives a recursive relation for the velocity between subsequent bounces. In problem a you also computed the time that it takes to carry out a single bounce. By assembling these results we can find a relation for the time $t_n$ for the $n$th bounce and the time $t_{n-1}$ for the previous bounce.

**Problem c:** Determine this relation. In addition, let us assume that the ball is thrown up the first time from $z = 0$ to reach a height $z = H$. Compute the time $t_0$ needed for the ball to make the first bounce and combine these results to show that

$$t_n = \sqrt{\frac{8H}{g}}(1-\gamma)^{n/2}, \qquad (2.30)$$

where $g$ is the acceleration of gravity.

We can now use this expression to determine the total time $T_N$ it takes to carry out $N$ bounces. This time is given by $T_N = \sum_{n=0}^{N} t_n$. By setting $N$ equal to infinity we can compute the time $T_\infty$ it takes to bounce infinitely often.

**Problem d:** Determine this time by carrying out the summation and show that it is given by:

$$T_\infty = \sqrt{\frac{8H}{g}} \frac{1}{1 - \sqrt{1-\gamma}}. \qquad (2.31)$$

Hint: write $(1-\gamma)^{n/2}$ as $(\sqrt{1-\gamma})^n$ and treat $\sqrt{1-\gamma}$ as the parameter $x$ in the appropriate Taylor series of section 2.1.

This result seems to suggest that the time it takes to bounce infinitely often is indeed finite.

**Problem e:** Show that this is indeed the case, except when the ball loses no energy between subsequent bounces. Hint: translate the condition that the ball loses no energy into one of the quantities in equation (2.31).

Expression (2.31) looks messy. It often happens in mathematical physics that the final expression resulting from a calculation is so complex that it is difficult to understand it. However, often we know that certain terms in an expression can be assumed to be very small (or very large). This may allow us to obtain an approximate expression that is of a simpler form. In this way we trade accuracy for simplicity and understanding. In practice, this often turns out to be a good deal! In our example of the bouncing ball we assume that the energy loss at each bounce is small, i.e. that $\gamma$ is small.

**Problem f:** Show that in this case $T_\infty \approx \sqrt{(8H/g)}2/\gamma$ by using the leading terms of the appropriate Taylor series of section 2.1.

This result is actually quite useful. It tells us *how* the total bounce time approaches infinity when the energy loss $\gamma$ goes to zero.

In this example we have solved the problem in little steps. In general we will take larger steps during this course, and you will have to discover how to divide a large step in smaller steps. The next problem is a 'large' problem; solve it by dividing it in smaller problems. First formulate the smaller problems as ingredients for the large problem before you actually start working on the smaller problems.

> *Make it a habit whenever you are solving problems to first formulate a strategy for how you are going to attack a problem before you actually start working on the sub-problems. Make a list if this helps you and don't be deterred if you cannot solve a particular sub-problem. Perhaps you can solve the other sub-problems and somebody else can help you with the one you cannot solve.*

Keeping this in mind, solve the following 'large' problem:

**Problem g:** Let the total distance travelled by the ball in the vertical direction during infinitely many bounces be denoted by $S$. Show that $S = 2H/\gamma$.

## 2.4 Reflection and transmission by a stack of layers

In 1917 Lord Rayleigh [71] addressed the question of why some birds and insects have beautiful iridescent colors. He explained this by studying the reflective properties of a stack of thin reflective layers. This problem is also of interest in geophysics; in exploration seismology one is also interested in the reflection and transmission properties of stacks of reflective layers in the Earth. Lord Rayleigh solved this problem in the following way.

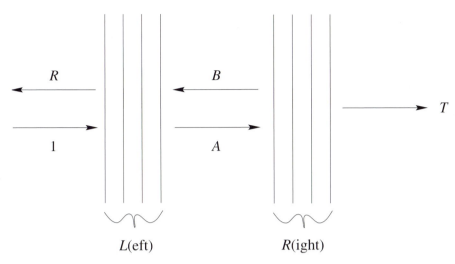

Fig. 2.4. Geometry of the problem where stacks of $n$ and $m$ reflective layers are combined. The notation of the strength of left- and right-going waves is indicated.

Suppose we have one stack of layers on the left with reflection coefficient $R_L$ and transmission coefficient $T_L$ and another stack of layers on the right with reflection coefficient $R_R$ and transmission coefficient $T_R$. If we add these two stacks together to obtain a larger stack of layers, what are the reflection coefficient $R$ and transmission coefficient $T$ of the total stack of layers? See figure 2.4 for the scheme of this problem. Note that the reflection coefficient is defined as the ratio of the strengths of the reflected and the incident waves, similarly the transmission coefficient is defined as the ratio of the strengths of the transmitted wave and the incident wave. To highlight the essential arguments we simplify the analysis and ignore that the reflection coefficient for waves incident from the left and the right are in general not the same. However, this simplification does not change the essence of the coming arguments.

Before we start solving the problem, let us speculate what the transmission coefficient of the combined stack is. It may seem natural to assume that the transmission coefficient of the combined stack is the product of the transmission coefficient of the individual stacks:

$$T \overset{?}{=} T_L T_R . \tag{2.32}$$

However, this result is wrong and we will discover why this is so. Consider figure 2.4 again. The unknown quantities are $R$, $T$ and the coefficients $A$ and $B$ for the right-going and left-going waves between the stacks. An incident wave with strength 1 impinges on the stack from the left. Let us first determine the coefficient $A$ of the right-going waves between the

stacks. The right-going wave between the stacks contains two contributions: the wave transmitted from the left (this contribution has a strength
$1 \times T_L$) and the wave reflected towards the right due the incident left-
going wave with strength $B$ (this contribution has a strength $B \times R_L$).
This implies that:

$$A = T_L + BR_L \ . \tag{2.33}$$

**Problem a:** Using similar arguments show that:

$$B = AR_R \ , \tag{2.34}$$

$$T = AT_R \ , \tag{2.35}$$

$$R = R_L + BT_L \ . \tag{2.36}$$

This is all we need to solve our problem. The system of equations (2.33)–
(2.36) consists of four linear equations with four unknowns $A$, $B$, $R$ and $T$.
We could solve this system of equations by brute force, but some thought
will make life easier for us. Note that the last two equations immediately
give $T$ and $R$ once $A$ and $B$ are known. The first two equations give $A$
and $B$.

**Problem b:** Show that

$$A = \frac{T_L}{(1 - R_L R_R)} \ , \tag{2.37}$$

$$B = \frac{T_L R_R}{(1 - R_L R_R)} . \tag{2.38}$$

This is a puzzling result, the right-going wave $A$ between the layers not
only contains the transmission coefficient of the left layer $T_L$ but also an
additional term $1/(1 - R_L R_R)$.

**Problem c:** Make a series expansion of $1/(1 - R_L R_R)$ in the quantity
$R_L R_R$ and show that this term accounts for the waves that bounce
back and forth between the two stacks. Hint: use that $R_L$ is the
reflection coefficient for a wave that reflects from the left stack and
$R_R$ is the reflection coefficient for one that reflects from the right
stack so that $R_L R_R$ is the total reflection coefficient for a wave that
bounces once between the left and the right stacks.

This implies that the term $1/(1 - R_L R_R)$ accounts for the waves that
bounce back and forth between the two stacks of layers. It is for this
reason that we call this a *reverberation* term. It plays an important role
in computing the response of layered media.

**Problem d:** Show that the reflection and transmission coefficients of the combined stack of layers are given by:

$$R = R_L + \frac{T_L^2 R_R}{(1 - R_L R_R)},\qquad (2.39)$$

$$T = \frac{T_L T_R}{(1 - R_L R_R)}.\qquad (2.40)$$

At the beginning of this section we conjectured that the transmission coefficient of the combined stacks is the product of the transmission coefficient of the separate stacks, see expression (2.32).

**Problem e:** Is this conjecture correct? Under which conditions is it approximately correct?

Equations (2.39) and (2.40) are very useful for computing the reflection and transmission coefficients of a large stack of layers. The reason for this is that it is extremely simple to determine the reflection and transmission coefficients of a very thin layer using the Born approximation. (The Born approximation is treated in section 22.2.) Let the reflection and transmission coefficients of a *single* thin layer $n$ be denoted by $r_n$ and $t_n$ respectively and let the reflection and transmission coefficients of a *stack* of $n$ layers be denoted by $R_n$ and $T_n$ respectively. Suppose that the left stack consists on $n$ layers and that we want to add an $(n+1)$th layer to the stack. In that case the right stack consists of a single $(n+1)$th layer so that $R_R = r_{n+1}$ and $T_R = t_{n+1}$ and the reflection and transmission coefficients of the left stack are given by $R_L = R_n$, $T_L = T_n$. Using this in expressions (2.39) and (2.40) yields

$$R_{n+1} = R_n + \frac{T_n^2 r_{n+1}}{(1 - R_n r_{n+1})},\qquad (2.41)$$

$$T_{n+1} = \frac{T_n t_{n+1}}{(1 - R_n r_{n+1})}.\qquad (2.42)$$

This means that given the known response of a stack of $n$ layers, one can easily compute the effect of adding the $(n+1)$th layer to this stack. In this way one can recursively build up the response of the complex reflector out of the known response of very thin reflectors. Computers are pretty stupid, but they are ideally suited for applying rules (2.41) and (2.42) a large number of times. Of course this process has to be begun with a medium in which no layers are present.

**Problem f:** What are the reflection coefficient $R_0$ and the transmission coefficient $T_0$ when there are no reflective layers present yet? Describe how one can compute the response of a thick stack of layers once we know the response of a very thin layer.

In developing this theory, Lord Rayleigh prepared the foundations for a theory that later became known as *invariant embedding* which turns out to be extremely useful for a number of scattering and diffusion problems [10] [90].

The main conclusion of the treatment of this section is that the transmission of a combination of two stacks of layers is not the product of the transmission coefficients of the two separate stacks because the waves that repeatedly reflect between the two stacks leave an imprint on the transmission coefficient as well. Paradoxically, Berry and Klein [12] showed in their analysis of 'transparent mirrors' that for the special case of a large stack of layers with random transmission coefficients the total transmission coefficients *is* the product of the transmission coefficients of the individual layers, despite the fact that multiple reflections play a crucial role in this process.

# 3

## Spherical and cylindrical coordinates

Many problems in mathematical physics exhibit a spherical or cylindrical symmetry. For example, the gravity field of the Earth is to first order spherically symmetric. Waves excited by a stone thrown into water are usually cylindrically symmetric. Although there is no reason why problems with such a symmetry cannot be analyzed using Cartesian coordinates (i.e. $(x, y, z)$-coordinates), it is usually not very convenient to use such a coordinate system. The reason for this is that the theory is usually much simpler when one selects a coordinate system with symmetry properties that are the same as the symmetry properties of the physical system that one wants to study. It is for this reason that spherical coordinates and cylindrical coordinates are introduced in this section. It takes a certain effort to become acquainted with these coordinate system, but this effort is well spent because it makes solving a large class of problems much easier.

### 3.1 Introducing spherical coordinates

In figure 3.1 a Cartesian coordinate system with its $x$-, $y$- and $z$-axes is shown as well as the location of a point $\mathbf{r}$. This point can be described either by its $x$-, $y$- and $z$-components or by the radius $r$ and the angles $\theta$ and $\varphi$ shown in figure 3.1. In the latter case one uses spherical coordinates. Comparing the angles $\theta$ and $\varphi$ with the geographical coordinates that define a point on the globe one sees that $\varphi$ can be compared with *longitude* and $\theta$ can be compared with *colatitude*, which is defined as (*latitude – 90 degrees*).

**Problem a:** The city of Utrecht in the Netherlands is located at 52 degrees north and 5 degrees east. Compute the angles $\theta$ and $\varphi$ (in radians) that correspond to this point on the sphere.

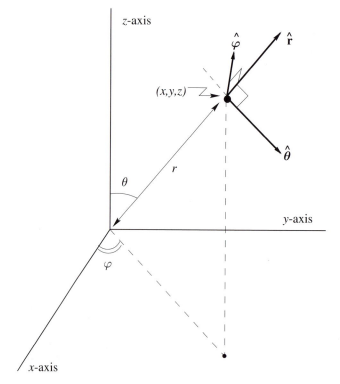

Fig. 3.1.   Definition of the angles used in the spherical coordinates.

The angle $\varphi$ runs from 0 to $2\pi$, while $\theta$ has values between 0 and $\pi$. In terms of Cartesian coordinates the position vector can be written as:

$$\mathbf{r} = x\hat{\mathbf{x}} + y\hat{\mathbf{y}} + z\hat{\mathbf{z}} , \tag{3.1}$$

where the caret (^) is used to denote a vector that is of unit length. An arbitrary vector can of course also be expressed in these basis vectors:

$$\mathbf{u} = u_x\hat{\mathbf{x}} + u_y\hat{\mathbf{y}} + u_z\hat{\mathbf{z}} . \tag{3.2}$$

We want also to express the same vector in basis vectors that are related to the spherical coordinate system. Before we can do so we must first establish the connection between the Cartesian coordinates $(x, y, z)$ and the spherical coordinates $(r, \theta, \varphi)$.

**Problem b:** Use figure 3.1 to show that the Cartesian coordinates are given by:

$$\left. \begin{array}{rcl} x & = & r\sin\theta\cos\varphi , \\ y & = & r\sin\theta\sin\varphi , \\ z & = & r\cos\theta . \end{array} \right\} \tag{3.3}$$

**Problem c:** Use these expressions to derive the following expression for the spherical coordinates in terms of the Cartesian coordinates:

$$\left.\begin{array}{rcl} r & = & \sqrt{x^2 + y^2 + z^2}\,, \\ \theta & = & \arccos\left(z/\sqrt{x^2 + y^2 + z^2}\right), \\ \varphi & = & \arctan\left(y/x\right). \end{array}\right\} \tag{3.4}$$

We now have obtained the relation between the Cartesian coordinates $(x, y, z)$ and the spherical coordinates $(r, \theta, \varphi)$. We want also to express the vector **u** of equation (3.2) in spherical coordinates:

$$\mathbf{u} = u_r \hat{\mathbf{r}} + u_\theta \hat{\boldsymbol{\theta}} + u_\varphi \hat{\boldsymbol{\varphi}}\,, \tag{3.5}$$

and we want to know the relation between the components $(u_x, u_y, u_z)$ in Cartesian coordinates and the components $(u_r, u_\theta, u_\varphi)$ of the same vector expressed in spherical coordinates. In order to do this we first need to determine the unit vectors $\hat{\mathbf{r}}$, $\hat{\boldsymbol{\theta}}$ and $\hat{\boldsymbol{\varphi}}$. In Cartesian coordinates, the unit vector $\hat{\mathbf{x}}$ points along the $x$-axis. This is a different way of saying that it is a unit vector pointing in the direction of increasing values of $x$ for constant values of $y$ and $z$; in other words, $\hat{\mathbf{x}}$ can be written as: $\hat{\mathbf{x}} = \partial\mathbf{r}/\partial x$.

**Problem d:** Verify this by showing that the differentiation $\hat{\mathbf{x}} = \partial\mathbf{r}/\partial x$

leads to the correct unit vector in the $x$-direction: $\hat{\mathbf{x}} = \begin{pmatrix} 1 \\ 0 \\ 0 \end{pmatrix}$.

Now consider the unit vector $\hat{\boldsymbol{\theta}}$. Using the same argument as for the unit vector $\hat{\mathbf{x}}$ we know that $\hat{\boldsymbol{\theta}}$ is directed towards increasing values of $\theta$ for constant values of $r$ and $\varphi$. This means that $\hat{\boldsymbol{\theta}}$ can be written as $\hat{\boldsymbol{\theta}} = C\partial\mathbf{r}/\partial\theta$. The constant $C$ follows from the requirement that $\hat{\boldsymbol{\theta}}$ is of unit length.

**Problem e:** Use this reasoning for all the unit vectors $\hat{\mathbf{r}}$, $\hat{\boldsymbol{\theta}}$ and $\hat{\boldsymbol{\varphi}}$ and expression (3.3) to show that:

$$\hat{\mathbf{r}} = \frac{\partial\mathbf{r}}{\partial r}, \quad \hat{\boldsymbol{\theta}} = \frac{1}{r}\frac{\partial\mathbf{r}}{\partial\theta}, \quad \hat{\boldsymbol{\varphi}} = \frac{1}{r\sin\theta}\frac{\partial\mathbf{r}}{\partial\varphi}\,, \tag{3.6}$$

and that this result can also be written as

$$\hat{\mathbf{r}} = \begin{pmatrix} \sin\theta\cos\varphi \\ \sin\theta\sin\varphi \\ \cos\theta \end{pmatrix}, \quad \hat{\boldsymbol{\theta}} = \begin{pmatrix} \cos\theta\cos\varphi \\ \cos\theta\sin\varphi \\ -\sin\theta \end{pmatrix}, \quad \hat{\boldsymbol{\varphi}} = \begin{pmatrix} -\sin\varphi \\ \cos\varphi \\ 0 \end{pmatrix}. \tag{3.7}$$

These equations give the unit vectors $\hat{\mathbf{r}}$, $\hat{\boldsymbol{\theta}}$ and $\hat{\boldsymbol{\varphi}}$ in Cartesian coordinates.

On the right hand side of (3.6) the derivatives of the position vector are divided by 1, $r$ and $r \sin \theta$ respectively. These factors are usually shown in the following notation:

$$h_r = 1, \quad h_\theta = r, \quad h_\varphi = r \sin \theta . \tag{3.8}$$

These scale factors play a very important role in the general theory of curvilinear coordinate systems, see Butkov [18] for details. The material presented in the remainder of this chapter as well as the derivation of vector calculus in spherical coordinates can be based on the scale factors given in (3.8). However, this approach will not be taken here.

**Problem f:** Verify explicitly that the vectors $\hat{\mathbf{r}}$, $\hat{\boldsymbol{\theta}}$ and $\hat{\boldsymbol{\varphi}}$ defined in this way form an orthonormal basis, i.e. they are of unit length and perpendicular to each other:

$$(\hat{\mathbf{r}} \cdot \hat{\mathbf{r}}) = \left(\hat{\boldsymbol{\theta}} \cdot \hat{\boldsymbol{\theta}}\right) = (\hat{\boldsymbol{\varphi}} \cdot \hat{\boldsymbol{\varphi}}) = 1 , \tag{3.9}$$

$$\left(\hat{\mathbf{r}} \cdot \hat{\boldsymbol{\theta}}\right) = (\hat{\mathbf{r}} \cdot \hat{\boldsymbol{\varphi}}) = \left(\hat{\boldsymbol{\theta}} \cdot \hat{\boldsymbol{\varphi}}\right) = 0 . \tag{3.10}$$

**Problem g:** Using expressions (3.7) for the unit vectors $\hat{\mathbf{r}}$, $\hat{\boldsymbol{\theta}}$ and $\hat{\boldsymbol{\varphi}}$ show by calculating the cross-products explicitly that

$$\hat{\mathbf{r}} \times \hat{\boldsymbol{\theta}} = \hat{\boldsymbol{\varphi}}, \quad \hat{\boldsymbol{\theta}} \times \hat{\boldsymbol{\varphi}} = \hat{\mathbf{r}}, \quad \hat{\boldsymbol{\varphi}} \times \hat{\mathbf{r}} = \hat{\boldsymbol{\theta}} . \tag{3.11}$$

The Cartesian basis vectors $\hat{\mathbf{x}}$, $\hat{\mathbf{y}}$ and $\hat{\mathbf{z}}$ point in the same direction at every point in space. This is not true for the spherical basis vectors $\hat{\mathbf{r}}$, $\hat{\boldsymbol{\theta}}$ and $\hat{\boldsymbol{\varphi}}$; for different values of the angles $\theta$ and $\varphi$ these vectors point in different directions. This implies that these unit vectors are functions of both $\theta$ and $\varphi$. For several applications it is necessary to know how the basis vectors change with $\theta$ and $\varphi$. This change is described by the derivative of the unit vectors with respect to the angles $\theta$ and $\varphi$.

**Problem h:** Show by direct differentiation of expressions (3.7) that the derivatives of the unit vectors with respect to the angles $\theta$ and $\varphi$ are given by:

$$\left.\begin{aligned}
\partial \hat{\mathbf{r}}/\partial \theta &= \hat{\boldsymbol{\theta}} , & \partial \hat{\mathbf{r}}/\partial \varphi &= \sin \theta \, \hat{\boldsymbol{\varphi}} , \\
\partial \hat{\boldsymbol{\theta}}/\partial \theta &= -\hat{\mathbf{r}} , & \partial \hat{\boldsymbol{\theta}}/\partial \varphi &= \cos \theta \, \hat{\boldsymbol{\varphi}} , \\
\partial \hat{\boldsymbol{\varphi}}/\partial \theta &= 0 , & \partial \hat{\boldsymbol{\varphi}}/\partial \varphi &= -\sin \theta \, \hat{\mathbf{r}} - \cos \theta \, \hat{\boldsymbol{\theta}} .
\end{aligned}\right\} \tag{3.12}$$

## 3.2 Changing coordinate systems

Now that we have derived the properties of the unit vectors $\hat{\mathbf{r}}$, $\hat{\boldsymbol{\theta}}$ and $\hat{\boldsymbol{\varphi}}$ we are in the position to derive how the components $(u_r, u_\theta, u_\varphi)$ of the vector $\mathbf{u}$ defined in equation (3.5) are related to the usual Cartesian coordinates $(u_x, u_y, u_z)$. This can most easily be achieved by writing expressions (3.7) in the following form:

$$\left.\begin{array}{l} \hat{\mathbf{r}} = \sin\theta\cos\varphi\,\hat{\mathbf{x}} + \sin\theta\sin\varphi\,\hat{\mathbf{y}} + \cos\theta\,\hat{\mathbf{z}}\,, \\ \hat{\boldsymbol{\theta}} = \cos\theta\cos\varphi\,\hat{\mathbf{x}} + \cos\theta\sin\varphi\,\hat{\mathbf{y}} - \sin\theta\,\hat{\mathbf{z}}\,, \\ \hat{\boldsymbol{\varphi}} = -\sin\varphi\,\hat{\mathbf{x}} + \cos\varphi\,\hat{\mathbf{y}}\,. \end{array}\right\} \qquad (3.13)$$

**Problem a:** Convince yourself that this expression can also be written in a symbolic form as

$$\begin{pmatrix} \hat{\mathbf{r}} \\ \hat{\boldsymbol{\theta}} \\ \hat{\boldsymbol{\varphi}} \end{pmatrix} = \mathbf{M} \begin{pmatrix} \hat{\mathbf{x}} \\ \hat{\mathbf{y}} \\ \hat{\mathbf{z}} \end{pmatrix}\,, \qquad (3.14)$$

with the matrix $\mathbf{M}$ given by

$$\mathbf{M} = \begin{pmatrix} \sin\theta\cos\varphi & \sin\theta\sin\varphi & \cos\theta \\ \cos\theta\cos\varphi & \cos\theta\sin\varphi & -\sin\theta \\ -\sin\varphi & \cos\varphi & 0 \end{pmatrix}\,. \qquad (3.15)$$

Of course expression (3.14) can only be considered to be a shorthand notation for equations (3.13) since the entries in (3.14) are vectors rather than single components. However, expression (3.14) is a convenient shorthand notation.

The relation between the spherical components $(u_r, u_\theta, u_\varphi)$ and the Cartesian components $(u_x, u_y, u_z)$ of the vector $\mathbf{u}$ can be obtained by inserting expressions (3.13) for the spherical coordinate unit vectors into the relation $\mathbf{u} = u_r\hat{\mathbf{r}} + u_\theta\hat{\boldsymbol{\theta}} + u_\varphi\hat{\boldsymbol{\varphi}}$.

**Problem b:** Do this and collect together all terms multiplying the unit vectors $\hat{\mathbf{x}}$, $\hat{\mathbf{y}}$ and $\hat{\mathbf{z}}$ to show that expression (3.5) for the vector $\mathbf{u}$ is equivalent to:

$$\begin{array}{rl} \mathbf{u} = & (u_r\sin\theta\cos\varphi + u_\theta\cos\theta\cos\varphi - u_\varphi\sin\varphi)\,\hat{\mathbf{x}} \\ + & (u_r\sin\theta\sin\varphi + u_\theta\cos\theta\sin\varphi + u_\varphi\cos\varphi)\,\hat{\mathbf{y}} \\ + & (u_r\cos\theta - u_\theta\sin\theta)\,\hat{\mathbf{z}}\,. \end{array} \qquad (3.16)$$

**Problem c:** Show that this relation can also be written as:

$$\begin{pmatrix} u_x \\ u_y \\ u_z \end{pmatrix} = \mathbf{M}^T \begin{pmatrix} u_r \\ u_\theta \\ u_\varphi \end{pmatrix}\,, \qquad (3.17)$$

where the matrix $\mathbf{M}$ is given by (3.15). In this expression, $\mathbf{M}^T$ is the transpose of the matrix $\mathbf{M}$; i.e. it is the matrix obtained by interchanging rows and columns of the matrix $M_{ij}^T = M_{ji}$.

We have not yet reached with equation (3.17) our goal of expressing the spherical coordinate components $(u_r, u_\theta, u_\varphi)$ of the vector $\mathbf{u}$ in the Cartesian components $(u_x, u_y, u_z)$. This is most easily achieved by multiplying (3.17) with the inverse matrix $(\mathbf{M}^T)^{-1}$, which gives:

$$\begin{pmatrix} u_r \\ u_\theta \\ u_\varphi \end{pmatrix} = \left( \mathbf{M}^T \right)^{-1} \begin{pmatrix} u_x \\ u_y \\ u_z \end{pmatrix}. \tag{3.18}$$

However, now we have only shifted the problem because we don't know the inverse $(\mathbf{M}^T)^{-1}$. One could of course painstakingly compute this inverse, but this would be a laborious process that we can avoid. It follows by inspection of (3.15) that all the columns of $\mathbf{M}$ are of unit length and that the columns are orthogonal. This implies that $\mathbf{M}$ is an orthogonal matrix. Orthogonal matrices have the useful property that the transpose of the matrix is identical to the inverse of the matrix: $\mathbf{M}^{-1} = \mathbf{M}^T$.

**Problem d:** The property $\mathbf{M}^{-1} = \mathbf{M}^T$ can be verified explicitly by showing that $\mathbf{M}\mathbf{M}^T$ and $\mathbf{M}^T\mathbf{M}$ are equal to the identity matrix, do this!

Note that we have obtained the inverse of the matrix by making a guess and by verifying that this guess indeed solves our problem. This approach is often very useful in solving mathematical problems; there is nothing wrong with making a guess as long as you check afterwards that your guess is indeed a solution to your problem. Since we know that $\mathbf{M}^{-1} = \mathbf{M}^T$, it follows that $(\mathbf{M}^T)^{-1} = (\mathbf{M}^{-1})^{-1} = \mathbf{M}$.

**Problem e:** Use these results to show that the spherical coordinate components of $\mathbf{u}$ are related to the Cartesian coordinates by the following transformation rule:

$$\begin{pmatrix} u_r \\ u_\theta \\ u_\varphi \end{pmatrix} = \begin{pmatrix} \sin\theta\cos\varphi & \sin\theta\sin\varphi & \cos\theta \\ \cos\theta\cos\varphi & \cos\theta\sin\varphi & -\sin\theta \\ -\sin\varphi & \cos\varphi & 0 \end{pmatrix} \begin{pmatrix} u_x \\ u_y \\ u_z \end{pmatrix}. \tag{3.19}$$

## 3.3 The acceleration in spherical coordinates

You may wonder whether we really need all these transformation rules between a Cartesian coordinate system and a system of spherical coordinates. The answer is yes! An important example can be found in meteorology where air moves around a sphere. The velocity $\mathbf{v}$ of the air can be expressed in spherical coordinates as:

$$\mathbf{v} = v_r\hat{\mathbf{r}} + v_\theta\hat{\boldsymbol{\theta}} + v_\varphi\hat{\boldsymbol{\varphi}} . \tag{3.20}$$

The motion of the air is governed by Newton's law, but when the velocity $\mathbf{v}$ and the force $\mathbf{F}$ are both expressed in spherical coordinates it would be wrong to express the $\theta$-component of Newton's law as: $\rho dv_\theta/dt = F_\theta$. The reason is that the basis vectors of the spherical coordinate system depend on the position. When a particle moves, the directions of the basis vectors change as well. This is a different way of saying that the spherical coordinate system is not a Cartesian system where the orientation of the coordinate axes is independent of the position. When computing the acceleration in such a system additional terms appear that account for the fact that the coordinate system is not an inertial system. The results of the section 3.1 contain all the ingredients we need.

Let us follow a particle moving over a sphere. The position vector $\mathbf{r}$ has an obvious expansion in spherical coordinates:

$$\mathbf{r} = r\hat{\mathbf{r}} . \tag{3.21}$$

The velocity is obtained by taking the time-derivative of this expression. However, the unit vector $\hat{\mathbf{r}}$ is a function of the angles $\theta$ and $\varphi$, see equation (3.7). This means that when we take the time-derivative of (3.21) to obtain the velocity we also need to differentiate $\hat{\mathbf{r}}$ with time. Note that this is not the case with the Cartesian expression $\mathbf{r} = x\hat{\mathbf{x}} + y\hat{\mathbf{y}} + z\hat{\mathbf{z}}$ because the unit vectors $\hat{\mathbf{x}}$, $\hat{\mathbf{y}}$ and $\hat{\mathbf{z}}$ are constant, hence they do not change when the particle moves and they thus have a vanishing time-derivative.

An as example, let us compute the time-derivative of $\hat{\mathbf{r}}$. This vector is a function of $\theta$ and $\varphi$, and these angles both change with time as the particle moves. Using the chain rule it thus follows that:

$$\frac{d\hat{\mathbf{r}}}{dt} = \frac{d\hat{\mathbf{r}}(\theta, \varphi)}{dt} = \frac{d\theta}{dt}\frac{\partial\hat{\mathbf{r}}}{\partial\theta} + \frac{d\varphi}{dt}\frac{\partial\hat{\mathbf{r}}}{\partial\varphi} . \tag{3.22}$$

The derivatives $\partial\hat{\mathbf{r}}/\partial\theta$ and $\partial\hat{\mathbf{r}}/\partial\varphi$ can be eliminated with (3.12).

**Problem a:** Use expressions (3.12) to eliminate the derivatives $\partial\hat{\mathbf{r}}/\partial\theta$ and $\partial\hat{\mathbf{r}}/\partial\varphi$ and carry out a similar analysis for the time-derivatives of the unit vectors $\hat{\boldsymbol{\theta}}$ and $\hat{\boldsymbol{\varphi}}$ to show that:

$$\left. \begin{array}{l} \dfrac{d\hat{\mathbf{r}}}{dt} = \dot{\theta}\,\hat{\boldsymbol{\theta}} + \sin\theta\,\dot{\varphi}\,\hat{\boldsymbol{\varphi}} , \\[2mm] \dfrac{d\hat{\boldsymbol{\theta}}}{dt} = -\dot{\theta}\,\hat{\mathbf{r}} + \cos\theta\,\dot{\varphi}\,\hat{\boldsymbol{\varphi}} , \\[2mm] \dfrac{d\hat{\boldsymbol{\varphi}}}{dt} = -\sin\theta\,\dot{\varphi}\,\hat{\mathbf{r}} - \cos\theta\,\dot{\varphi}\,\hat{\boldsymbol{\theta}} . \end{array} \right\} \tag{3.23}$$

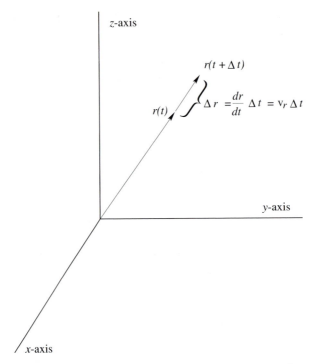

Fig. 3.2.    Definition of the geometric variables used to derive the radial component of the velocity.

In these and other expressions in this section a dot is used to denote the
time-derivative: $\dot{F} \equiv dF/dt$.

**Problem b:** Use the first line of (3.23) and the definition $\mathbf{v} = d\mathbf{r}/dt$ to
show that in spherical coordinates:

$$\mathbf{v} = \dot{r}\hat{\mathbf{r}} + r\dot{\theta}\hat{\boldsymbol{\theta}} + r\sin\theta\,\dot{\varphi}\,\hat{\boldsymbol{\varphi}} \; . \tag{3.24}$$

In spherical coordinates the components of the velocity are thus given by:

$$\left.\begin{array}{l} v_r = \dot{r} \; , \\ v_\theta = r\dot{\theta} \; , \\ v_\varphi = r\sin\theta\,\dot{\varphi} \; . \end{array}\right\} \tag{3.25}$$

This result can be interpreted geometrically. As an example, let us consider the radial component of the velocity, see figure 3.2. To obtain the
radial component of the velocity we keep the angles $\theta$ and $\varphi$ fixed and
let the radius $r(t)$ change to $r(t + \Delta t)$ over a time $\Delta t$. The particle has
moved a distance $r(t + \Delta t) - r(t) = (dr/dt)/\Delta t$ in a time $\Delta t$, so that the
radial component of the velocity is given by $v_r = dr/dt = \dot{r}$. This is the
result given by the first line of (3.25).

**Problem c:** Use similar geometric arguments to explain the form of the velocity components $v_\theta$ and $v_\varphi$ given in (3.25).

**Problem d:** We are now in the position to compute the acceleration in spherical coordinates. To do this differentiate (3.24) with respect to time and use expression (3.23) to eliminate the time-derivatives of the basis vectors. Use this to show that the acceleration $\mathbf{a}$ is given by:

$$
\begin{aligned}
\mathbf{a} =\ & \left( \dot{v}_r - \dot{\theta} v_\theta - \sin\theta\ \dot{\varphi} v_\varphi \right) \hat{\mathbf{r}} \\
& + \left( \dot{v}_\theta + \dot{\theta} v_r - \cos\theta\ \dot{\varphi} v_\varphi \right) \hat{\boldsymbol{\theta}} \\
& + \left( \dot{v}_\varphi + \sin\theta\ \dot{\varphi} v_r + \cos\theta\ \dot{\varphi} v_\theta \right) \hat{\boldsymbol{\varphi}}\ .
\end{aligned} \tag{3.26}
$$

**Problem e:** This expression is not quite satisfactory because it contains both the components of the velocity as well as the time-derivatives $\dot{\theta}$ and $\dot{\varphi}$ of the angles. Eliminate the time-derivatives with respect to the angles in favor of the components of the velocity using expressions (3.25) to show that the components of the acceleration in spherical coordinates are given by:

$$
\left.
\begin{aligned}
a_r &= \dot{v}_r - \frac{v_\theta^2 + v_\varphi^2}{r}\ , \\[2mm]
a_\theta &= \dot{v}_\theta + \frac{v_r v_\theta}{r} - \frac{v_\varphi^2}{r \tan\theta}\ , \\[2mm]
a_\varphi &= \dot{v}_\varphi + \frac{v_r v_\varphi}{r} + \frac{v_\theta v_\varphi}{r \tan\theta}\ .
\end{aligned}
\right\} \tag{3.27}
$$

It thus follows that the components of the acceleration in a spherical coordinate system are not simply the time-derivatives of the components of the velocity in that system. The reason for this is that the spherical coordinate system uses basis vectors that change when the particle moves. Expressions (3.27) play a crucial role in meteorology and oceanography where one describes the motion of the atmosphere or ocean[39]. Of course, in that application one should account for the Earth's rotation as well so that terms accounting for the Coriolis force and the centrifugal force need to be added, see section 12.3. It should also be noted that the analysis of this section has been oversimplified when applied to the ocean or atmosphere. The reason for this is that the time-derivative of a moving parcel of air or water contains the explicit time-derivative $\partial/\partial t$ as well as a term $\mathbf{v} \cdot \nabla$ that accounts for the fact that the properties change because the parcel moves to another location in space. The difference between the partial derivative $\partial/\partial t$ and the total derivative $d/dt$ is treated in section

4.5; this distinction has not been taken into account in the analysis in this section. A complete treatment is given by Holton [39].

## 3.4 Volume integration in spherical coordinates

Carrying out a volume integration in Cartesian coordinates involves multiplying the function to be integrated by an infinitesimal volume element $dxdydz$ and integrating over all volume elements:

$$\iiint FdV = \iiint F(x,y,z)dxdydz \ .$$

Although this seems to be a simple procedure, it can be quite complex when the function $F$ depends in a complex way on the coordinates $(x, y, z)$ or when the limits of integration are not simple functions of $x$, $y$ and $z$.

**Problem a:** Compute the volume of a sphere of radius $R$ by taking $F = 1$ and integrating the volume integral in Cartesian coordinates over the volume of the sphere. Show first that in Cartesian coordinates the volume of the sphere can be written as

$$volume = \int_{-R}^{R} \int_{-\sqrt{R^2-x^2}}^{\sqrt{R^2-x^2}} \int_{-\sqrt{R^2-x^2-y^2}}^{\sqrt{R^2-x^2-y^2}} dzdydx \ , \qquad (3.28)$$

and carry out the integrations next.

After carrying out this exercise you have probably become convinced that using Cartesian coordinates is not the most efficient way to derive that the volume of a sphere with radius $R$ is given by $4\pi R^3/3$. Using spherical coordinates appears to be the way to go, but for this one needs to be able to express an infinitesimal volume element $dV$ in spherical coordinates. In doing this we will use that the volume spanned by three vectors $\mathbf{a}$, $\mathbf{b}$ and $\mathbf{c}$ is given by

$$volume = \det(\ \mathbf{a}\ ,\ \mathbf{b}\ ,\ \mathbf{c}\ ) = \begin{vmatrix} a_x & b_x & c_x \\ a_y & b_y & c_y \\ a_z & b_z & c_z \end{vmatrix} \ . \qquad (3.29)$$

If we change the spherical coordinate $\theta$ by an increment $d\theta$, the position vector will change from $\mathbf{r}(r, \theta, \varphi)$ to $\mathbf{r}(r, \theta+d\theta, \varphi)$, and this corresponds to a change $\mathbf{r}(r, \theta+d\theta, \varphi)-\mathbf{r}(r, \theta, \varphi) = \partial\mathbf{r}/\partial\theta\, d\theta$ in the position vector. Using the same reasoning for the variation of the position vector with $r$ and $\varphi$ it follows that the infinitesimal volume $dV$ corresponding to increments $dr$, $d\theta$ and $d\varphi$ is given by

$$dV = \det\left(\frac{\partial\mathbf{r}}{\partial r}dr\ ,\ \frac{\partial\mathbf{r}}{\partial\theta}d\theta\ ,\ \frac{\partial\mathbf{r}}{\partial\varphi}d\varphi\right). \qquad (3.30)$$

**Problem b:** Show that this can be written as:

$$dV = \begin{vmatrix} \dfrac{\partial x}{\partial r} & \dfrac{\partial x}{\partial \theta} & \dfrac{\partial x}{\partial \varphi} \\[2mm] \dfrac{\partial y}{\partial r} & \dfrac{\partial y}{\partial \theta} & \dfrac{\partial y}{\partial \varphi} \\[2mm] \dfrac{\partial z}{\partial r} & \dfrac{\partial z}{\partial \theta} & \dfrac{\partial z}{\partial \varphi} \end{vmatrix} \underbrace{\phantom{xxxxxxxxxxxx}}_{J} dr d\theta d\varphi = J dr d\theta d\varphi . \tag{3.31}$$

The determinant $J$ is called the *Jacobian*, which is also sometimes written as:

$$J = \frac{\partial(x, y, z)}{\partial(r, \theta, \varphi)} , \tag{3.32}$$

but it should be kept in mind that this is nothing more than a new notation for the determinant in (3.31).

**Problem c:** Use expressions (3.3) and (3.31) to show that

$$J = r^2 \sin\theta . \tag{3.33}$$

Note that the Jacobian $J$ in (3.33) is the product of the scale factors defined in equation (3.8): $J = h_r h_\theta h_\varphi$. This is not a coincidence; in general the scale factors contain all the information needed to compute the Jacobian for an orthogonal curvilinear coordinate system, see Butkov [18] for details.

**Problem d:** A volume element $dV$ is thus given in spherical coordinates by $dV = r^2 \sin\theta\, dr d\theta d\varphi$. Consider the volume element $dV$ in figure 3.3 that is defined by infinitesimal increments $dr$, $d\theta$ and $d\varphi$. Give an alternative derivation of this expression for $dV$ that is based on geometric arguments only.

In some applications one wants to integrate over the surface of a sphere rather than over a volume. For example, if one wants to compute the cooling of the Earth, one needs to integrate the heat flow over the Earth's surface. The treatment used for deriving the volume integral in spherical coordinates can also be used to derive the surface integral. A key element in the analysis is that the surface spanned by two vectors $\mathbf{a}$ and $\mathbf{b}$ is given by $|\mathbf{a} \times \mathbf{b}|$. Again, an increment $d\theta$ of the angle $\theta$ corresponds to a change $(\partial \mathbf{r}/\partial \theta)\, d\theta$ of the position vector. A similar result holds when the angle $\varphi$ is changed.

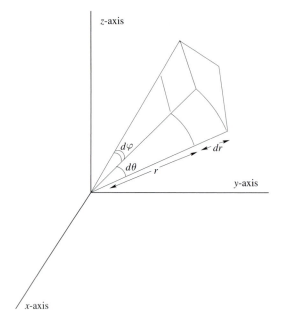

Fig. 3.3.   Definition of the geometric variables for an infinitesimal volume element $dV$.

**Problem e:** Use these results to show that the surface element $dS$ corresponding to infinitesimal changes $d\theta$ and $d\varphi$ is given by

$$dS = \left| \frac{\partial \mathbf{r}}{\partial \theta} \times \frac{\partial \mathbf{r}}{\partial \varphi} \right| d\theta d\varphi \ . \tag{3.34}$$

**Problem f:** Use expression (3.3) to compute the vectors in the cross product and use this to derive that

$$dS = r^2 \sin \theta \ d\theta d\varphi \ . \tag{3.35}$$

**Problem g:** Using the geometric variables in figure 3.3 give an alternative derivation of this expression for a surface element that is based on geometric arguments only.

**Problem h:** Compute the volume of a sphere with radius $R$ using spherical coordinates. Pay special attention to the range of integration for the angles $\theta$ and $\varphi$, see section 3.1.

## 3.5  Cylindrical coordinates

Cylindrical coordinates are useful in problems that exhibit cylindrical symmetry rather than spherical symmetry. An example is the generation

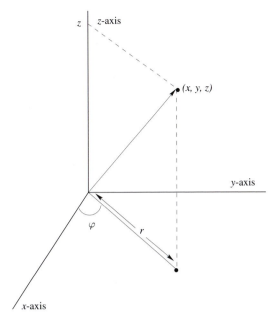

Fig. 3.4. Definition of the geometric variables used in cylindrical coordinates.

of water waves when a stone is thrown into a pond, or more importantly when an earthquake excites a tsunami in the ocean. In cylindrical coordinates a point is specified by giving its distance $r = \sqrt{x^2 + y^2}$ to the $z$-axis, the angle $\varphi$ and the $z$-coordinate, see figure 3.4 for the definition of the variables. All the results we need could be derived using an analysis like those shown in the previous sections. However, in such an approach we would do a large amount of unnecessary work. The key is to realize that at the equator of a spherical coordinate system (i.e. at the locations where $\theta = \pi/2$) the spherical coordinate system and the cylindrical coordinate system are identical, see figure 3.5. An inspection of this figure shows that all results obtained for spherical coordinates can be used for cylindrical coordinates by making the following substitutions:

$$\left.\begin{array}{l} r = \sqrt{x^2 + y^2 + z^2} \rightarrow \sqrt{x^2 + y^2}\ , \\ \theta \rightarrow \pi/2\ , \\ \hat{\boldsymbol{\theta}} \rightarrow -\hat{\mathbf{z}}\ , \\ r d\theta \rightarrow -dz\ . \end{array}\right\} \tag{3.36}$$

**Problem a:** Convince yourself of this. To derive the third line consider the unit vectors pointing in the direction of increasing values of $\theta$ and $z$ at the equator.

**Problem b:** Use the results of the previous sections and the substitutions (3.36) to show the following properties for a system of

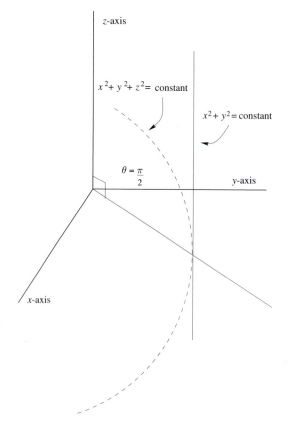

**Fig. 3.5.** At the equator the spherical coordinate system has the same properties as a system of cylindrical coordinates.

cylindrical coordinates:

$$\left. \begin{array}{l} x = r\cos\varphi \,, \\ y = r\sin\varphi \,, \\ z = z \,. \end{array} \right\} \qquad (3.37)$$

$$\hat{\mathbf{r}} = \begin{pmatrix} \cos\varphi \\ \sin\varphi \\ 0 \end{pmatrix} \,, \quad \hat{\boldsymbol{\varphi}} = \begin{pmatrix} -\sin\varphi \\ \cos\varphi \\ 0 \end{pmatrix} \,, \quad \hat{\mathbf{z}} = \begin{pmatrix} 0 \\ 0 \\ 1 \end{pmatrix} \,, \qquad (3.38)$$

$$dV = r\,dr\,d\varphi\,dz \,, \qquad (3.39)$$

$$dS = r\,dz\,d\varphi \,. \qquad (3.40)$$

**Problem c:** Derive these properties directly using geometric arguments.

# 4

## The gradient

In this chapter the gradient of a function is introduced. The gradient plays an important role in the differentiation and integration of functions in more than one dimension (section 4.3). Newton's law is derived in section 4.4 from the concept of energy conservation. As a by-product of this derivation it follows that the force is the (negative) gradient of the potential energy. The gradient plays a crucial role in the difference between the partial time derivatives and the total time derivatives. This resulting distinction between a Eulerian and Lagrangian formulation of problems involving fluid flow is shown in section 4.5. In section 4.6 expressions for the gradient in spherical coordinates and cylindrical coordinates are derived.

### 4.1 Properties of the gradient vector

Let us consider a function $f$ that depends on the variables $x$ and $y$ in a plane. We want to describe how this function changes when we move from point $A$ in the plane to point $C$ as shown in figure 4.1. The resulting change in the function $f$ is denoted by $\delta f = f_C - f_A$, where $f_A$ denotes, for example, the function $f$ at point $A$. It follows from figure 4.1 that

$$\left.\begin{array}{l} f_A = f(x, y) \, , \\ f_B = f(x + \delta x, y) \, , \\ f_C = f(x + \delta x, y + \delta y) \, . \end{array}\right\} \tag{4.1}$$

**Problem a:** Show that

$$\delta f = f_C - f_A = f_B - f_A + f_C - f_B \, . \tag{4.2}$$

33

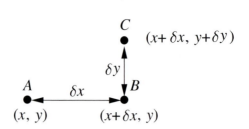

Fig. 4.1.   Definition of the points $A$, $B$ and $C$.

**Problem b:** Use (4.1) and expression (2.18) to derive that for small values of $\delta x$ and $\delta y$:

$$\left. \begin{aligned} f_B - f_A &= \frac{\partial f}{\partial x}(x,y)\,\delta x \ , \\[2mm] f_C - f_B &= \frac{\partial f}{\partial y}(x+\delta x,y)\,\delta y \ . \end{aligned} \right\} \tag{4.3}$$

**Problem c:** Insert this result in (4.2) and derive that to leading order in $\delta x$ and $\delta y$ the result can be written as:

$$\delta f = \frac{\partial f}{\partial x}(x,y)\,\delta x + \frac{\partial f}{\partial y}(x,y)\,\delta y \ . \tag{4.4}$$

This equation has the same form as the inner product between two vectors $\mathbf{a}$ and $\mathbf{b}$ in two dimensions: $(\mathbf{a}\cdot\mathbf{b}) = a_x b_x + a_y b_y$. Let us define $\delta \mathbf{r}$ as the difference in the location vector of the points $C$ and $A$: $\delta \mathbf{r} \equiv \mathbf{r}_C - \mathbf{r}_A$. This vector has components $\delta x$ and $\delta y$ so that in two dimensions:

$$\delta \mathbf{r} = \begin{pmatrix} \delta x \\ \delta y \end{pmatrix} . \tag{4.5}$$

Similarly we define a vector that contains as the $x$-component the partial $x$-derivative of the function $f$ and as the $y$-component the partial $y$-derivative:

$$\nabla f \equiv \begin{pmatrix} \partial f/\partial x \\ \partial f/\partial y \end{pmatrix} . \tag{4.6}$$

This vector is called the *gradient of $f$*. It is customary to denote the gradient of a function $f$ by the symbol $\nabla f$, but another notation you may find in the literature is grad $f$.

**Problem d:** Show that the increment $\delta f$ of expression (4.4) can be expressed as the inner product of two vectors:

$$\delta f = (\nabla f \cdot \delta \mathbf{r}) \ . \tag{4.7}$$

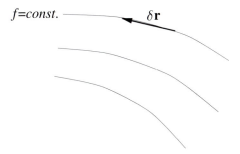

Fig. 4.2. Contour lines defined by the condition $f = const.$ and a perturbation $\delta\mathbf{r}$ in the position vector along a contour line.

This expression shows directly why the gradient is such a useful vector. Once we know the gradient $\nabla f$, we can use (4.7) to compute the change in the function $f$ when we change the point of evaluation over an arbitrary step $\delta\mathbf{r}$. Note that this expression holds for *any* direction in which we can take this step. It should be noted that (4.7) is based on the first order Taylor solutions in expression (4.3). This means that (4.7) only holds in the limit $\delta\mathbf{r} \to 0$. However, this expression is still extremely useful because it forms the basis of the rules for differentiation and integration in more than one dimension. We will return to this issue in section 4.3.

**Problem e:** The derivation up to this point has been for two space dimensions only. Repeat this derivation for three space dimensions and derive that (4.7) still holds when the gradient in three dimensions is defined as

$$
\nabla f \equiv \begin{pmatrix} \partial f/\partial x \\ \partial f/\partial y \\ \partial f/\partial z \end{pmatrix} \quad \text{and} \quad \delta\mathbf{r} \equiv \begin{pmatrix} \delta x \\ \delta y \\ \delta z \end{pmatrix}. \qquad (4.8)
$$

**Problem f:** Compute $\nabla f$ when $f(x, y, z) = x\, e^{-y} \sin z$.

The gradient is a vector and it therefore has a direction and a magnitude. The direction of the gradient can be obtained from expression (4.7). Let us consider a change $\delta\mathbf{r}$ such that the function $f$ does not change in this direction. This means that $\delta\mathbf{r}$ is chosen in such a way that the displacement is along a plane where $f = const.$ as depicted in figure 4.2. For such a change $\delta\mathbf{r}$ the corresponding change $\delta f$ is by definition equal to zero, hence $(\nabla f \cdot \delta\mathbf{r}) = 0$.

**Problem g:** Use the properties of the inner product of two vectors to show that this identity implies that the vectors $\nabla f$ and $\delta\mathbf{r}$ are for this special perturbation perpendicular: $\nabla f \perp \delta\mathbf{r}$.

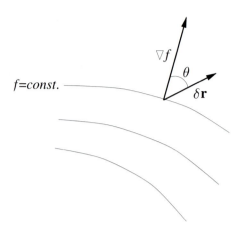

Fig. 4.3.   Contour lines defined by the condition $f = const.$ and a perturbation $\delta \mathbf{r}$ in the position vector in an arbitrary direction.

This last identity of course only holds when the step $\delta \mathbf{r}$ is taken along the surface where the function $f$ is constant. However, it does imply that the gradient vector is perpendicular to the surface $f = const.$ Now that we know this, the gradient can still point in two directions because one can move in two directions perpendicular to a surface. Consider figure 4.3 where we consider a step $\delta \mathbf{r}$ in an arbitrary direction.

**Problem h:** Use (4.7) to show that

$$\delta f = |\nabla f| \, |\delta \mathbf{r}| \cos \theta \,, \tag{4.9}$$

where $\theta$ is the angle between $\nabla f$ and $\delta \mathbf{r}$.

The change $\delta f$ is largest when the vectors $\nabla f$ and $\delta \mathbf{r}$ point in the same direction because in that case $\cos \theta$ is equal to its maximum value of 1, i.e. when $\theta = 0$. We also know that $\delta f$ increases most rapidly when $\delta \mathbf{r}$ is directed in such a way that one moves from smaller values of $f$ towards larger values of $f$. Since $\delta f$ is largest when $\theta = 0$ this means the gradient also points from small values of $f$ towards high values of $f$.

The magnitude of the gradient vector can also be obtained from (4.9). Let us consider a step $\delta \mathbf{r}$ in the direction of the gradient vector, hence a step in the direction of increasing values of $f$. In that case the vectors $\nabla f$ and $\delta \mathbf{r}$ are parallel and $\cos \theta = 1$. This means that (4.9) implies that

$$|\nabla f| = \frac{\delta f}{|\delta \mathbf{r}|} \,, \tag{4.10}$$

where $\delta \mathbf{r}$ is a step in the direction of increasing values of $f$. Summarizing this we obtain the following properties of the gradient vector $\nabla f$:

1. The gradient of a function $f$ is perpendicular to the surface $f = const.$

2. The gradient points in the direction of increasing values of $f$.

3. The magnitude of the gradient is the change $\delta f$ in the function in the direction of the largest increase divided by the distance $|\delta \mathbf{r}|$ in that direction.

## 4.2 The pressure force

An example of a two-dimensional function is shown in figure 4.4 where the atmospheric pressure on the Earth's surface is shown in units of millibars.

**Problem a:** Draw the gradient vector of the pressure $p$ in a number at locations on this map. When doing so draw larger arrows where the gradient is greater.

**Problem b:** What is the approximate gradient $|\nabla p|$ over Ireland? Is the gradient vector larger over Ireland than over Iceland?

There is a very good reason why a map of the pressure is used to illustrate the concept of the gradient. When the air pressure is not constant, a parcel of air will experience a force that pushes it from a region of high pressure towards a region of lower air pressure. This force increases when the pressure varies more rapidly with distance. This can be described by the pressure force $\mathbf{F}_p$ that is given by:

$$\mathbf{F}_p = -\nabla p . \tag{4.11}$$

**Problem c:** The pressure in the weather map of figure 4.4 is shown in unit of millibars. If you have worked out problem b properly you will have deduced that the units of the pressure force are in millibars per kilometre. A pressure is a force per unit area because the pressure times the area gives the total force that acts on this area. Deduce from this result that the pressure force has the dimensions *force/volume*.

**Problem d:** Use the results of this section to show that this force indeed points from regions of high pressure towards regions of lower pressure. Draw the direction of the pressure force at the high-pressure area in figure 4.4 over Ireland.

You may have noted that we have not really derived expression (4.11) for the pressure force. Sometimes only physical arguments are used to state a physical law. However, one often needs to verify that the arguments

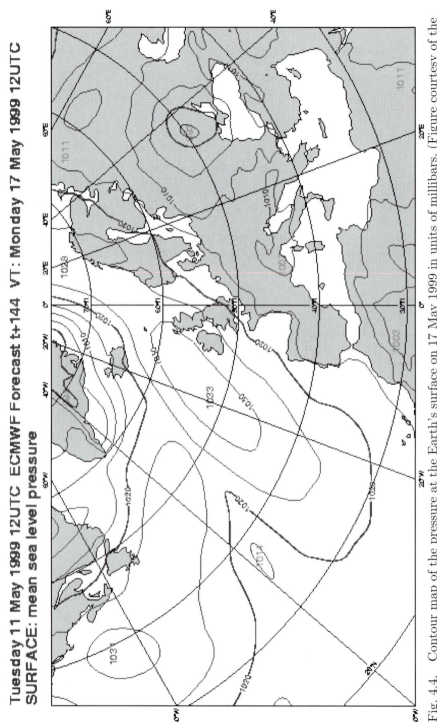

Fig. 4.4. Contour map of the pressure at the Earth's surface on 17 May 1999 in units of millibars. (Figure courtesy of the European Centre for Medium-Range Weather Forecasts (ECMWF).)

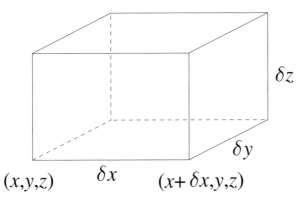

Fig. 4.5. Definition of geometric variables for the derivation of the pressure force.

employed have a proper mathematical basis. It is possible to derive the pressure force from the fact that the force that acts in a gas or a fluid on a hypothetical surface with that medium is perpendicular to that surface and its strength is given by the pressure times the surface area.

**Problem e:** Consider a small volume element with lengths $\delta x$, $\delta y$ and $\delta z$ in the three coordinate directions as shown in figure 4.5. Let us consider the net force in the $x$-direction. Use the concept of pressure as described above to show that the force acting on the left-hand plane of the volume is given by $p(x, y, z)\delta y\delta z\hat{\mathbf{x}}$ and that the force acting on the right-hand plane of the volume is given by $-p(x + \delta x, y, z)\delta y\delta z\hat{\mathbf{x}}$, where $\hat{\mathbf{x}}$ is the unit vector in the $x$-direction. Explain the minus sign in the second term.

The top, bottom, front and back surfaces do not contribute to the $x$-component of the pressure force because the pressure force acting on these surfaces is directed in the $y$- or $z$-direction. This implies that the $x$-component of the total force is given by

$$f_x = -\left[p(x + \delta x, y, z) - p(x, y, z)\right]\delta y\delta z\ \hat{\mathbf{x}}\ . \qquad (4.12)$$

**Problem f:** Use expression (2.18) to show that this can also be written as

$$f_x = -\frac{\partial p}{\partial x}\delta V\ \hat{\mathbf{x}}\ , \qquad (4.13)$$

where the volume $\delta V$ is equal to $\delta x\delta y\delta z$.

**Problem g:** Apply the same reasoning to obtain the $y$- and $z$-components of the force and show that the net force felt by the volume is given by

$$\mathbf{f} = -\nabla p\ \delta V\ . \qquad (4.14)$$

You may have been puzzled by the fact that in problem c you deduced that the pressure force $\mathbf{F}_p$ has the dimensions $force/volume$. This means it is not really a force. In fact, we see in (4.14) that the net force that acts of the volume is given by $-\nabla p \delta V$. You have to keep in mind that the volume $\delta V$ is not a physical entity, instead it is a mathematical volume used in our reasoning. The net force is proportional to $\delta V$, and since this volume is physically meaningless the net force has no physical meaning either. However, when $\rho$ is the mass-density of the gas or fluid then $\delta m = \rho \, \delta V$ is the mass of the volume. This means that both the mass of the volume and the net force that acts on the volume are proportional to $\delta V$. When we apply Newton's law $\delta m \mathbf{a} = \mathbf{f}$ to this volume, we can divide both the left hand side and the right hand side by the arbitrary volume $\delta V$ and Newton's law then takes the form $\rho \mathbf{a} = \mathbf{F}_p$, with the pressure force given by (4.11).

**Problem h:** Verify this statement.

This means that the pressure force $\mathbf{F}_p$ should be seen as a force per unit volume, just as the density $\rho$ is the mass per unit volume. In fluid mechanics one always works with physical quantities per unit volume, for the simple reason that a gas or fluid is not composed of physical small volumes. We will return to a more rigorous treatment of Newton's law in fluid dynamics in section 10.3 where the conservation of momentum in a continuous medium is treated.

The pressure force is one of the most important forces in fluid mechanics, meteorology and oceanography because it is the variation in the pressure that underlies the motion of fluids and gases. Physically it states the simple fact that a gas or fluid is pushed away from regions of higher pressure to regions of lower pressure.

## 4.3 Differentiation and integration

The results of the previous section hold for infinitesimal changes $\delta \mathbf{r}$ in the position only. However, a change over a finite distance can be thought of as being built up from many infinitesimal steps. Let us consider two points $A$ and $B$ that may be far apart as shown in figure 4.6. Between the points $A$ and $B$ we can insert a large number of points $P_1, P_2, \ldots, P_N$. The difference in the function values at the points $A$ and $B$ can then be written as

$$f_B - f_A = (f_B - f_N) + (f_N - f_{N-1}) + \cdots + (f_2 - f_1) + (f_1 - f_A) \, , \quad (4.15)$$

where $f_j$ denotes the function evaluated at point $P_j$. What we are really doing is dividing the large distance from $A$ to $B$ in infinitesimally smaller

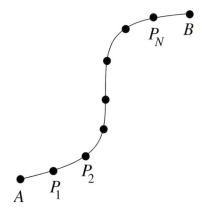

Fig. 4.6. The definition of the points $A$ and $B$ at the end of an interval and the intermediary points $P_1, P_2, \ldots, P_N$.

intervals. If we take enough of these subintervals we can apply (4.7) to each of these sub-intervals.

**Problem a:** Show that

$$f_B - f_A = \sum (\nabla f \cdot \delta \mathbf{r}) \, , \qquad (4.16)$$

where the sum is over the sub-intervals used in (4.15) and where $\delta \mathbf{r}$ is the increment in the position vector in each interval.

The analysis in this section only holds on the limit $N \to \infty$ where the interval is divided in infinitesimal intervals. In that case the summation in (4.16) is replaced by an integration and the notation $d\mathbf{r}$ is used rather than $\delta \mathbf{r}$:

$$f_B - f_A = \int_A^B (\nabla f \cdot d\mathbf{r}) \, . \qquad (4.17)$$

This expression is extremely useful because it makes it possible to compute the change in a function between two points once one knows the gradient of that function. It should be noted that (4.17) holds for *any* path that joins the points $A$ and $B$. This can easily be seen from the derivation of this section, because the points $P_1, P_2, \ldots, P_N$ can be arbitrarily chosen as long as they form a continuous path that joins the points $A$ and $B$. This property can sometimes be exploited by choosing the path for which the calculation of the integral is easiest.

**Problem b:** Use these results to show that the line integral of $\nabla f$ along any closed contour is equal to zero:

$$\oint (\nabla f \cdot d\mathbf{r}) = 0 \, . \qquad (4.18)$$

The gradient can also be used to determine the derivative of a function in a given direction. The *directional derivative* $df/ds$ of the function in the direction of a unit vector $\hat{\mathbf{n}}$ is defined as the change of the function in the direction of $\hat{\mathbf{n}}$ normalized per unit distance:

$$\frac{df}{ds}(\mathbf{r}) = \lim_{\delta s \to 0} \frac{f(\mathbf{r} + \hat{\mathbf{n}}\delta s) - f(\mathbf{r})}{\delta s} . \tag{4.19}$$

**Problem c:** Use (4.7) in the numerator of this expression to derive that

$$\frac{df}{ds}(\mathbf{r}) = (\hat{\mathbf{n}} \cdot \nabla f) . \tag{4.20}$$

This expression is important because it allows us to compute the derivative of a function in *any* arbitrary direction once the gradient is known.

Equations (4.17) and (4.20) generalize the rules for the integration and differentiation of functions of one variable to more space dimensions.

**Problem d:** To see this, let the function $f$ depend on the variable $x$ only and let the points $A$ and $B$ be located on the $x$-axis. Let $\hat{\mathbf{x}}$ denote the unit vector in the direction of the $x$-axis. Use the relation $d\mathbf{r} = \hat{\mathbf{x}}dx$ to show that in that case (4.17) reduces to the well-known integration rule of a function that depends only on one variable:

$$f_B - f_A = \int_A^B \frac{\partial f}{\partial x} dx . \tag{4.21}$$

**Problem e:** Let the vector $\hat{\mathbf{n}}$ be directed along the $x$-axis so that $\hat{\mathbf{n}} = \hat{\mathbf{x}}$. Show that in that case the directional derivative along the $x$-axis in (4.20) is given by $\partial f/\partial x$.

## 4.4 Newton's law from energy conservation

In classical mechanics one can start from Newton's law and derive that the total energy of a mechanical system without friction is conserved. You will probably have noted by now that in physics there is often no proof of the basic laws that form the starting point of the analysis. For example, there is no 'proof' of Newton's law. Its use is justified by the observation that it describes the motion of the Sun, Moon and planets very accurately. To a certain extent, it is arbitrary which physical law one uses as a starting point. In this section we use the concept of energy conservation as a starting point and then derive Newton's law.

Let us consider a mechanical system without friction. The total energy $E$ is the sum of the kinetic energy $\frac{1}{2}mv^2$ and the potential energy $V(\mathbf{r})$:

$$\frac{1}{2}mv^2 + V(\mathbf{r}) = E . \tag{4.22}$$

We assume that in such a system the total energy $E$ is conserved; this means that the time derivative of this quantity is equal to zero

$$\frac{dE}{dt} = 0 \, . \tag{4.23}$$

In order to derive Newton's law from this expression we need to take the time derivative of both the kinetic and the potential energy.

**Problem a:** Show that the time derivative of the kinetic energy is given by

$$\frac{d}{dt}\left(\frac{1}{2}mv^2\right) = m\left(\mathbf{v}\cdot\frac{d\mathbf{v}}{dt}\right) \, . \tag{4.24}$$

Hint: write $v^2 = v_x^2 + v_y^2 + v_z^2$ and differentiate each term.

We also need to compute the time derivative of the potential energy. At first you may be tempted to conclude that this time derivative is equal to zero because the potential depends on the position $\mathbf{r}$ but not on time. However, as the particle moves through space it is at different positions at different times. Therefore, the position of the particle is a function of time: $\mathbf{r} = \mathbf{r}(t)$. For this reason, the potential should be written as $V(\mathbf{r}(t))$. The time derivative of the potential then follows from the usual rule for taking a derivative:

$$\frac{dV(\mathbf{r})}{dt} = \lim_{\delta t \to 0} \frac{V(\mathbf{r}(t+\delta t)) - V(\mathbf{r}(t))}{\delta t} \, . \tag{4.25}$$

**Problem b:** Use (4.7) to show that this time derivative can be written as

$$\frac{dV(\mathbf{r})}{dt} = \lim_{\delta t \to 0} \frac{(\nabla V \cdot \delta\mathbf{r})}{\delta t} \, , \tag{4.26}$$

with $\delta\mathbf{r} = \mathbf{r}(t+\delta t) - \mathbf{r}(t)$.

**Problem c:** Use the definition of the velocity $\mathbf{v} = \lim_{\delta t \to 0} \delta\mathbf{r}/\delta t$ to show that

$$\frac{dV(\mathbf{r})}{dt} = (\mathbf{v}\cdot\nabla V) \, . \tag{4.27}$$

**Problem d:** Use these results to show that the law of energy conservation (4.23) can be written as

$$\mathbf{v}\cdot\left(m\frac{d\mathbf{v}}{dt} + \nabla V\right) = 0 \, . \tag{4.28}$$

Since $\mathbf{v}$ is arbitrary, (4.28) is valid for any velocity vector $\mathbf{v}$. This can only be the case when the term in brackets is equal to zero.

**Problem e:** Show that this implies that

$$m\frac{d\mathbf{v}}{dt} = \mathbf{F} \ , \tag{4.29}$$

with

$$\mathbf{F} = -\nabla V \ . \tag{4.30}$$

Equation (4.29) is Newton's law; we have derived it here from the requirement of energy conservation. As a by-product we have shown that
the force in Newton's law is equal to $-\nabla V$. Note that the force (4.30)
and the pressure force (4.11) share the common property that they follow
from the gradient of a scalar function. You may find this analogy useful.
We have argued in section 4.2 that the pressure force pushes air from
a region of higher pressure to a region of lower pressure. By the same
reasoning the force $\mathbf{F} = -\nabla V$ pushes a particle from regions of higher
potential energy to regions of lower potential energy. A simple example of
this is gravity. The gravitational potential energy increases with the distance to the attracting body. This means that according to the reasoning
given here the gravitational force forces a particle closer to the attracting
body. This is an observation that may be obvious to you when you see an
apple fall from a tree. However, you may want to keep the analogy of the
pressure force and the force associated with a general potential energy in
mind because it helps to understand the concept of potential energy.

With all the elements we have assembled here, we can now derive the
concept of power. This quantity is defined as the energy delivered to the
particle per unit time. This means that the power is defined as the time
derivative of the kinetic energy.

**Problem f:** Use (4.23), (4.27) and (4.30) to derive that the power is
given by

$$\frac{d}{dt}\left(\frac{1}{2}mv^2\right) = (\mathbf{F} \cdot \mathbf{v}) \ . \tag{4.31}$$

How can we understand this result? Let the particle be displaced over a
distance $\delta\mathbf{r}$ in a time increment $\delta t$. The work done by the force is given by
$(\mathbf{F} \cdot \delta\mathbf{r})$. This means that the work per unit time is given by $(\mathbf{F} \cdot \delta\mathbf{r})/\delta t$.
In the limit $\delta t \to 0$ the quantity $\delta\mathbf{r}/\delta t$ is the velocity $\mathbf{v}$, so that the power
is given by $(\mathbf{F} \cdot \mathbf{v})$ as stated in (4.31). In section 10.3 we will derive the
time derivative of the kinetic energy of a gas or fluid, and it will be shown
in (10.22) that for such a system the energy delivered by the force per
unit time is also given by $(\mathbf{F} \cdot \mathbf{v})$, with $\mathbf{F}$ the force per unit volume.

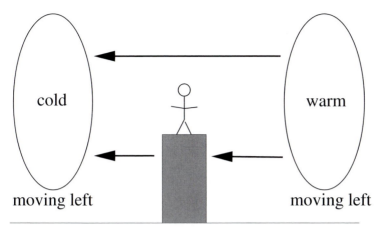

Fig. 4.7. An observer standing on a fixed tower experiences an increase in temperature because warm air moves towards the observer.

## 4.5 Total and partial time derivatives

We have seen in the previous section that the potential that acts on a particle changes with time when the particle moves to a different location. This principle of a temporal change that is caused by a motion in the system is very general. As an example, consider the situation in figure 4.7 where an observer measures the temperature in an observational tower. The motion of the wind is towards the left, and a region of warm air is transported leftward with the wind. The observer detects an increase of the temperature with time, so that $\partial T/\partial t > 0$. In this expression the partial derivative symbol is used deliberately. In general, the temperature field is a function of both time and the space coordinates: $T = T(\mathbf{r}, t)$. For the observer in the observational tower the space coordinates are fixed, which means that any change detected by the observer is due to the change of the local temperature only but that the location is fixed. This is expressed by the definition of the partial derivative:

$$\frac{\partial T(\mathbf{r}, t)}{\partial t} \equiv \lim_{\delta t \to 0} \frac{T(\mathbf{r}, t + \delta t) - T(\mathbf{r}, t)}{\delta t} . \qquad (4.32)$$

In this derivative the time is varied but all other variables are kept fixed.

In contrast to the previous example let us now consider an observer that is carried along in a balloon by the wind, see figure 4.8. We assume in this situation that the temperature field is fixed in time, so that $\partial T/\partial t = 0$. The observer is carried by the wind from a cold region to a warmer region. This means that just like the observer in figure 4.7 this second observer experiences an increase in the temperature. The rate of change of the temperature will be denoted by the total time derivative $dT/dt$. We can

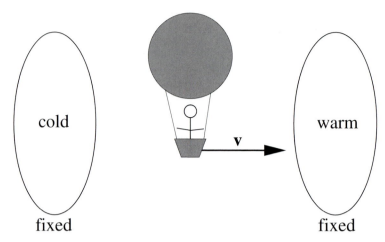

Fig. 4.8.   An observer in a balloon flies through a temperature field that is constant at each location. Because the balloon flies to a warmer region the observer detects an increase in temperature.

see from this example that $dT/dt$ can be larger than zero while $\partial T/\partial t = 0$. The total time derivative is defined in the following way:

$$\frac{dT(\mathbf{r}, t)}{dt} = \lim_{\delta t \to 0} \frac{T(\mathbf{r}(t + \delta), t + \delta t) - T(\mathbf{r}(t), t)}{\delta t} . \qquad (4.33)$$

The only difference between this and the partial time derivative (4.32) is that in (4.33) the position of the observation point changes, whereas in the partial time derivative (4.32) the position is fixed.

**Problem a:** Assuming that the temperature field in figure 4.8 depends on position but not on time, show that $dT/dt$ has the same form as expression (4.25) and use the results of section 4.4 to show that for the observer in figure 4.8

$$\frac{dT(\mathbf{r})}{dt} = (\mathbf{v} \cdot \nabla T) \qquad \text{(in the special case of figure 4.8)} . \quad (4.34)$$

**Problem b:** Draw the gradient vector $\nabla T$ in figure 4.8 at the location of the observer and show from the geometry of this vector and the velocity vector that $dT/dt > 0$.

This means that the observers in both figures 4.7 and 4.8 detect a rise in temperature, but their description of this temperature change is completely different. The first observer feels an increase in temperature because the local temperature changes at her location; this is described by the partial time derivative $\partial T/\partial t$. The second observer detects an increase in temperature because she is transported to a warmer region in a fixed temperature field; she describes this by the total time derivative that is for this special case given by $dT/dt = (\mathbf{v} \cdot \nabla T)$.

In general the temperature field may change because of a combination of a change in the temperature at a fixed location and a movement of the observer. We will assume now that the temperature field depends on all three space coordinates and on time: $T = T(x(t), y(t), z(t), t)$. In this notation it is explicit that the space coordinates of the observation point change with time as well. The total time derivative is given by (4.33) and can be written as:

$$\frac{dT}{dt} = \lim_{\delta t \to 0} \frac{T(x(t + \delta t), y(t + \delta t), z(t + \delta t), t + \delta t) - T(x(t), y(t), z(t), t)}{\delta t} .$$

(4.35)

Let the $x$-coordinate over the time interval change from $x$ to $x + \delta x$, and the $y$- and $z$-coordinates change in a similar way.

**Problem c:** Use the first order Taylor series to show that

$$\delta x = x(t + \delta t) - x(t) \approx \frac{\partial x}{\partial t} \delta t = v_x \delta t .$$

(4.36)

In the limit $\delta t \to 0$ the approximation becomes an identity. Explain that $v_x$ is the $x$-component of the velocity vector.

**Problem d:** Generalize (2.22) for the case of a function that depends both on $x$, $y$, $z$ and on $t$ to the following first order Taylor series:

$$T(x + \delta x, y + \delta y, z + \delta z, t + \delta t) \approx T(z, y, z, t)$$
$$+ \frac{\partial T}{\partial x} \delta x + \frac{\partial T}{\partial y} \delta y + \frac{\partial T}{\partial z} \delta z + \frac{\partial T}{\partial t} \delta t . \quad (4.37)$$

**Problem e:** Insert (4.36) and (4.37) in (4.35) and take the limit $\delta t \to 0$ to derive that

$$\frac{dT}{dt} = \frac{\partial T}{\partial x} v_x + \frac{\partial T}{\partial y} v_y + \frac{\partial T}{\partial z} v_z + \frac{\partial T}{\partial t} .$$

(4.38)

**Problem f:** Use the definition of the gradient vector to rewrite this as

$$\frac{dT}{dt} = \frac{\partial T}{\partial t} + (\mathbf{v} \cdot \nabla T) .$$

(4.39)

**Problem g:** This is the general expression of the total time derivative. Show that the time derivatives (4.32) and (4.34) seen by the observers in figures 4.7 and 4.8 can both be obtained from this general expression for the total time derivative.

The temperature field was used in this section only for illustrative purposes. The analysis is of course applicable to any function that depends on both time and the space coordinates. Also, in the analysis we used 'observers' to fix our minds. However, in general there are no observers and it is not essential that there is anybody present to 'observe' the change in temperature. The total time derivative is always related to the movement of a certain quantity. In section 4.4 this was the movement of a particle that is subjected to a potential $V(\mathbf{r})$. In the description of a gas it may concern the motion of a parcel of material in the gas that is carried around by the flow.

Whenever one describes a continuous system such as the motion in the atmosphere one has a choice in the way that one sets up this description. One option is to consider every quantity at a fixed location in space, and to specify how these quantities change with time. In that case the change of the properties with time is described by the partial time derivative $\partial/\partial t$. This is called a *Eulerian* description. As an alternative one may follow a certain property while it is being carried around by the flow. In that case the change of this property with time is given by the total time derivative $d/dt$. This is called a *Lagrangian* description. It depends on the problem which description is most convenient. In numerical applications the Eulerian description is usually most convenient because one can work with a fixed system of space coordinates and one only needs to specify how a property changes with time in that coordinate system. On the other hand, if one wants to study, for example, the spreading of pollution by wind one aims at tracking particles, and for this the Lagrangian formulation is most convenient. For this reason it is important that one can transform the physical laws back and forth between a Eulerian and a Lagrangian formulation. We return to this issue in chapter 10.

## 4.6 The gradient in spherical coordinates

The gradient vector is defined in (4.8) using a system of Cartesian coordinates. In many applications it is much more convenient to use a system of curvilinear coordinates, especially spherical coordinates and cylindrical coordinates which are of great importance. It is therefore useful to obtain the form of the gradient in these coordinate systems as well. In this section we consider the expression of the gradient in spherical coordinates. Before we do this we will first rewrite the expression of the gradient in Cartesian coordinates.

**Problem a:** Show that the gradient in Cartesian coordinates can also be written as:

$$\nabla f = \hat{\mathbf{x}}\frac{\partial f}{\partial x} + \hat{\mathbf{y}}\frac{\partial f}{\partial y} + \hat{\mathbf{z}}\frac{\partial f}{\partial z} . \qquad (4.40)$$

**Problem b:** Take the inner product of this expression with the unit vector in the $x$-direction to show that

$$\frac{\partial f}{\partial x} = (\hat{\mathbf{x}} \cdot \nabla f) \ . \tag{4.41}$$

Expression (3.4) gives the spherical coordinates $r$, $\theta$ and $\varphi$ in terms of the Cartesian coordinates $x$, $y$ and $z$. Using the chain rule of differentiation one can then use for example that

$$\frac{\partial}{\partial x} = \frac{\partial r}{\partial x} \frac{\partial}{\partial r} + \frac{\partial \theta}{\partial x} \frac{\partial}{\partial \theta} + \frac{\partial \varphi}{\partial x} \frac{\partial}{\partial \varphi} \ .$$

Together with expression (3.7) for the unit vectors $\hat{\mathbf{r}}$, $\hat{\boldsymbol{\theta}}$ and $\hat{\boldsymbol{\varphi}}$ this could be used to derive the expression for the gradient in spherical coordinates. However, this approach is algebraically very complex and does not give much insight. Here we derive the gradient vector in spherical coordinates using (4.7). The idea is that the component of the gradient along a certain coordinate axis is simply given by the rate of change of the function in that direction: $\delta f = (\nabla f \cdot \delta \mathbf{s})$. Suppose the change in position $\delta \mathbf{s}$ is in the direction of a unit vector $\hat{\mathbf{e}}$, then the change in position is given by $\delta \mathbf{s} = \hat{\mathbf{e}} \, \delta s$, where $\delta s$ is given by: $\delta s \equiv |\delta \mathbf{s}|$. This means that $\delta f = (\nabla f \cdot \hat{\mathbf{e}}) \, \delta s$. However, as $(\nabla f \cdot \hat{\mathbf{e}})$ is the component of the gradient vector in the direction of the unit vector $\hat{\mathbf{e}}$, this means that the component $\nabla_e f$ of the gradient vector in the direction of $\hat{\mathbf{e}}$ is given by:

$$\nabla_e f \equiv (\nabla f \cdot \hat{\mathbf{e}}) = \frac{\delta f}{\delta s} = \lim_{\delta s \to 0} \frac{f(\mathbf{r} + \hat{\mathbf{e}}\delta s) - f(\mathbf{r})}{\delta s} \ . \tag{4.42}$$

Consider the spherical coordinate system shown in figure 4.9. Let us move in the direction of the unit vector $\hat{\boldsymbol{\theta}}$ from the point $A$ with spherical coordinates $(r, \theta, \varphi)$ to point $B$ with spherical coordinates $(r, \theta + \delta\theta, \varphi)$.

**Problem c:** Derive from figure 4.9 that the distance $\delta s$ between the points $A$ and $B$ is related to the change in the angle $\delta\theta$ by the relation $\delta s = r\delta\theta$.

**Problem d:** Use this in (4.42) to derive that the $\theta$-component of the gradient vector is given by

$$\nabla_\theta f = \lim_{\delta\theta \to 0} \frac{f(r, \theta + \delta\theta, \varphi) - f(r, \theta, \varphi)}{r\delta\theta} = \frac{1}{r} \frac{\partial f}{\partial \theta} \ . \tag{4.43}$$

**Problem e:** For a change from point $A$ to point $C$ in the $\hat{\mathbf{r}}$ direction over an increment $\delta r$ in the radius the change in the position has

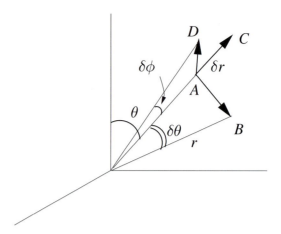

Fig. 4.9.  Definition of the geometric variables and the points $A$, $B$, $C$ and $D$ in spherical coordinates.

magnitude $\delta s = \delta r$. Derive from this that the radial component is given by

$$\nabla_r f = \frac{\partial f}{\partial r} \ . \tag{4.44}$$

**Problem f:** Use the geometry of figure 4.9 to show that for a change from point $A$ to point $D$ in the $\hat{\varphi}$-direction over an increment $\delta\varphi$ in the azimuth $\varphi$ the change in the position has magnitude $\delta s = r\,\sin\theta\,\delta\varphi$. Derive from this that the azimuthal component of the gradient is given by

$$\nabla_\varphi f = \frac{1}{r \sin\theta} \frac{\partial f}{\partial \varphi} \ . \tag{4.45}$$

Combining these result we obtain the expression for the gradient in spherical coordinates:

$$\nabla f = \hat{\mathbf{r}} \frac{\partial f}{\partial r} + \hat{\boldsymbol{\theta}} \frac{1}{r} \frac{\partial f}{\partial \theta} + \hat{\boldsymbol{\varphi}} \frac{1}{r \sin\theta} \frac{\partial f}{\partial \varphi} \ . \tag{4.46}$$

Note that the gradient in spherical coordinates is *not* given by $\nabla f = \hat{\mathbf{r}}\partial f/\partial r + \hat{\boldsymbol{\theta}}\partial f/\partial\theta + \hat{\boldsymbol{\varphi}}\partial f/\partial\varphi$. There is a simple reason why this expression must be wrong. The first term on the right hand side has the dimension $f/length$ because the radius is a length, while the second term has the dimension $f$ because the angle $\theta$ is expressed in radians which is a dimensionless property. This means that this expression must be wrong. The factors $1/r$ and $1/r\sin\theta$ in (4.46) account for the scaling of the distance $\delta s$ with the change in the angles $\delta\theta$ and $\delta\varphi$, respectively. These terms therefore account for the fact that the system is curvilinear.

**Problem g:** Show that each of the terms in (4.46) has the dimension *f/length*.

**Problem h:** Do the same analysis for cylindrical coordinates and show that the gradient in cylindrical coordinates is given by

$$\nabla f = \hat{\mathbf{r}}\frac{\partial f}{\partial r} + \hat{\boldsymbol{\varphi}}\frac{1}{r}\frac{\partial f}{\partial \varphi} + \hat{\mathbf{z}}\frac{\partial f}{\partial z}\ . \tag{4.47}$$

# 5

---

# The divergence of a vector field

The physical meaning of the divergence cannot be understood without understanding what the flux of a vector field is, and what the sources and sinks of a vector field are.

## 5.1 The flux of a vector field

To fix our mind, let us consider a vector field $\mathbf{v}(\mathbf{r})$ that represents the flow of a fluid that has a constant density. We define a surface $S$ in this fluid. Of course the surface has an orientation in space, and the unit vector perpendicular to $S$ is denoted by $\hat{\mathbf{n}}$. Infinitesimal elements of this surface are denoted with $d\mathbf{S} \equiv \hat{\mathbf{n}} dS$. Now suppose we are interested in the volume of fluid that flows per unit time through the surface $S$; this quantity is called $\Phi$. When we want to know the flow through the surface, we only need to consider the component of $\mathbf{v}$ perpendicular to the surface, the flow along the surface is not relevant.

**Problem a:** Show that the component of the flow *across* the surface is given by $(\mathbf{v} \cdot \hat{\mathbf{n}})\hat{\mathbf{n}}$ and that the flow *along* the surface is given by $\mathbf{v} - (\mathbf{v} \cdot \hat{\mathbf{n}})\hat{\mathbf{n}}$. If you find this problem difficult you may want to look ahead to section 12.1.

Using this result the volume of the flow through the surface per unit time is given by:

$$\Phi = \iint (\mathbf{v} \cdot \hat{\mathbf{n}}) dS = \iint \mathbf{v} \cdot d\mathbf{S} \; ; \tag{5.1}$$

this expression defines the flux $\Phi$ of the vector field $\mathbf{v}$ through the surface $S$. The definition of a flux is not restricted to the flow of fluids: a flux can be computed for any vector field. However, the analogy of fluid flow often is very useful for understanding the meaning of the flux and divergence.

**Problem b:** The electric field generated by a point charge $q$ at the origin is given by

$$\mathbf{E}(\mathbf{r}) = \frac{q\hat{\mathbf{r}}}{4\pi\varepsilon_0 r^2} \ . \tag{5.2}$$

In this expression $\hat{\mathbf{r}}$ is the unit vector in the radial direction and $\varepsilon_0$ is the permittivity. Compute the flux of the electric field through a spherical surface of radius $R$ with the point charge at its center. Show that this flux is independent of the radius $R$ and find its relation to the charge $q$ and the permittivity $\varepsilon_0$. Choose the coordinate system you use for the integration carefully.

**Problem c:** To first order the magnetic field of the Earth is a dipole field. (This is the field generated by a magnetic north pole and magnetic south pole that are very close together.) The dipole vector $\mathbf{m}$ points from the south pole of the dipole to the north pole and its size is given by the strength of the dipole. The magnetic field $\mathbf{B}(\mathbf{r})$ is given by (ref. [42], p. 182):

$$\mathbf{B}(\mathbf{r}) = \frac{3\hat{\mathbf{r}}(\hat{\mathbf{r}} \cdot \mathbf{m}) - \mathbf{m}}{r^3} \ . \tag{5.3}$$

Compute the flux of the magnetic field through the surface of the Earth (assume that the Earth is a sphere of radius $R$). Hint: when you select a coordinate system, think not only about the geometry of the coordinate system (i.e. Cartesian or spherical coordinates), but also choose the *direction* of the axes of your coordinate system with care.

## 5.2 Introduction of the divergence

In order to introduce the divergence, consider an infinitesimal rectangular volume with sides $dx$, $dy$ and $dz$ (see fig 5.1 for the definition of the geometric variables). The outward flux through the right hand surface perpendicular through the $x$-axis is given by $v_x(x + dx, y, z)dydz$, because $v_x(x + dx, y, z)$ is the component of the flow perpendicular to that surface and $dydz$ is the area of the surface. By the same token, the flux through the left hand surface perpendicular through the $x$-axis is given by $-v_x(x, y, z)dydz$, the $-$ sign is due to the fact the component of $\mathbf{v}$ in the direction *outward* from the cube is given by $-v_x$. (Alternatively one can say that for this surface the unit vector perpendicular to the surface and pointing outwards is given by $\hat{\mathbf{n}} = -\hat{\mathbf{x}}$.) This means that the total outward flux through the two surfaces is given by $v_x(x + dx, y, z)dydz - v_x(x, y, z)dydz = (\partial v_x/\partial x) dxdydz$. The same reasoning applies to the surfaces perpendicular to the $y$- and $z$-axes. This

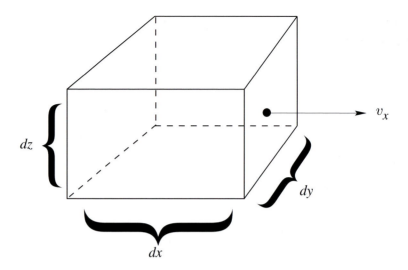

Fig. 5.1.   Definition of the geometric variables in the calculation of the flux of a vector field through an infinitesimal rectangular volume.

means that the total outward flux through the sides of the cubes is:

$$d\Phi = \left( \frac{\partial v_x}{\partial x} + \frac{\partial v_y}{\partial y} + \frac{\partial v_z}{\partial z} \right) dV = (\nabla \cdot \mathbf{v}) \, dV \; , \tag{5.4}$$

where $dV$ is the volume $dxdydz$ of the cube and $(\nabla \cdot \mathbf{v})$ is the *divergence* of the vector field $\mathbf{v}$. The above definition does not really tell us yet what the divergence really is. Dividing (5.4) by $dV$ one obtains $(\nabla \cdot \mathbf{v}) = d\Phi/dV$. This allows us to state in words what the divergence is:

> *The divergence of a vector field is the outward flux of the vector field per unit volume.*

To fix our minds again let us consider a physical example in two dimensions in which a fluid is pumped into this two-dimensional space at the origin $\mathbf{r} = 0$. For simplicity we assume that the fluid is incompressible, which means that the mass-density is constant. We do not know yet what the resulting flow field is, but we know two things. Away from the source at $\mathbf{r} = 0$ there are no sources or sinks of fluid flow. This means that the flux of the flow through any closed surface $S$ that does not enclose any sources or sinks must be zero: 'what goes in must come out.' This in turn means that the divergence of the flow is zero, except possibly near the source at $\mathbf{r} = 0$:

$$(\nabla \cdot \mathbf{v}) = 0 \qquad \text{for} \qquad \mathbf{r} \neq 0 \; . \tag{5.5}$$

In addition we know that due to the symmetry of the problem the flow is directed in the radial direction and depends on the radius $r$ only:

$$\mathbf{v}(\mathbf{r}) = f(r)\mathbf{r} . \tag{5.6}$$

**Problem a:** Show this.

This is enough information to determine the flow field. Of course, it is a problem that we cannot immediately insert (5.6) in (5.5) because we have not yet derived an expression for the divergence in cylindrical coordinates. However, there is another way to determine the flow from the expression above.

**Problem b:** Using that $r = \sqrt{x^2 + y^2}$ show that

$$\frac{\partial r}{\partial x} = \frac{x}{r} , \tag{5.7}$$

and derive the corresponding equation for $y$. Using expressions (5.6), (5.7) and the chain rule for differentiation show that

$$(\nabla \cdot \mathbf{v}) = 2f(r) + r\frac{df}{dr} \qquad \text{(cylindrical coordinates) .} \tag{5.8}$$

**Problem c:** Insert this result in (5.5) and show that the flow field is given by $\mathbf{v}(\mathbf{r}) = A\mathbf{r}/r^2$. Make a sketch of the flow field.

The constant $A$ is yet to be determined. Let a volume of liquid $V$ be injected per unit time at the source $\mathbf{r} = 0$.

**Problem d:** Show that $V = \int \mathbf{v} \cdot d\mathbf{S}$ (where the integration is over an arbitrary surface around the source at $\mathbf{r} = 0$). Choosing a suitable surface, derive that

$$\mathbf{v}(\mathbf{r}) = \frac{V}{2\pi} \frac{\hat{\mathbf{r}}}{r} . \tag{5.9}$$

Note that the unit vector $\hat{\mathbf{r}}$ is now used rather than the position vector $\mathbf{r}$.

From this simple example of a single source at $\mathbf{r} = 0$ more complex examples can be obtained. Suppose we have a source at $\mathbf{r}_+ = (L, 0)$, where a volume $V$ is injected per unit time, and a sink at $\mathbf{r}_- = (-L, 0)$, where a volume $-V$ is removed per unit time. The total flow field can be obtained by superposition of flow fields of the form (5.9) for the source and the sink.

**Problem e:** Show that the $x$- and $y$-components of the flow field in this case are given by:

$$v_x(x, y) = \frac{V}{2\pi} \left[ \frac{x - L}{(x - L)^2 + y^2} - \frac{x + L}{(x + L)^2 + y^2} \right], \qquad (5.10)$$

$$v_y(x, y) = \frac{V}{2\pi} \left[ \frac{y}{(x - L)^2 + y^2} - \frac{y}{(x + L)^2 + y^2} \right], \qquad (5.11)$$

and sketch the resulting flow field. This is most easily accomplished by determining from these expressions the flow field at some selected lines such as the $x$- and $y$-axes.

One may also be interested in computing the *streamlines* of the flow. These are the lines along which material particles flow. The streamlines can be found by using the fact that the time derivative of the position of a material particle is the velocity:

$$d\mathbf{r}/dt = \mathbf{v}(\mathbf{r}). \qquad (5.12)$$

Inserting expressions (5.10) and (5.11) into equation (5.12) leads to two coupled differential equations for $x(t)$ and $y(t)$ which are difficult to solve. Fortunately, there are more intelligent ways of retrieving the streamlines. We will return to this issue in section 15.3.

## 5.3 Sources and sinks

In the example of the fluid flow given above the fluid flow moves away from the source and converges on the sink. The terms 'source' and 'sink' have a clear physical meaning since they are directly related to the 'source' of water as from a tap, and a 'sink' as the sink in a bathtub. The flow lines of the water flow diverge from the source while they converge towards the sink. This explains the term 'divergence', because this quantity simply indicates to what extent flow lines originate (in case of a source) or end (in case of a sink).

This definition of sources and sinks is not restricted to fluid flow. For example, for the electric field the term 'fluid flow' should be replaced by the term 'field lines.' Electrical field lines originate at positive charges and end at negative charges.

**Problem a:** To verify this, show that the divergence of the electrical field (5.2) for a point charge in three dimensions vanishes except near the point charge at $\mathbf{r} = 0$. Show also that the net flux through a small sphere surrounding the charge is positive (negative) when the charge $q$ is positive (negative).

What we have just discovered is that the electric charge is the source of the electric field. This is reflected in the Maxwell equation for the divergence of the electric field:

$$(\nabla \cdot \mathbf{E}) = \rho(\mathbf{r})/\varepsilon_0 . \tag{5.13}$$

In this expression $\rho(\mathbf{r})$ is the charge density, which is simply the electric charge per unit volume just as the mass-density denotes the mass per unit volume. In addition, expression (5.13) contains the permittivity $\varepsilon_0$. This term serves as a *coupling constant* since it describes how 'much' electrical field is generated by a given electrical charge density. It is obvious that a constant is needed here because the charge density and the electrical field have different physical dimensions, hence a proportionality factor must be present. However, the physical meaning of a coupling constant goes much deeper, because it prescribes how strong the field is that is generated by a given source. This constant describes how strongly cause (the source) and effect (the field) are coupled.

**Problem b:** Show that the divergence of the magnetic field (5.3) for a dipole $\mathbf{m}$ at the origin is zero everywhere, *including* the location of the dipole.

By analogy with (5.13) one might expect that the divergence of the magnetic field is related to a magnetic charge density:

$$(\nabla \cdot \mathbf{B}) = coupling\ const.\ \rho_B(\mathbf{r}) ,$$

where $\rho_B$ would be the 'density of magnetic charge.' However, particles with a magnetic charge (usually called 'magnetic monopoles') have not been found in nature despite extensive searches. Therefore the Maxwell equation for the divergence of the magnetic field is:

$$(\nabla \cdot \mathbf{B}) = 0 , \tag{5.14}$$

but we should remember that this divergence is zero because of the observational absence of magnetic monopoles rather than a vanishing coupling constant.

## 5.4 The divergence in cylindrical coordinates

In the previous analysis we have only expressed the divergence in Cartesian coordinates: $\nabla \cdot \mathbf{v} = \partial_x v_x + \partial_y v_y + \partial_z v_z$. As you have (hopefully) discovered, the use of other coordinate systems such as cylindrical coordinates or spherical coordinates can often make life much simpler. Here we derive an expression for the divergence in cylindrical coordinates. In this system, the distance $r = \sqrt{x^2 + y^2}$ of a point to the $z$-axis, the azimuth

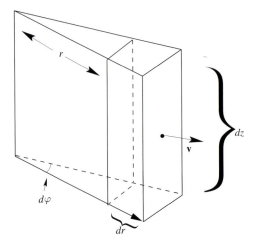

Fig. 5.2.    Definition of the geometric variables for the computation of the divergence in cylindrical coordinates.

$\varphi(= \arctan(y/x))$ and $z$ are used as coordinates, see section 3.5. A vector
$\mathbf{v}$ can be decomposed into components in this coordinate system:

$$\mathbf{v} = v_r\hat{\mathbf{r}} + v_\varphi\hat{\boldsymbol{\varphi}} + v_z\hat{\mathbf{z}} , \tag{5.15}$$

where $\hat{\mathbf{r}}$, $\hat{\boldsymbol{\varphi}}$ and $\hat{\mathbf{z}}$ are unit vectors in the direction of increasing values of
$r$, $\varphi$ and $z$, respectively. As shown in section 5.2 the divergence is the flux
per unit volume. Let us consider the infinitesimal volume corresponding
to increments $dr$, $d\varphi$ and $dz$ shown in figure 5.2. Let us first consider the
flux of $\mathbf{v}$ through the surface elements perpendicular to $\hat{\mathbf{r}}$. The size of this
surface is $rd\varphi dz$ and $(r+dr)d\varphi dz$ at $r$ and $r+dr$ respectively. The normal
components of $\mathbf{v}$ through these surfaces are $v_r(r, \varphi, z)$ and $v_r(r + dr, \varphi, z)$
respectively. Hence the total flux through these two surface is given by
$v_r(r + dr, \varphi, z)(r + dr)d\varphi dz - v_r(r, \varphi, z)(r)d\varphi dz.$

**Problem a:** Show that to first order in $dr$ this quantity is equal to

$$\frac{\partial}{\partial r}(rv_r)\, dr d\varphi dz .$$

Hint: use a first order Taylor expansion for $v_r(r + dr, \varphi, z)$ in the
quantity $dr$.

**Problem b:** Show that the flux through the surfaces perpendicular to $\hat{\boldsymbol{\varphi}}$
is to first order in $d\varphi$ given by $(\partial v_\varphi/\partial\varphi)dr d\varphi dz$.

**Problem c:** Show that the flux through the surfaces perpendicular to $\hat{\mathbf{z}}$
is to first order in $dz$ given by $(\partial v_z/\partial z)rdr d\varphi dz$.

The volume of the infinitesimal part of space shown in figure 5.2 is given by $r dr d\varphi dz$.

**Problem d:** Use the fact that the divergence is the flux per unit volume to show that in cylindrical coordinates:

$$(\nabla \cdot \mathbf{v}) = \frac{1}{r}\frac{\partial}{\partial r}(r v_r) + \frac{1}{r}\frac{\partial v_\varphi}{\partial \varphi} + \frac{\partial v_z}{\partial z} . \qquad (5.16)$$

**Problem e:** Use this result to rederive (5.8) without using Cartesian coordinates as an intermediary.

In spherical coordinates a vector $\mathbf{v}$ can be expanded in the components $v_r$, $v_\theta$ and $v_\varphi$ in the directions of increasing values of $r$, $\theta$ and $\varphi$ respectively. In this coordinate system $r$ has a different meaning than in cylindrical coordinates because in spherical coordinates $r = \sqrt{x^2 + y^2 + z^2}$.

**Problem f:** Use the same reasoning as for the cylindrical coordinates to show that in spherical coordinates

$$\nabla \cdot \mathbf{v} = \frac{1}{r^2}\frac{\partial}{\partial r}\left(r^2 v_r\right) + \frac{1}{r \sin\theta}\frac{\partial}{\partial \theta}(\sin\theta\, v_\theta) + \frac{1}{r \sin\theta}\frac{\partial v_\varphi}{\partial \varphi} . \qquad (5.17)$$

## 5.5 Is life possible in a five-dimensional world?

In this section we will investigate whether the motion of the Earth around the Sun is stable or not. This means that we ask ourselves whether, when the position of the Earth is perturbed (for example by the gravitational attraction of the other planets or by a passing asteroid), the gravitational force causes the Earth to return to its original position (stability) or spiral away from (or towards) the Sun. It turns out that these stability properties depend on the spatial dimension! We know that we live in a world of three spatial dimensions, but it is interesting to investigate whether the orbit of the Earth would also be stable in a world with a different number of spatial dimensions.

In the Newtonian theory the gravitational field $\mathbf{g}(\mathbf{r})$ satisfies (see ref. [62]):

$$(\nabla \cdot \mathbf{g}) = -4\pi G \rho , \qquad (5.18)$$

where $\rho(\mathbf{r})$ is the mass-density and $G$ is the gravitational constant which has a value of $6.67 \times 10^{-8}$ cm$^3$ g$^{-1}$ s$^{-2}$. The term $G$ plays the role of a coupling constant, just like the $1/$permittivity in (5.13). Note that the right hand side of the gravitational field equation (5.18) has the opposite sign to the right hand side of the electric field equation (5.13). This is due to the fact that two electric charges of equal sign repel each other, while two masses of equal sign (mass being positive) attract each other. If the sign of the right hand side of (5.18) were positive, masses would repel each other and structures such as planets, the solar system and stellar systems would not exist.

**Problem a:** We argued in section 5.3 that electric field lines start at positive charges and end at negative charges. By analogy we expect that gravitational field lines end at the (positive) masses that generate the field. However, where do the gravitational field lines start?

Let us first determine the gravitational field of the Sun in $N$ dimensions. Outside the Sun the mass-density vanishes; this means that $(\nabla \cdot \mathbf{g}) = 0$. We assume that the mass-density in the Sun is spherically symmetric, then the gravitational field must be spherically symmetric too and is thus of the form:

$$\mathbf{g}(\mathbf{r}) = f(r)\mathbf{r} . \tag{5.19}$$

In order to make further progress we must derive the divergence of a spherically symmetric vector field in $N$ dimensions. Generalizing expression (5.17) to an arbitrary number of dimensions is not trivial, but fortunately this is not needed. We will make use of the property that in $N$ dimensions: $r = \sqrt{\sum_{i=1}^{N} x_i^2}$.

**Problem b:** Derive from this expression that

$$\partial r/\partial x_j = x_j/r . \tag{5.20}$$

Use this result to derive that for a vector field of the form (5.19):

$$(\nabla \cdot \mathbf{g}) = Nf(r) + r\frac{\partial f}{\partial r} . \tag{5.21}$$

Outside the Sun, where the mass-density vanishes and $(\nabla \cdot \mathbf{g}) = 0$ we can use this result to solve for the gravitational field.

**Problem c:** Derive that

$$\mathbf{g}(\mathbf{r}) = -\frac{A}{r^{N-1}}\hat{\mathbf{r}} , \tag{5.22}$$

and check this result for three spatial dimensions.

At this point the constant $A$ is not determined, but this is not important for the coming arguments. The minus sign has been added for convenience: the gravitational field points towards the Sun and hence $A > 0$.

Associated with the gravitational field is a gravitational force that attracts the Earth towards the Sun. If the mass of the Earth is denoted by $m$, this force is given by

$$\mathbf{F}_{grav} = -\frac{Am}{r^{N-1}}\hat{\mathbf{r}} \ , \qquad (5.23)$$

and is directed towards the Sun. For simplicity we assume that the Earth is in a circular orbit. This means that the attractive gravitational force is balanced by the repulsive centrifugal force which is given by

$$\mathbf{F}_{cent} = \frac{mv^2}{r}\hat{\mathbf{r}} \ . \qquad (5.24)$$

In equilibrium these forces balance: $\mathbf{F}_{grav} + \mathbf{F}_{cent} = 0$.

**Problem d:** Derive the velocity $v$ from this requirement.

We now assume that the distance to the Sun changes from $r$ to $r + \delta r$: The perturbation in the position is therefore $\delta\mathbf{r} = \delta r \ \hat{\mathbf{r}}$. Because of this perturbation, the gravitational force and the centrifugal force are perturbed too. These quantities will be denoted by $\delta\mathbf{F}_{grav}$ and $\delta\mathbf{F}_{cent}$ respectively, see figure 5.3.

**Problem e:** Show that the Earth moves back to its original position when:

$$(\delta\mathbf{F}_{grav} + \delta\mathbf{F}_{cent}) \cdot \delta\mathbf{r} < 0 \qquad \text{(stability)} \ . \qquad (5.25)$$

Hint: consider the cases in which the radius is increased ($\delta r > 0$) and decreased ($\delta r < 0$) separately.

Hence the orbital motion is stable for perturbations when the gravitational field satisfies the criterion (5.25). In order to compute the change in the centrifugal force we use that angular momentum is conserved, i.e. $mrv = m(r + \delta r)(v + \delta v)$. In what follows we will consider small perturbations and will retain only terms of first order in the perturbation. This means that we will ignore higher order terms such as the product $\delta r \delta v$.

**Problem f:** Determine $\delta v$ from (5.24) and the condition that angular momentum is conserved and derive that

$$\delta\mathbf{F}_{cent} = -\frac{3mv^2}{r^2}\delta\mathbf{r} \ , \qquad (5.26)$$

then use (5.23) to show that

$$\delta\mathbf{F}_{grav} = (N-1)\frac{Am}{r^N}\delta\mathbf{r} \ . \qquad (5.27)$$

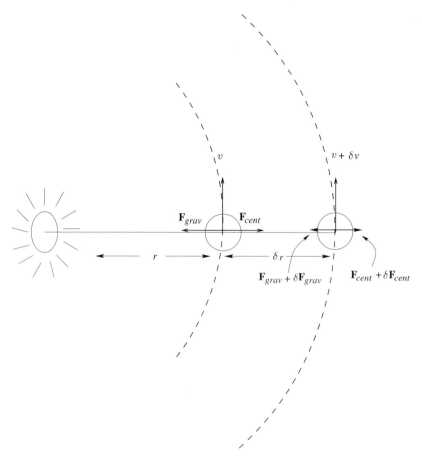

Fig. 5.3.   Definition of variables for the perturbed orbit of the Earth.

Note that the perturbation of the centrifugal force does not depend on the number of spatial dimensions, but that the perturbation of the gravitational force does depend on $N$.

**Problem g:** Using the value of the velocity derived in problem d and expressions (5.26)–(5.27) show that according to the criterion (5.25) the orbital motion is stable in less than four spatial dimensions. Show also that the requirement for stability is independent of the original distance $r$.

This is a very interesting result. It implies that orbital motion is unstable in more than four spatial dimensions. This means that in a world with five spatial dimensions the solar system would not be stable. Life seems to be tied to planetary systems with a central star which supplies the energy to sustain life on the orbiting planet(s). This implies that life would be impossible in a five-dimensional world! Note also that the stability requirement is independent of $r$, i.e. the stability properties of orbital motion do not depend on the size of the orbit. This implies that the gravitational field does not have 'stable regions' and 'unstable regions', the stability property depends only on the number of spatial dimensions.

# 6

---

# The curl of a vector field

## 6.1 Introduction of the curl

We will introduce the *curl* of a vector field $\mathbf{v}$ by its formal definition in terms of Cartesian coordinates $(x, y, z)$ and unit vectors $\hat{\mathbf{x}}$, $\hat{\mathbf{y}}$ and $\hat{\mathbf{z}}$ in the $x$-, $y$- and $z$-directions respectively:

$$curl\ \mathbf{v} \equiv \begin{vmatrix} \hat{\mathbf{x}} & \hat{\mathbf{y}} & \hat{\mathbf{z}} \\ \partial_x & \partial_y & \partial_z \\ v_x & v_y & v_z \end{vmatrix} = \begin{pmatrix} \partial_y v_z - \partial_z v_y \\ \partial_z v_x - \partial_x v_z \\ \partial_x v_y - \partial_y v_x \end{pmatrix}. \tag{6.1}$$

It can be seen that the curl is a *vector*, in contrast to the divergence which is a scalar. The notation with the determinant is of course incorrect because the entries in a determinant should be numbers rather than vectors such as $\hat{\mathbf{x}}$ or differentiation operators such as $\partial_y = \partial/\partial y$. However, the notation in terms of a determinant is a simple way to remember the definition of the curl in Cartesian coordinates. We will write the curl of a vector field also as: curl $\mathbf{v} = \nabla \times \mathbf{v}$.

**Problem a:** Verify that this notation with the curl expressed as the outer product of the operator $\nabla$ and the vector $\mathbf{v}$ is consistent with the definition (6.1).

In general the curl is a three-dimensional vector. To see the physical interpretation of the curl, we will make life easy for ourselves by choosing a Cartesian coordinate system in which the $z$-axis is aligned with curl $\mathbf{v}$. In that coordinate system the curl is given by: curl $\mathbf{v} = (\partial_x v_y - \partial_y v_x)\hat{\mathbf{z}}$. Consider a little rectangular surface element oriented perpendicular to the $z$-axis with sides $dx$ and $dy$ respectively, see figure 6.1. We consider the line integral $\oint_{dxdy} \mathbf{v} \cdot d\mathbf{r}$ along a closed loop defined by the sides of this surface element, integrating in the counterclockwise direction. This line integral can be written as the sum of the integral over the four sides of the surface element.

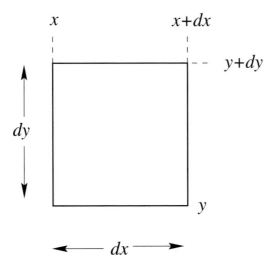

Fig. 6.1.   Definition of the geometric variables for the interpretation of the curl.

**Problem b:** Show that the line integral is given by $\oint_{dxdy} \mathbf{v} \cdot d\mathbf{r} = v_x(x,y)dx + v_y(x+dx,y)dy - v_x(x,y+dy)dx - v_y(x,y)dy$, and use a first order Taylor expansion to write this as

$$\oint_{dxdy} \mathbf{v} \cdot d\mathbf{r} = (\partial_x v_y - \partial_y v_x)dxdy .$$  (6.2)

This expression can be rewritten as:

$$(\text{curl } \mathbf{v})_z = (\partial_x v_y - \partial_y v_x) = \frac{\oint_{dxdy} \mathbf{v} \cdot d\mathbf{r}}{dxdy} .$$  (6.3)

In this form we can express the meaning of the *curl* in words:

> *The component of curl* **v** *in a certain direction is the closed line integral of* **v** *along a closed path perpendicular to this direction, normalized per unit surface area.*

Note that this interpretation is similar to the interpretation of the divergence given in section 5.2. There is, however, one major difference. The *curl* is a vector while the divergence is a scalar. This is reflected in our interpretation of the curl because a surface has an orientation defined by its normal vector, hence the curl is a vector too.

## 6.2 What is the curl of the vector field?

In order to discover the meaning of the curl, we consider again an incompressible fluid and study the curl of the velocity vector **v**, because this

will allow us to discover when the curl is nonzero. It is not only for a di-
dactic purpose that we consider the curl of fluid flow. In fluid mechanics
this quantity plays such a crucial role that it is given a special name, the
*vorticity* $\boldsymbol{\omega}$:

$$\boldsymbol{\omega} \equiv \nabla \times \mathbf{v} \ . \tag{6.4}$$

To simplify things further we assume that the fluid moves in the $(x, y)$-
plane only (i.e. $v_z = 0$) and that the flow depends only on $x$ and $y$:
$\mathbf{v} = \mathbf{v}(x, y)$.

**Problem a:** Show that for such a flow

$$\boldsymbol{\omega} = \nabla \times \mathbf{v} = (\partial_x v_y - \partial_y v_x)\hat{\mathbf{z}} \ . \tag{6.5}$$

We first consider an axisymmetric flow field. Such a flow field has rotational symmetry around one axis, and we will take this as the $z$-axis.
Because of the cylindrical symmetry and the fact that it is assumed that
the flow does not depend on $z$, the components $v_r$, $v_\varphi$ and $v_z$ depend
neither on the azimuth $\varphi$ ($= \arctan y/x$) used in the cylinder coordinates
nor on $z$ but only on the distance $r = \sqrt{x^2 + y^2}$ to the $z$-axis.

**Problem b:** Show that it follows from expression (5.16) for the divergence in cylindrical coordinates that for an axisymmetric flow field
for an incompressible fluid (where $(\nabla \cdot \mathbf{v}) = 0$ everywhere including
the $z$-axis where $r = \sqrt{x^2 + y^2} = 0$) the radial component of the
velocity must vanish: $v_r = 0$.

This result simply reflects that for an incompressible flow with cylindrical
symmetry there can be no net flow towards (or away from) the symmetry
axis. The only nonzero component of the flow is therefore in the direction
of $\hat{\varphi}$. This implies that the velocity field must be of the form:

$$\mathbf{v} = \hat{\varphi} v(r) \ ; \tag{6.6}$$

see figure 6.2 for a sketch of this flow field. The problem we now face is that
definition (6.1) is expressed in Cartesian coordinates while the velocity in
equation (6.6) is expressed in cylindrical coordinates. In section 6.6 an
expression for the curl in cylindrical coordinates will be derived. As an
alternative, one can express the unit vector $\hat{\varphi}$ in Cartesian coordinates.

**Problem c:** Verify that:

$$\hat{\varphi} = \begin{pmatrix} -y/r \\ x/r \\ 0 \end{pmatrix} \ . \tag{6.7}$$

Hints: draw this vector in the $(x, y)$-plane, verify that this vector is
perpendicular to the position vector $\mathbf{r}$ and that it is of unit length.
Alternatively you can use expression (3.38).

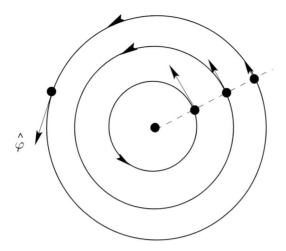

Fig. 6.2.   Sketch of an axisymmetric source-free flow in the $(x, y)$-plane.

**Problem d:** Use (6.5), (6.7) and the chain rule for differentiation to show that for the flow field (6.6):

$$(\nabla \times \mathbf{v})_z = \frac{\partial v}{\partial r} + \frac{v}{r} \ . \tag{6.8}$$

Hint: you have to use the derivatives $\partial r/\partial x$ and $\partial r/\partial y$ again. You determined these in section 5.2.

## 6.3 The first source of vorticity: rigid rotation

In general, a nonzero curl of a vector field can have two origins: rigid rotation and shear. In this section we will treat the effect of rigid rotation. Because we will use fluid flow as an example we will speak about the vorticity, but keep in mind that the results of this section (and the next) apply to any vector field. We consider a velocity field that describes a rigid rotation with the $z$-axis as the axis of rotation and an angular velocity $\Omega$.

**Problem a:** Show that the associated velocity field is of the form (6.6) with $v(r) = \Omega r$. Verify explicitly that every particle in the flow makes one revolution in a time $T = 2\pi/\Omega$ and that this time does not depend on the position of the particle.

**Problem b:** Show that for this velocity field: $\nabla \times \mathbf{v} = 2\Omega\hat{\mathbf{z}}$.

This means that the vorticity is twice the rotational vector $\Omega\hat{\mathbf{z}}$. This result is derived here for the special case that the $z$-axis is the axis of rotation. (This can always be achieved because one is free in the choice of the

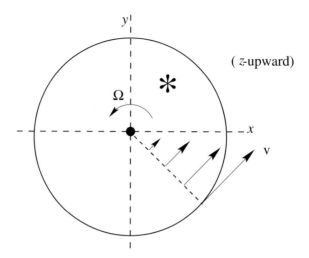

Fig. 6.3.   The vorticity for a rigid rotation.

orientation of the coordinate system.) In section 6.11 of Boas [15] it is
shown with a very different derivation that the vorticity for rigid rotation
is given by $\boldsymbol{\omega} = \nabla \times \mathbf{v} = 2\boldsymbol{\Omega}$, where $\boldsymbol{\Omega}$ is the rotation vector. (Beware, the
notation used by Boas [15] is different from ours in a deceptive way!)

We see that rigid rotation leads to a vorticity that is twice the rotation
rate. Imagine we place a paddle-wheel in the flow field that is associated
with the rigid rotation, see figure 6.3. This paddle-wheel moves with the
flow and makes one revolution about its axis in a time $2\pi/\Omega$. Note also
that for the sense of rotation shown in figure 6.3 the paddle-wheel moves in
the counterclockwise direction and that the curl points along the positive
$z$-axis. This implies that the rotation of the paddle-wheel not only denotes
that the curl is nonzero, but also that the rotation vector of the paddle
is directed along the curl! This actually explains the origin of the word
*vorticity*. In a vortex, the flow rotates around a rotational axis. The curl
increases with the rotation rate, hence it increases with the strength of
the vortex. This strength of the vortex has been dubbed *vorticity*, and
this term therefore reflects the fact that the curl of velocity denotes the
(local) intensity of rotation in the flow.

## 6.4 The second source of vorticity: shear

In addition to rigid rotation, shear is another cause of vorticity. In order to
see this we consider a fluid in which the flow is only in the $x$-direction and
where the flow depends on the $y$-coordinate only: $v_y = v_z = 0$, $v_x = f(y)$.

**Problem a:** Show that this flow does not describe a rigid rotation. Hint:
how long does it take before a fluid particle returns to its original
position?

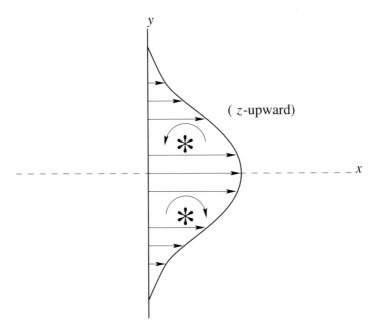

Fig. 6.4.   Sketch of the flow field for a shear flow.

**Problem b:** Show that for this flow

$$\nabla \times \mathbf{v} = -\frac{\partial f}{\partial y}\hat{\mathbf{z}} \ . \tag{6.9}$$

As a special example consider the velocity given by:

$$v_x = f(y) = v_0 e^{-y^2/L^2} \ . \tag{6.10}$$

This flow field is sketched in figure 6.4.

**Problem c:** Verify for yourself that paddle-wheels placed in the flow rotate in the sense indicated in figure 6.4.

**Problem d:** Compute $\nabla \times \mathbf{v}$ for this flow field and verify that both the curl and the rotational vector of the paddle wheels are aligned with the $z$-axis. Show that the vorticity is positive where the paddle-wheels rotate in the counterclockwise direction and that the vorticity is negative where the paddle-wheels rotate in the clockwise direction.

It follows from the example in this section and the example in section 6.3 that both rotation and shear cause a nonzero vorticity. Both phenomena lead to the rotation of imaginary paddle-wheels embedded in the vector field. Therefore, the curl of a vector field measures the local rotation of the vector field (in a literal sense). This explains why in some languages (i.e. Dutch) the notation *rot* **v** is used rather than *curl* **v**. Note that this interpretation of the *curl* as a measure of (local) rotation is consistent with (6.3) in which the curl is related to the value of the line integral along the small contour. If the flow (locally) rotates and if we integrate along the fluid flow, the line integral $\oint \mathbf{v} \cdot d\mathbf{r}$ will be relatively large, so that this line integral indeed measures the local rotation.

Rotation and shear each contribute to the *curl* of a vector field. Let us consider once again a vector field of the form (6.6) which is axially symmetric around the $z$-axis. In the following we don't require the rotation around the $z$-axis to be rigid, so that $v(r)$ in (6.6) is still arbitrary. We know that both the rotation around the $z$-axis and the shear are a source of vorticity.

**Problem e:** Show that for the flow

$$v(r) = \frac{A}{r} \tag{6.11}$$

the vorticity vanishes, with $A$ a constant that is not yet determined. Make a sketch of this flow field.

The vorticity of this flow vanishes despite the fact that the flow rotates around the $z$-axis (but not in rigid rotation) and that the flow has a nonzero shear. The reason that the vorticity vanishes is that the contribution of the rotation around the $z$-axis to the vorticity is equal but of opposite sign to the contribution of the shear, so that the total vorticity vanishes. Note that this implies that a paddle-wheel does not change its orientation as it moves with this flow!

## 6.5 The magnetic field induced by a straight current

At this point you may have the impression that the flow field (6.11) has been contrived in an artificial way. However, keep in mind that all the arguments of the previous section apply to any vector field and that fluid flow was used only as an example to fix our mind. As an example we consider the generation of the magnetic field **B** by an electrical current **J** that is independent of time. The Maxwell equation for the curl of the magnetic field in vacuum is for time-independent fields given by:

$$\nabla \times \mathbf{B} = \mu_0 \mathbf{J} , \tag{6.12}$$

see equation (5.22) in ref. [42]. In this expression $\mu_0$ is the magnetic permeability of vacuum. It plays the role of a coupling constant since it governs the strength of the magnetic field that is generated by a given current. It plays the same role as 1/permittivity in (5.13) or the gravitational constant $G$ in (5.18). The vector $\mathbf{J}$ denotes the electric current per unit volume (properly called the electric current density).

For simplicity we consider an electric current running through an infinite straight wire along the $z$-axis. Because of rotational symmetry around the $z$-axis and because of translational invariance along the $z$-axis the magnetic field depends neither on $\varphi$ nor on $z$ and must be of the form (6.6). Away from the wire the electrical current $\mathbf{J}$ vanishes.

**Problem a:** Show that

$$\mathbf{B} = \frac{A}{r}\hat{\varphi} \; . \tag{6.13}$$

A comparison with equation (6.11) shows that for this magnetic field the contribution of the 'rotation' around the $z$-axis to $\nabla \times \mathbf{B}$ is exactly balanced by the contribution of the 'magnetic shear' to $\nabla \times \mathbf{B}$. It should be noted that the magnetic field derived in this section is of great importance because this field has been used to define the unit of electrical current, the ampère. However, this can only be done when the constant $A$ in expression (6.13) is known.

**Problem b:** Why does the treatment of this section not tell us what the relation is between the constant $A$ and the current $\mathbf{J}$ in the wire?

We will return to this issue in section 8.3.

## 6.6 Spherical coordinates and cylindrical coordinates

In section 5.4, expressions for the divergence in spherical coordinates and cylindrical coordinates were derived. Here we will do the same for the curl because these expressions are frequently very useful. It is possible to derive the curl in curvilinear coordinates by systematically transforming all the elements of the involved vectors and all the differentiations from Cartesian to curvilinear coordinates. As an alternative, we use the physical interpretation of the *curl* given by expression (6.3) to derive the *curl* in spherical coordinates. This expression simply states that a certain component of the *curl* of a vector field $\mathbf{v}$ is the line integral $\oint \mathbf{v} \cdot d\mathbf{r}$ along a contour perpendicular to the component of the *curl* that we are considering, normalized by the surface area bounded by that contour. As an example we will derive for a system of spherical coordinates the $\varphi$-component of the *curl*; see figure 6.5 for the definition of the geometric variables.

Consider the grey infinitesimal surface in figure 6.5. When we carry out the line integration along the surface we integrate in the direction shown in the figure. The reason for this is that the azimuth $\varphi$ increases

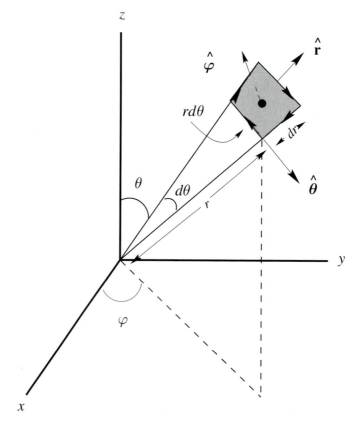

Fig. 6.5. Definition of the geometric variables for the computation of the *curl* in spherical coordinates.

when we move into the figure, hence $\hat{\varphi}$ points into the figure. Following the rules of a right handed screw, this corresponds to the indicated sense of integration. The area enclosed by the contour is given by $rd\theta dr$. By summing the contributions of the four sides of the contour we find, using expression (6.3), that the $\varphi$-component of $\nabla \times \mathbf{v}$ is given by:

$$(\nabla \times \mathbf{v})_{\varphi} = \frac{1}{rd\theta dr}[v_{\theta}(r + dr, \theta)(r + dr)d\theta - v_r(r, \theta + d\theta)dr$$

$$-v_{\theta}(r, \theta)rd\theta + v_r(r, \theta)dr] . \quad (6.14)$$

In this expression $v_r$ and $v_{\theta}$ denote the components of $\mathbf{v}$ in the radial direction and in the direction of $\hat{\boldsymbol{\theta}}$ respectively.

**Problem a:** Verify expression (6.14).

This result can be simplified by Taylor expanding the components of $\mathbf{v}$ in $dr$ and $d\theta$ and linearizing the resulting expression in the infinitesimal increments $dr$ and $d\theta$.

**Problem b:** Do this and show that the final result does not depend on $dr$ and $d\theta$ and is given by:

$$(\nabla \times \mathbf{v})_\varphi = \frac{1}{r}\frac{\partial}{\partial r}(rv_\theta) - \frac{1}{r}\frac{\partial v_r}{\partial \theta} \, . \tag{6.15}$$

The same treatment can be applied to the other components of the curl. This leads to the following expression for the curl in spherical coordinates:

$$\nabla \times \mathbf{v} = \hat{\mathbf{r}}\frac{1}{r\sin\theta}\left[\frac{\partial}{\partial\theta}(\sin\theta v_\varphi) - \frac{\partial v_\theta}{\partial\varphi}\right] + \hat{\boldsymbol{\theta}}\frac{1}{r}\left[\frac{1}{\sin\theta}\frac{\partial v_r}{\partial\varphi} - \frac{\partial}{\partial r}(rv_\varphi)\right]$$
$$+ \hat{\boldsymbol{\varphi}}\frac{1}{r}\left[\frac{\partial}{\partial r}(rv_\theta) - \frac{\partial v_r}{\partial\theta}\right]. \tag{6.16}$$

**Problem c:** Use a similar analysis to show that in cylindrical coordinates $(r, \varphi, z)$ the curl is given by:

$$\nabla \times \mathbf{v} = \hat{\mathbf{r}}\left[\frac{1}{r}\frac{\partial v_z}{\partial\varphi} - \frac{\partial v_\varphi}{\partial z}\right] + \hat{\boldsymbol{\varphi}}\left[\frac{\partial v_r}{\partial z} - \frac{\partial v_z}{\partial r}\right] + \hat{\mathbf{z}}\frac{1}{r}\left[\frac{\partial}{\partial r}(rv_\varphi) - \frac{\partial v_r}{\partial\varphi}\right], \tag{6.17}$$

with $r = \sqrt{x^2 + y^2}$.

**Problem d:** Use this result to rederive (6.8) for vector fields of the form $\mathbf{v} = \mathbf{v}(r)\hat{\boldsymbol{\varphi}}$. Hint: use the same method as used in the derivation of (6.14) and treat the three components of the curl separately.

# 7

---

# The theorem of Gauss

In section 5.5 we determined the gravitational field in $N$ dimensions using as the only ingredient that in free space, where the mass-density vanishes, the divergence of the gravitational field vanishes: $(\nabla \cdot \mathbf{g}) = 0$. This was sufficient to determine the gravitational field in expression (5.22). However, that expression is not quite satisfactory because it contains a constant $A$ that is unknown. In fact, at this point we have no idea how this constant is related to the mass $M$ that causes the gravitational field! The reason for this is simple: in order to derive the gravitational field in (5.22) we have only used the field equation (5.18) for free space (where $\rho = 0$). However, if we want to find the relation between the mass and the resulting gravitational field we must also use the field equation $(\nabla \cdot \mathbf{g}) = -4\pi G\rho$ at places where the mass is present. More specifically, we have to *integrate* the field equation in order to find the total effect of the mass. The theorem of Gauss gives us an expression for the volume integral of the divergence of a vector field.

## 7.1 Statement of Gauss's law

In section 5.2 it was shown that the divergence is the flux per unit volume. In fact, (5.4) gives us the outward flux $d\Phi$ through an infinitesimal volume $dV$: $d\Phi = (\nabla \cdot \mathbf{v})dV$. We can immediately integrate this expression to find the total flux through the surface $S$ which encloses the total volume $V$:

$$\oint_S \mathbf{v} \cdot d\mathbf{S} = \int_V (\nabla \cdot \mathbf{v})dV \ . \tag{7.1}$$

In deriving this, equation (5.4) has been used to express the total flux on the left hand side. This expression is called the theorem of Gauss.

Note that in the derivation of (7.1) we did not use the dimensionality of the space: this relation holds in any number of dimensions. You may

74

recognize the one-dimensional version of (7.1). In one dimension the vector **v** has only one component $v_x$, hence $(\nabla \cdot \mathbf{v}) = \partial_x v_x$. A 'volume' in one dimension is simply a line. Let this line run from $x = a$ to $x = b$. The 'surface' of a one-dimensional volume consists of the endpoints of this line, so that the left hand side of (7.1) is the difference of the function $v_x$ at its endpoints. This implies that the theorem of Gauss in one dimension is:

$$v_x(b) - v_x(a) = \int_a^b \frac{\partial v_x}{\partial x} dx \; . \tag{7.2}$$

This expression will be familiar to you. We will use the two-dimensional version of the theorem of Gauss in section 8.2 to derive the theorem of Stokes.

**Problem a:** Compute the flux of the vector field $\mathbf{v}(x, y, z) = (x+y+z)\hat{\mathbf{z}}$ through a sphere of radius $R$ centered on the origin by explicitly computing the integral that defines the flux.

**Problem b:** Show that the total flux of the magnetic field of the Earth through your skin is zero.

**Problem c:** Solve problem a without carrying out any integration explicitly.

## 7.2 The gravitational field of a spherically symmetric mass

In this section we use Gauss's law (7.1) to show that the gravitational field of a body with a spherically symmetric mass-density $\rho$ depends only on the total mass and not on the distribution of the mass over that body. For a spherically symmetric body the mass-density depends only on radius: $\rho = \rho(r)$. Because of the spherical symmetry of the mass, the gravitational field is spherically symmetric and points in the radial direction:

$$\mathbf{g}(\mathbf{r}) = g(r)\hat{\mathbf{r}} \; . \tag{7.3}$$

**Problem a:** Use the field equation (5.18) for the gravitational field and Gauss's law applied to a surface that completely encloses the mass to show that

$$\oint_S \mathbf{g} \cdot d\mathbf{S} = -4\pi GM \; , \tag{7.4}$$

where $M$ is the total mass of the body.

**Problem b:** Use a sphere of radius $r$ as the surface in (7.4) to show that the gravitational field is given in three dimensions by

$$\mathbf{g}(\mathbf{r}) = -\frac{GM}{r^2}\hat{\mathbf{r}} \; . \tag{7.5}$$

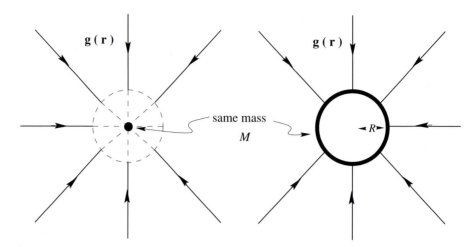

Fig. 7.1.  Two different bodies with different mass distributions that generate the same gravitational field for distances larger than the radius of the body on the right.

This is an intriguing result. What we have shown here is that the gravitational field depends only the total mass of the spherically symmetric body, but not on the distribution of the mass within that body. As an example consider two bodies with the same mass. One body has all its mass located in a small ball near the origin and the other body has all its mass distributed on a thin spherical shell of radius $R$, see figure 7.1. According to (7.5) these bodies generate exactly the same gravitational field outside the body. This implies that gravitational measurements taken outside the two bodies cannot be used to distinguish between them. The nonunique relation between the gravity field and the underlying mass distribution is of importance for the interpretation of gravitational measurements taken in geophysical surveys.

**Problem c:** Let us assume that the mass is located within a sphere of radius $R$, and that the mass-density within that sphere is constant. Integrate (5.18) over a sphere of radius $r < R$ to show that the gravitational field within the sphere is given by:

$$\mathbf{g}(\mathbf{r}) = - \frac{MGr}{R^3} \hat{\mathbf{r}} \ . \tag{7.6}$$

Plot the gravitational field as a function of $r$ when the distance increases from zero to a distance larger than the radius $R$. Verify explicitly that the gravitational field is continuous at the radius $R$ of the sphere.

Note that all these conclusions hold identically for the electrical field when we replace the mass-density by the charge-density, because (5.13) for the divergence of the electric field has the same form as (5.18) for the gravitational field. As an example we will consider a hollow spherical shell of radius $R$. On the spherical shell electrical charge is distributed with a constant charge density: $\rho = const.$

**Problem d:** Use expression (5.13) for the electric field and Gauss's law to show that within the hollow sphere the electric field vanishes: $\mathbf{E}(\mathbf{r}) = 0$ for $r < R$.

This result implies that when a charge is placed within such a spherical shell the electrical field generated by the charge distribution on the shell exerts no net force on this charge: the charge will not move. Since the electrical potential satisfies $\mathbf{E} = -\nabla V$, the result derived in problem d implies that the potential is constant within the sphere. This property has actually been used to determine experimentally whether the electric field indeed satisfies (5.13) (which implies that the field of point charge decays as $1/r^2$). Measurements of the potential differences within a hollow spherical shell as described in problem d can be carried out with very great sensitivity. Experiments based on this principle (usually in a more elaborated form) have been used to ascertain the decay of the electric field of a point charge with distance. Writing the field strength as $1/r^{2+\varepsilon}$ it has been shown that $\varepsilon = (2.7 \pm 3.1) \times 10^{-16}$, see section I.2 of Jackson [42] for a discussion. The small value of $\varepsilon$ is a remarkable experimental confirmation of (5.13) for the electric field.

## 7.3 A representation theorem for acoustic waves

Acoustic waves are waves that propagate through a gas or fluid. You can hear the voice of others because acoustic waves propagate from their vocal tract to your ear. Acoustic waves are frequently used to describe the propagation of waves through the Earth. Since the Earth is a solid body, this is strictly speaking not correct, but under certain conditions (small scattering angles) the errors can be acceptable. The pressure field $p(\mathbf{r})$ of acoustic waves satisfies the following partial differential equation in the frequency domain:

$$\nabla \cdot \left( \frac{1}{\rho} \nabla p \right) + \frac{\omega^2}{\kappa} p = f . \tag{7.7}$$

In this expression $\rho(\mathbf{r})$ is the mass-density of the medium, $\omega$ is the angular frequency and $\kappa(\mathbf{r})$ is the compressibility (a factor that describes how strongly the medium resists changes in its volume). The right hand side

$f(\mathbf{r})$ describes the source of the acoustic wave. This term accounts, for example, for the action of your voice.

We now consider two pressure fields $p_1(\mathbf{r})$ and $p_2(\mathbf{r})$ that both satisfy (7.7) with sources $f_1(\mathbf{r})$ and $f_2(\mathbf{r})$ on the right hand side of the equation.

**Problem a:** Multiply equation (7.7) for $p_1$ with $p_2$, multiply equation (7.7) for $p_2$ with $p_1$ and subtract the resulting expressions. Integrate the result over a volume $V$ to show that:

$$\int_V \left[ p_2 \nabla \cdot \left( \frac{1}{\rho} \nabla p_1 \right) - p_1 \nabla \cdot \left( \frac{1}{\rho} \nabla p_2 \right) \right] dV = \int_V (p_2 f_1 - p_1 f_2)\, dV \ .$$
$$(7.8)$$

Ultimately we want to relate the wave field at the surface $S$ that encloses the volume $V$ to the wave field within the volume. Obviously, Gauss's law is the tool for doing this. The problem we face is that Gauss's law holds for the volume integral of the divergence, whereas in (7.8) we have the *product* of a divergence (such as $\nabla \cdot (\rho^{-1} \nabla p_1)$) with another function (such as $p_2$). This means we have to 'make' a divergence.

**Problem b:** First derive that

$$\nabla \cdot (f\mathbf{v}) = f\,(\nabla \cdot \mathbf{v}) + \mathbf{v} \cdot \nabla f \ .$$
$$(7.9)$$

Hint: write the left hand side in its components and use the product rule for differentiation.

**Problem c:** Use $f = p_2$ and $\mathbf{v} = \rho^{-1} \nabla p_1$ in (7.9) to show that

$$p_2 \nabla \cdot \left( \frac{1}{\rho} \nabla p_1 \right) = \nabla \cdot \left( \frac{1}{\rho} p_2 \nabla p_1 \right) - \frac{1}{\rho}(\nabla p_1 \cdot \nabla p_2) \ .$$
$$(7.10)$$

What we are doing here is similar to the standard derivation of integration by parts. The easiest way to show that $\int_a^b f(\partial g / \partial x) dx = [f(x)g(x)]_a^b - \int_a^b (\partial f / \partial x) g dx$ is to integrate the identity $f(\partial g / \partial x) = \partial (fg)/dx - (\partial f / \partial x)g$ from $x = a$ to $x = b$. This last equation has exactly the same structure as (7.10).

**Problem d:** Use (7.8), (7.10) and Gauss's law to derive that

$$\oint_S \frac{1}{\rho} (p_2 \nabla p_1 - p_1 \nabla p_2) \cdot d\mathbf{S} = \int_V (p_2 f_1 - p_1 f_2)\, dV \ .$$
$$(7.11)$$

This expression forms the basis for the proof that reciprocity holds for acoustic waves. (Reciprocity means that the wavefield propagating from point $A$ to point $B$ is identical to the wavefield that propagates in the reverse direction from point $B$ to point $A$.) To see the power of expression (7.11), consider the special case that the source $f_2$ of $p_2$ is of unit strength and that this source is localized in a very small volume around a point $\mathbf{r}_0$ within the volume. This means that $f_2$ on the right hand side of (7.11) is only nonzero at $\mathbf{r}_0$. The corresponding volume integral $\int_V p_1 f_2 dV$ is in that case given by $p_1(\mathbf{r}_0)$. The wavefield $p_2(\mathbf{r})$ generated by this point source is called the *Green's function,* and this special solution is denoted by $G(\mathbf{r}, \mathbf{r}_0)$. (The concept of the Green's function is introduced in great detail in chapter 17.) The argument $\mathbf{r}_0$ is added to indicate that this is the wavefield at location $\mathbf{r}$ due to a unit source at location $\mathbf{r}_0$. We now consider a solution $p_1$ that has no sources within the volume $V$ (i.e. $f_1 = 0$). Let us simplify the notation further by dropping the subscript '1' in $p_1$.

**Problem e:** Show by making all these changes that (7.11) can be written as:

$$p(\mathbf{r}_0) = \oint_S \frac{1}{\rho} [p(\mathbf{r})\nabla G(\mathbf{r}, \mathbf{r}_0) - G(\mathbf{r}, \mathbf{r}_0)\nabla p(\mathbf{r})] \cdot d\mathbf{S} . \qquad (7.12)$$

This result is called the 'representation theorem' because it gives the wavefield inside the volume when the wavefield (and its gradient) are specified on the surface that bounds this volume. Expression (7.12) can be used to formally derive Huygens's principle which states that every point on a wavefront acts as a source for other waves and that interference between these waves determines the propagation of the wavefront. Equation (7.12) also forms the basis for imaging techniques for seismic data [78]. In seismic exploration one records the wavefield at the Earth's surface. This can be used by taking the Earth's surface as the surface $S$ over which the integration is carried out. If the Green's function $G(\mathbf{r}, \mathbf{r}_0)$ is known, one can use expression (7.12) to compute the wavefield in the interior of the Earth. Once the wavefield in the interior of the Earth is known, one can deduce some of the properties of the material in the Earth. In this way, equation (7.12) (or its elastic generalization) forms the basis of seismic imaging techniques.

   This treatment suggests that if one knows the wavefield at a surface, one can determine the wavefield anywhere within the surface, and hence one can determine the properties of the medium within the surface. This in turn suggests that the problem of seismic imaging has been solved by the treatment of this section. There is, however, a catch. In order to apply (7.12) one must know the Green's function $G$. To know this function one must know the medium. We thus have the interesting situation that once we know the Green's function, we can deduce the properties of the medium from (7.12), but that in order to use the Green's function we must know the medium. This suggests that seismic imaging cannot be carried out.

Fortunately, it turns out that in practice it suffices to use a reasonable *estimate* of the Green's function in (7.12). Such an estimated Green's function can be computed once the velocity in the medium is reasonably well known. It is for this reason that *velocity estimation* is such a crucial step in seismic data processing [20] [103].

## 7.4 Flowing probability

In classical mechanics, the motion of a particle with mass $m$ is governed by Newton's law: $m\ddot{\mathbf{r}} = \mathbf{F}$. When the force $\mathbf{F}$ is associated with a potential $V(\mathbf{r})$ the motion of the particle satisfies:

$$m\frac{d^2\mathbf{r}}{dt^2} = -\nabla V(\mathbf{r}) . \tag{7.13}$$

However, this law does not hold for particles that are very small. Atomic particles such as electrons are not described accurately by (7.13). One of the outstanding features of quantum mechanics is that an atomic particle is treated as a wave that describes the properties of that particle. (This rather vague statement reflects the wave-particle duality that forms the basis of quantum mechanics.) The wavefunction $\psi(\mathbf{r},t)$ that describes a particle that moves under the influence of a potential $V(\mathbf{r})$ satisfies the Schrödinger equation:*

$$i\hbar\frac{\partial\psi(\mathbf{r}, t)}{\partial t} = -\frac{\hbar^2}{2m}\nabla^2\psi(\mathbf{r}, t) + V(\mathbf{r})\psi(\mathbf{r}, t) . \tag{7.14}$$

In this expression, $\hbar$ is Planck's constant $h$ divided by $2\pi$.

**Problem a:** Analyse the physical dimension of the different terms to verify that Planck's constant has the dimension of angular momentum.

Planck's constant has the numerical value $h = 6.626 \times 10^{-34}$ kg m$^2$/s. Suppose we are willing to accept that the motion of an electron is described by the Schrödinger equation, then the following question arises: What is the position of the electron as a function of time? According to the Copenhagen interpretation of quantum mechanics, this is a meaningless question because the electron behaves like a wave and does not have a definite location. Instead, the wavefunction $\psi(\mathbf{r}, t)$ dictates how likely it is that the particle is at location $\mathbf{r}$ at time $t$. Specifically, the quantity $|\psi(\mathbf{r}, t)|^2$ is the probability-density of finding the particle at location $\mathbf{r}$ at time $t$. This implies that the probability $P_V$ that the particle is located

---

* In this expression $\nabla^2$ stands for the Laplacian which is treated in chapter 9. At this point you only need to know that the Laplacian of a function is the divergence of the gradient of that function: $\nabla^2\psi = \text{div grad } \psi = \nabla \cdot \nabla\psi$.

within the volume $V$ is given by $P_V = \int_V |\psi|^2 dV$. (Take care not to confuse the volume with the potential, because they are both indicated with the same symbol $V$.) This implies that the wavefunction is related to a probability. Instead of the motion of the electron, Schrödinger's equation dictates how the probability-density of the particle moves through space as time progresses. One expects that a 'probability current' is associated with this movement. In this section we determine this current using the theorem of Gauss.

**Problem b:** In the following we need the time derivative of $\psi^*(\mathbf{r}, t)$, where the asterisk denotes the complex conjugate. Derive the differential equation that $\psi^*(\mathbf{r}, t)$ obeys by taking the complex conjugate of Schrödinger's equation (7.14).

**Problem c:** Use this result to derive that for a volume $V$ that is fixed in time:

$$\frac{\partial}{\partial t} \int_V |\psi|^2 dV = \frac{i\hbar}{2m} \int_V (\psi^* \nabla^2 \psi - \psi \nabla^2 \psi^*) dV . \tag{7.15}$$

**Problem d:** Use Gauss's law to rewrite this expression as:

$$\frac{\partial}{\partial t} \int_V |\psi|^2 dV = \frac{i\hbar}{2m} \oint (\psi^* \nabla \psi - \psi \nabla \psi^*) \cdot d\mathbf{S} . \tag{7.16}$$

Hint: spot the divergence in (7.15) first.

The left hand side of this expression gives the time derivative of the probability that the particle is within the volume $V$. The only way the particle can enter or leave the volume is through the enclosing surface $S$. The right hand side therefore describes the 'flow' of probability through the surface $S$. More accurately, one can formulate this as the flux of the probability-density current.

**Problem e:** Show from (7.16) that the probability-density current $\mathbf{J}$ is given by:

$$\mathbf{J} = \frac{i\hbar}{2m} (\psi \nabla \psi^* - \psi^* \nabla \psi) . \tag{7.17}$$

Pay particular attention to the signs of the terms in this expression.

As an example let us consider a plane wave:

$$\psi(\mathbf{r}, t) = A e^{i(\mathbf{k} \cdot \mathbf{r} - \omega t)} , \tag{7.18}$$

where $\mathbf{k}$ is the wavevector and $A$ an unspecified constant.

**Problem f:** Show that the wavelength $\lambda$ is related to the wavevector by the relation $\lambda = 2\pi/|\mathbf{k}|$. In which direction does the wave propagate?

**Problem g:** Show that the probability-density current $\mathbf{J}$ for this wavefunction satisfies:

$$\mathbf{J} = \frac{\hbar\mathbf{k}}{m}|\psi|^2 \, . \tag{7.19}$$

This is a very interesting expression. The term $|\psi|^2$ gives the probability-density of the particle, while the probability-density current $\mathbf{J}$ physically describes the current of this probability-density. Since the probability-density current moves with the velocity of the particle (why?), the remaining terms on the right hand side of (7.19) must denote the velocity of the particle:

$$\mathbf{v} = \frac{\hbar\mathbf{k}}{m} \, . \tag{7.20}$$

Since the momentum $\mathbf{p}$ is the mass times the velocity, (7.20) can also be written as $\mathbf{p} = \hbar\mathbf{k}$. This relation was proposed by de Broglie in 1924 using completely different arguments than we have used here [17]. Its discovery was a major step in the development of quantum mechanics.

**Problem h:** Use this expression and the result of problem f to compute your own wavelength while you are riding your bicycle. Are quantum-mechanical phenomena important when you ride you bicycle? Use your wavelength as an argument. Did you know you possessed a wavelength?

# 8

---

# The theorem of Stokes

In chapter 7 we noted that in order to find the gravitational field of a mass we have to integrate the field equation (5.18) over the mass. Gauss's theorem can then be used to compute the integral of the divergence of the gravitational field. For the curl the situation is similar. In section 6.5 we computed the magnetic field generated by a current in a straight infinite wire. The field equation

$$\nabla \times \mathbf{B} = \mu_0 \mathbf{J} \qquad (6.12)$$

was used to compute the field away from the wire. However, the solution (6.13) contained an unknown constant $A$. The reason for this is that the field equation (6.12) was only used outside the wire, where $\mathbf{J} = 0$. The treatment of section 6.5 therefore did not provide us with the relation between the field $\mathbf{B}$ and its source $\mathbf{J}$. The only way to obtain this relation is to integrate the field equation. This implies we have to compute the integral of the curl of a vector field. The theorem of Stokes tells us how to do this.

## 8.1 Statement of Stokes's law

The theorem of Stokes is based on the principle that the curl of a vector field is the closed line integral of the vector field per unit surface area, see section 6.1. Mathematically this statement is expressed by (6.2) which we write in a slightly different form as:

$$\oint_{d\mathbf{S}} \mathbf{v} \cdot d\mathbf{r} = (\nabla \times \mathbf{v}) \cdot \hat{\mathbf{n}} \, dS = (\nabla \times \mathbf{v}) \cdot d\mathbf{S} . \qquad (8.1)$$

The only difference with (6.2) is that in the expression above we have not aligned the $z$-axis with the vector $\nabla \times \mathbf{v}$. The infinitesimal surface is therefore not necessarily confined to the $(x, y)$-plane and the $z$-component

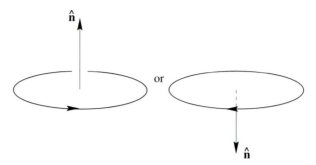

Fig. 8.1.  The relation between the sense of integration and the orientation of the surface.

of the curl is replaced by the component of the *curl* normal to the surface, hence the occurrence of the terms $\hat{\mathbf{n}}dS$ in (8.1). Expression (8.1) holds for an infinitesimal surface area. However, this expression can immediately be integrated to give the surface integral of the curl over a finite surface $S$ that is bounded by the curve $C$:

$$\oint_C \mathbf{v} \cdot d\mathbf{r} = \int_S (\nabla \times \mathbf{v}) \cdot d\mathbf{S} \ . \tag{8.2}$$

This result is known as the theorem of Stokes (or Stokes's law). The line integral on the left hand side is over the curve that bounds the surface $S$. A proper derivation of Stokes's law can be found in ref. [54].

Note that a line integration along a closed surface can be carried out in two directions. What is the direction of the line integral on the left hand side of Stokes's law (8.2)? To see this, we have to realize that Stokes's law was ultimately based on (6.2). The orientation of the line integration used in that expression is defined in figure 6.1, where it can be seen that the line integration is in the counterclockwise direction. In figure 6.1 the $z$-axis points out of the paper, which implies that the vector $d\mathbf{S}$ also points out of the paper. *This means that in Stokes's law the sense of the line integration and the direction of the surface vector $d\mathbf{S}$ are related through the rule for a right handed screw. This orientation is indicated in figure 8.1.*

There is something strange about Stokes's law. If we define a curve $C$ over which we carry out the line integration, we can define many different surfaces $S$ that are bounded by the same curve $C$. Apparently, the surface integral on the right hand side of Stokes's law does not depend on the specific choice of the surface $S$ as long as it is bounded by the curve $C$.

**Problem a:** Let us verify this property for an example. Consider the vector field $\mathbf{v} = r\hat{\varphi}$. Let the curve $C$ used for the line integral be a circle in the $(x, y)$-plane of radius $R$, see figure 8.2 for the geometry

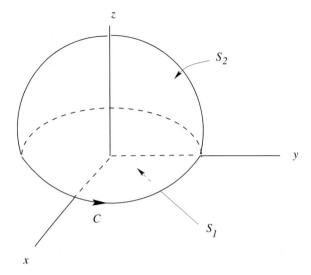

Fig. 8.2.   Definition of the geometric variables for problem a.

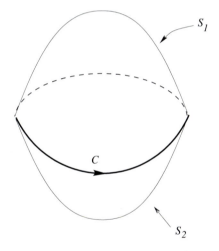

Fig. 8.3.   Two surfaces that are bounded by the same contour C.

of the problem. (i) Compute the line integral on the left hand side of (8.2) by direct integration. Compute the surface integral on the right hand side of (8.2) by (ii) integrating over a disc of radius $R$ in the $(x, y)$-plane (the surface $S_1$ in figure 8.2) and by (iii) integrating over the upper half of a sphere with radius $R$ (the surface $S_2$ in figure 8.2). Verify that the three integrals are identical.

It is actually not difficult to prove that the surface integral in Stokes's law is independent of the specific choice of the surface $S$ as long as it is bounded by the same contour $C$. Consider figure 8.3 where the two surfaces $S_1$ and $S_2$ are bounded by the same contour $C$. We want to show that the surface integral of $\nabla \times \mathbf{v}$ is the same for the two surfaces, i.e. that:

$$\int_{S_1} (\nabla \times \mathbf{v}) \cdot d\mathbf{S} = \int_{S_2} (\nabla \times \mathbf{v}) \cdot d\mathbf{S} . \tag{8.3}$$

We can form a closed surface $S$ by combining the surfaces $S_1$ and $S_2$.

**Problem b:** Show that (8.3) is equivalent to the condition

$$\oint_S (\nabla \times \mathbf{v}) \cdot d\mathbf{S} = 0 , \tag{8.4}$$

where the integration is over the closed surfaces defined by the combination of $S_1$ and $S_2$. Pay particular attention to the signs of the different terms.

**Problem c:** Use Gauss's law to convert (8.4) to a volume integral and show that the integral is indeed identical to zero. In doing so you need to use the identity $\nabla \cdot (\nabla \times \mathbf{v}) = 0$ (or in another notation div curl $\mathbf{v} = 0$). Make sure you can derive this identity.

The result you obtained in problem c implies that the condition (8.3) is indeed satisfied and that in the application of Stokes's law you can choose *any* surface as long as it is bounded by the contour over which the line integration is carried out. This is a very useful result because often the surface integration can be simplified by choosing the surface carefully.

## 8.2 Stokes's theorem from the theorem of Gauss

Stokes's law is concerned with surface integrations. Since the curl is intrinsically a three-dimensional vector, Stokes's law is inherently related to three space dimensions. However, if we consider a vector field that depends only on the coordinates $x$ and $y$ $(\mathbf{v} = \mathbf{v}(x, y))$ and that has a vanishing component in the $z$-direction $(v_z = 0)$, then $\nabla \times \mathbf{v}$ points along the $z$-axis. If we consider a contour $C$ that is confined to the $(x, y)$-plane, for such a vector field Stokes's law takes the form

$$\oint_C (v_x dx + v_y dy) = \int_S (\partial_x v_y - \partial_y v_x) \, dx dy . \tag{8.5}$$

**Problem a:** Verify this.

This result can be derived from the theorem of Gauss in two dimensions.

**Problem b:** Show that Gauss's law (7.1) for a vector field $\mathbf{u}$ in two dimensions can be written as

$$\oint_C (\mathbf{u} \cdot \hat{\mathbf{n}}) ds = \int_S (\partial_x u_x + \partial_y u_y) dx dy , \tag{8.6}$$

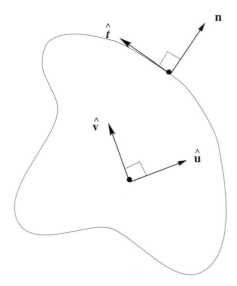

Fig. 8.4.   Definition of the geometric variables for the derivation of Stokes's law from the theorem of Gauss.

where the unit vector $\hat{\mathbf{n}}$ is perpendicular to the curve $C$ (see figure 8.4) and where $ds$ denotes the integration over the arclength of the curve $C$.

In order to derive the special form of Stokes's law (8.5) from Gauss's law (8.6) we have to define the relation between the vectors $\mathbf{u}$ and $\mathbf{v}$. Let the vector $\mathbf{u}$ follow from $\mathbf{v}$ by a clockwise rotation over 90 degrees, see figure 8.4.

**Problem c:** Show that:

$$v_x = -u_y \qquad \text{and} \qquad v_y = u_x . \tag{8.7}$$

We now define the unit vector $\hat{\mathbf{t}}$ to be directed along the curve $C$, see figure 8.4. Since a rotation is an orthonormal transformation, the inner product of two vectors is invariant for a rotation over 90 degrees so that $(\mathbf{u} \cdot \hat{\mathbf{n}}) = (\mathbf{v} \cdot \hat{\mathbf{t}})$.

**Problem d:** Verify this by expressing the components of $\hat{\mathbf{t}}$ in the components of $\hat{\mathbf{n}}$ and by using (8.7). The change in the position vector along the curve $C$ is given by

$$d\mathbf{r} = \hat{\mathbf{t}} \, ds = \begin{pmatrix} dx \\ dy \end{pmatrix} . \tag{8.8}$$

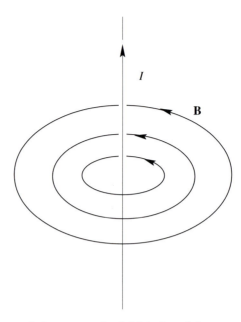

Fig. 8.5. Geometry of the magnetic field induced by a current in a straight infinite wire.

**Problem e:** Use these results to show that (8.5) follows from (8.6).

What you have shown here is that Stokes's law for the special case considered in this section is identical to the theorem of Gauss for two spatial dimensions.

## 8.3 The magnetic field of a current in a straight wire

We now return to the problem of the generation of the magnetic field induced by a current in an infinite straight wire that was discussed in section 6.5. Because of the cylindrical symmetry of the problem, we know that the magnetic field is oriented in the direction of the unit vector $\hat{\varphi}$ and that the field only depends on the distance $r = \sqrt{x^2 + y^2}$ to the wire:

$$\mathbf{B} = B(r)\hat{\varphi} \ . \tag{8.9}$$

The field can be found by integrating the field equation $\nabla \times \mathbf{B} = \mu_0 \mathbf{J}$ over a disc of radius $r$ perpendicular to the wire, see figure 8.5. When the disc is larger than the thickness of the wire the surface integral of $\mathbf{J}$ gives the electric current $I$ through the wire: $I = \int \mathbf{J} \cdot d\mathbf{S}$.

**Problem a:** Use these results and Stokes's law to show that:

$$\mathbf{B} = \frac{\mu_0 I}{2\pi r} \hat{\varphi} \ . \tag{8.10}$$

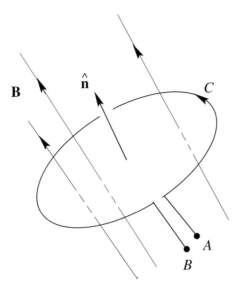

Fig. 8.6. A wire-loop in a time-dependent magnetic field.

We now have a relation between the magnetic field and the current that generates the field, hence the constant $A$ in expression (6.13) is now determined. Note that the magnetic field depends only on the total current through the wire: it does not depend on the distribution of the electric current density $\mathbf{J}$ within the wire as long as the electric current-density exhibits cylindrical symmetry. Compare this with the result you obtained in problem b of section 7.2!

## 8.4 Magnetic induction and Lenz's law

The theory in the previous section deals with the generation of a magnetic field by a current. A magnet placed in this field will experience a force exerted by the magnetic field. This force is essentially the driving force in electric motors; using an electrical current that changes with time a time-dependent magnetic field is generated that exerts a force on magnets attached to a rotation axis.

In this section we study the reverse effect: what is the electrical field generated by a magnetic field that changes with time? In a dynamo, a moving part (e.g. your bicycle wheel) moves a magnet. This creates a time-dependent electric field. This process is called magnetic induction and is described by the following Maxwell equation (see ref. [42]):

$$\nabla \times \mathbf{E} = -\frac{\partial \mathbf{B}}{\partial t} . \tag{8.11}$$

To fix our mind, let us consider a wire with endpoints $A$ and $B$, see figure 8.6. The direction of the magnetic field is indicated in this figure. In order to find the electric field induced in the wire, integrate (8.11) over the surface enclosed by the wire*

$$\int_S (\nabla \times \mathbf{E}) \cdot d\mathbf{S} = -\int_S \frac{\partial \mathbf{B}}{\partial t} \cdot d\mathbf{S} \ . \tag{8.12}$$

**Problem a:** Show that the right hand side of (8.12) is given by $-\partial\Phi/\partial t$, where $\Phi$ is the magnetic flux through the wire. (See section 5.1 for the definition of the flux.)

We have discovered that a change in the magnetic flux is the source of an electric field. The resulting field can be characterized by the *electromotive force* $F_{AB}$, which is a measure of the work done by the electric field on a unit charge when it moves from point $A$ to point $B$, see figure 8.6:

$$F_{AB} \equiv \int_A^B \mathbf{E} \cdot d\mathbf{r} \ . \tag{8.13}$$

**Problem b:** Show that the electromotive force satisfies

$$F_{AB} = -\frac{\partial\Phi}{\partial t} \ . \tag{8.14}$$

**Problem c:** Because of the electromotive force an electric current will flow through the wire. Determine the direction of the electric current in the wire. Show that this current generates a magnetic field that opposes the change in the magnetic field that generates this current. You learned in section 8.3 the direction of the magnetic field that is generated by an electric current in a wire.

What we have discovered in problem c is Lenz's law, which states that induction currents lead to a secondary magnetic field which opposes the change in the primary magnetic field that generates the electric current. This implies that coils in electrical systems exhibit a certain inertia in the sense that they resist changes in the magnetic field that passes through the coil. The amount of inertia is described by a quantity called the inductance $L$. This quantity plays a role similar to mass in classical mechanics because

---

* You may feel uncomfortable applying Stokes's law to the open curve $AB$. However, remember that the electric field associated with the varying magnetic field is a continuous function of the space variables. The electric field is therefore the same at the points $A$ and $B$ where the wire is open. The line integral over the open curve $C$ is therefore identical to the line integral along the closed contour.

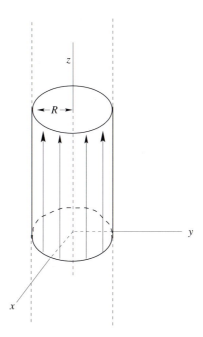

Fig. 8.7.    Geometry of the magnetic field.

the mass of a body also describes how strongly a body resists changing its velocity when an external force is applied.

## 8.5 The Aharonov–Bohm effect

It was shown in section 5.3 that because of the absence of magnetic monopoles the magnetic field is source-free: $(\nabla \cdot \mathbf{B}) = 0$. In electromagnetism one often expresses the magnetic field as the curl of a vector field $\mathbf{A}$:

$$\mathbf{B} = \nabla \times \mathbf{A} . \tag{8.15}$$

The advantage of writing the magnetic field in this way is that for *any* field $\mathbf{A}$ the magnetic field satisfies $(\nabla \cdot \mathbf{B}) = 0$ because $\nabla \cdot (\nabla \times \mathbf{A}) = 0$.

**Problem a:** Give a proof of this last identity.

The vector field $\mathbf{A}$ is called the *vector potential*. The reason for this name is that it plays a role similar to the electric potential $V$. The electric and the magnetic fields follow from $V$ and $\mathbf{A}$ respectively by differentiation: $\mathbf{E} = -\nabla V$ and $\mathbf{B} = \nabla \times \mathbf{A}$. The vector potential has the strange property that it can be nonzero (and variable) in parts of space where the magnetic field vanishes. As an example, consider a magnetic field with cylindrical symmetry along the $z$-axis which is constant for $r < R$ and which vanishes for $r > R$:

$$\mathbf{B} = \begin{cases} B_0 \hat{\mathbf{z}} & \text{for} \quad r < R \\ 0 & \text{for} \quad r > R \end{cases} ; \tag{8.16}$$

see figure 8.7 for a sketch of the magnetic field. Because of the cylindrical symmetry the vector potential is a function of the distance $r$ to the $z$-axis only and does not depend on $z$ or $\varphi$.

**Problem b:** Show that a vector potential of the form

$$\mathbf{A} = f(r)\hat{\varphi} \tag{8.17}$$

gives a magnetic field in the required direction. Give a derivation that $f(r)$ satisfies the following differential equation:

$$\frac{1}{r}\frac{\partial}{\partial r}(rf(r)) = \begin{cases} B_0 & \text{for} \quad r < R \\ 0 & \text{for} \quad r > R \end{cases} . \tag{8.18}$$

This differential equation for $f(r)$ can be immediately integrated. After integration two integration constants are present. These constants follow from the requirements that the vector potential is continuous at $r = R$ and that $f(r = 0) = 0$. (This latter requirement is needed because the direction of the unit vector $\hat{\varphi}$ is undefined on the $z$-axis, where $r = 0$. The vector potential therefore only has a unique value at the $z$-axis when $f(r = 0) = 0$.)

**Problem c:** Integrate the differential equation (8.18) subject to the boundary condition for $f(r = 0)$ to derive that the vector potential is given by

$$\mathbf{A} = \begin{cases} \frac{1}{2}B_0 r\hat{\varphi} & \text{for} \quad r < R \\ \frac{1}{2}B_0 \dfrac{R^2}{r}\hat{\varphi} & \text{for} \quad r > R \end{cases} . \tag{8.19}$$

The importance of this expression is that although the magnetic field is only nonzero for $r < R$, the vector potential (and its gradient) is nonzero everywhere in space! The vector potential is thus much more nonlocal than the magnetic field. This leads to a very interesting effect in quantum mechanics, called the Aharonov–Bohm effect.

Before introducing this effect we need to know more about quantum mechanics. As you have seen in section 7.4, the behavior of atomic 'particles' such as electrons is paradoxically described by a wave. The properties of this wave are described by Schrödinger's equation (7.14). When different waves propagate in the same region of space, interference can occur.

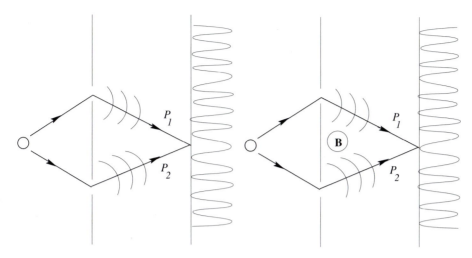

Fig. 8.8.   Experiment in which electrons travel through two slits and are detected on a screen behind the slits. The resulting interference pattern is sketched. The experiment without a magnetic field is shown on the left, the experiment with a magnetic field is shown on the right. Note the shift in the maxima and minima of the interference pattern between the two experiments.

In some parts of space the waves may enhance each other (constructive interference) while in other parts the waves cancel each other (destructive interference). This is observed for 'particle waves' when electrons are sent through two slits and then detected on a screen behind these slits, see the left hand panel of figure 8.8. You might expect the electrons to propagate like bullets along straight lines and only be detected in the two regions behind the two slits that are hit by particles that move along straight lines through the slits. However, this is not the case: in experiments one observes a pattern of fringes on the screen that are caused by the constructive and destructive interference of the electron waves. This interference pattern is sketched in figure 8.8 on the right hand side of the screens. This remarkable confirmation of the wave property of particles is described clearly in ref. [30]. (The situation is even more remarkable when one sends the electrons through the slits 'one-by-one' so that only one electron passes through the slits at a time, and one sees a dot at the detector for each electron. However, after many particles have arrived at the detector this pattern of dots forms the interference pattern of the waves, see ref. [79].)

Let us now consider the same experiment, but with a magnetic field given by (8.16) placed between the two slits. When the electrons propagate along the paths $P_1$ or $P_2$ they do not pass through this field, hence one would expect that the electrons would not be influenced by this field and that the magnetic field would not change the observed interference pattern at the detector. However, it is an observational fact that the magnetic field *does* change the interference pattern at the detector, see ref. [79] for examples. This surprising effect is called the Aharonov–Bohm effect [2].

In order to understand this effect, we should note that a magnetic field in quantum mechanics leads to a phase shift of the wavefunction. If the wavefunction in the absence of a magnetic field is given by $\psi(\mathbf{r})$, the wavefunction in the presence of the magnetic field is given by $\psi(\mathbf{r}) \times \exp[(ie/\hbar c) \int_P \mathbf{A} \cdot d\mathbf{r}]$, see ref. [76]. In this expression $\hbar$ is Planck's constant (divided by $2\pi$), $c$ is the speed of light and $\mathbf{A}$ is the vector potential associated with the magnetic field. The integration is over the path $P$ from the source of the particles to the detector. Consider now the waves that interfere in the two-slit experiment in the right hand panel of figure 8.8. The wave that travels through the upper slit experiences a phase shift $\exp[(ie/\hbar c) \int_{P_1} \mathbf{A} \cdot d\mathbf{r}]$, where the integration is over the path $P_1$ through the upper slit. The wave that travels through the lower slit obtains a phase shift $\exp[(ie/\hbar c) \int_{P_2} \mathbf{A} \cdot d\mathbf{r}]$ where the path $P_2$ runs through the lower slit.

**Problem d:** Show that the phase difference $\delta\varphi$ between the two waves due to the presence of the magnetic field is given by

$$\delta\varphi = \frac{e}{\hbar c} \oint_P \mathbf{A} \cdot d\mathbf{r} \; , \qquad (8.20)$$

where the path $P$ is the closed path from the source through the upper slit to the detector and back through the lower slit to the source.

This phase difference affects the interference pattern because it is the *relative* phase between interfering waves that determines whether the interference is constructive or destructive.

**Problem e:** Show that the phase difference can be written as

$$\delta\varphi = \frac{e\Phi}{\hbar c} \; , \qquad (8.21)$$

where $\Phi$ is the magnetic flux through the area enclosed by the path $P$.

This expression shows that the phase shift between the interfering waves is proportional to the magnetic field enclosed by the paths of the interfering waves, *despite the fact that the electrons never move through the magnetic field* $\mathbf{B}$. Mathematically the reason for this surprising effect is that the vector potential is nonzero throughout space even when the magnetic field is confined to a small region of space, see (8.19) as an example. However, this explanation is purely mathematical and does not seem to

Fig. 8.9. Vortices trailing form the wingtips of a Boeing 727. Figure courtesy of NASA.

agree with common sense. This has led to speculations that the vector potential is actually a more 'fundamental' quantity than the magnetic field [79].

## 8.6 Wingtips vortices

If you have watched aircraft closely, you may have noticed that sometimes a little stream of condensation is left behind by the wingtips, see figure 8.9. This is a different condensation trail than the thick contrails created by the engines. The condensation trails that start at the wingtips are due to a vortex (a spinning motion of the air) that is generated at the wingtips. This vortex is called the wingtip vortex. In this section we use Stokes's law to see that this wingtip vortex is closely related to the lift that is generated by the airflow along a wing.

Let us first consider the flow along a wing, see figure 8.10. A wing can only generate lift when it is curved. In figure 8.10 the air traverses a longer path along the upper part of the wing than along the lower part. Because of the curved upper side of the wing, the velocity of the airstream along the upper part of the wing is larger than the velocity along the lower part. From Bernoulli's law, this is the reason that a wing generates lift. (For

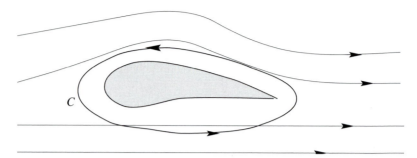

Fig. 8.10.   Sketch of the flow along an airfoil. The wing is shown in grey, the contour $C$ is shown by the thick solid line.

details of Bernoulli's law and other aspects of the flow along wings see ref. [89].)

**Problem a:** The *circulation* is defined as the line integral $\oint_C \mathbf{v} \cdot d\mathbf{r}$ of the velocity along a curve. Is the circulation positive or negative for the curve $C$ in figure 8.10 for the indicated sense of integration?

**Problem b:** Consider now the surface $S$ shown in figure 8.11. Show that the circulation satisfies

$$\oint_C \mathbf{v} \cdot d\mathbf{r} = \int_S \boldsymbol{\omega} \cdot d\mathbf{S} , \qquad (8.22)$$

where $\boldsymbol{\omega}$ is the vorticity. (See the sections 6.2–6.4.)

This expression implies that whenever lift is generated by the circulation along the contour $C$ around the wing, the integral of the vorticity over a surface that envelopes the wingtip is nonzero. The vorticity depends on the derivative of the velocity. Since the flow is relatively smooth along the wing, the derivative of the velocity field is largest near the wingtips. Therefore, expression (8.22) implies that vorticity is generated at the wingtips. As shown in section 6.3 the vorticity is a measure of the local vortex strength. A wing can only produce lift when the circulation along the curve $C$ is nonzero. The above reasoning implies that wingtip vortices are unavoidably associated with the lift produced by an airfoil.

**Problem c:** Consider the wingtip vortex shown in figure 8.11. You obtained the sign of the circulation $\oint_C \mathbf{v} \cdot d\mathbf{r}$ in problem a. Does this imply that the wingtip vortex rotates in the clockwise direction $A$ of figure 8.11 or in the counterclockwise direction $B$? Use equation (8.22) in your argumentation. You may assume that the vorticity is mostly concentrated at the trailing edge of the wingtips, see figure 8.11.

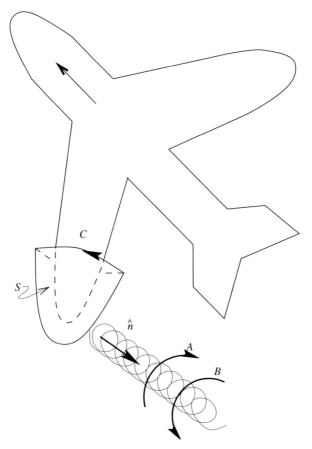

Fig. 8.11. Geometry of the surface $S$ and the wingtip vortex for an aircraft seen from above. The surface $S$ encloses the wingtip of the aircraft. The edge of this surface is the same contour $C$ as drawn in the previous figure.

The wingtip vortex carries kinetic energy. Modern aircraft such as the Boeing 747-400, the MD-11 and the Gulfstream 5 have wingtips that are turned upwards. These 'winglets' modify the vorticity at the wingtip in such a way that the induced drag on the aircraft is reduced.

Just like aircraft, sailing boats suffer from energy loss due a vortex that is generated at the upper part of the sail, see the discussion of Marchaj [53]. (A sail can be considered to be a 'vertical wing.') Consider the two boats shown in figure 8.12 that have sails with the same surface area but with different aspect ratios. The boat on the left will, in general, sail faster. The reason for this is that for the two boats the difference in the wind speed between the two sides of the sail will be roughly identical. This means that for the boat on the right the circulation $\oint_C \mathbf{v} \cdot d\mathbf{r}$ will be larger than that for the boat on the left, simply because the integration

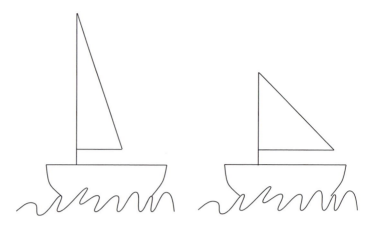

Fig. 8.12.   Two boats carrying sails with very different aspect ratios.

contour $C$ is longer. The vorticity generated at the top of the sail of the boat on the right is therefore larger than for the boat on the left. Since this resulting 'sailtip' vortex leads to a dissipation of energy, the sail of the boat on the left has a higher efficiency. For the same reason, planes that have to fly with a minimal energy loss (such as gliders) have thin and long wings. In contrast to this, planes that may waste energy in order to fly at a very high speed (such as the Concorde) have wings of a very different shape. Birds follow the same rule: birds that fly relatively slowly but that can glide efficiently over long distances (such as the albatross) have long and thin wings, whereas birds that do not need to fly efficiently (such as a crow) have shorter and thicker wings.

# 9

## The Laplacian

The Laplacian of a function consists of a special combination of the second partial derivatives of that function. Before we introduce this quantity, the relation between the second derivative and the curvature of a function is established in section 9.1. The Laplacian is introduced in section 9.3 using the physical example of a soap film that minimizes its surface area. The analysis used for this example is introduced in section 9.2 where a proof is given that the shortest distance between two points is a straight line. The concept of the Laplacian is used in section 9.5 to study the stability of matter, while in section 9.6 the implications for the initiation of lightning is considered. Finally, the Laplacian in cylindrical and spherical coordinates is derived, and this is used in section 9.8 to derive an averaging integral for harmonic functions.

### 9.1 The curvature of a function

Let us consider a function $f(x)$ that has an extremum. We are free to choose the origin of the $x$-axis, and the origin is chosen here at the location of the extremum. This means that the function $f(x)$ is stationary at the location $x = 0$, see figure 9.1. The behavior of the function near its maximum can be studied using the Taylor series (2.11)

$$f(x) = f(0) + \frac{df}{dx}(x = 0)\ x + \frac{1}{2}\frac{d^2 f}{dx^2}(x = 0)\ x^2 + \cdots \ . \qquad (2.11)$$

The point $x = 0$ is a maximum or a minimum which implies by definition that the first derivative vanishes at $x = 0$, so that the function behaves near the extremum as

$$f(x) = f(0) + \frac{1}{2}\frac{d^2 f}{dx^2}(x = 0)\ x^2 + \cdots \ . \qquad (9.1)$$

99

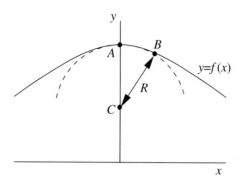

Fig. 9.1.   Definition of the geometric variables in the computation of the radius
of curvature of a function.

When the first derivative vanishes, the function can have either a maximum or a minimum. In more dimensions, the function can have a minimum in one direction and a maximum in another direction. For this reason
the terminology extremum is not very accurate. Instead, one refers to a
point at which the first derivative vanishes as a *stationary* point. This
term is chosen because at such a point, the function according to (9.1)
does not vary to first order with the independent variable. In more than
one dimension a stationary point is defined by the requirement that the
first partial derivatives with respect to *all* variables vanish.

The property of stationary points that the first derivative vanishes corresponds with the fact that at an extremum the slope of the function
vanishes. The dominant behavior of the function near its stationary point
is given by its *curvature* that is described by the $x^2$ term in (9.1). In order
to characterize this curvature one can define a circle as shown in figure 9.1
that touches the function $f(x)$ near its extremum. This *tangent-circle* is
shown by a dashed line in figure 9.1. The radius $R$ of this circle measures
the curvature of the function at its extremum. For this reason, $R$ is called
the *radius of curvature*.

**Problem a:** Convince yourself that a small radius of curvature $R$ implies
a large curvature of $f(x)$ and a large radius $R$ corresponds to a small
curvature of $f(x)$. What is the shape of $f(x)$ when the radius of
curvature tends to infinity ($R \to \infty$)? Hint: draw this situation.

The radius of curvature can be determined using the points $A$, $B$ and $C$
in figure 9.1.

**Problem b:** Use figure 9.1 to show that these points have the following
coordinates in the $(x, y)$-plane

$$\mathbf{r}_A = \begin{pmatrix} 0 \\ f(0) \end{pmatrix}, \quad \mathbf{r}_B = \begin{pmatrix} x \\ f(x) \end{pmatrix}, \quad \mathbf{r}_C = \begin{pmatrix} 0 \\ f(0) - R \end{pmatrix}. \quad (9.2)$$

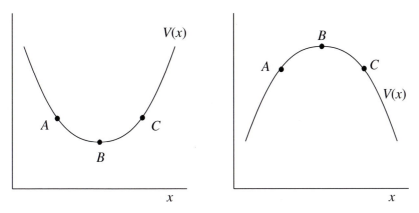

Fig. 9.2. The behavior of a function near a minimum and a maximum.

**Problem c:** Since $A$ and $B$ are located on the same circle with $C$ as center, the distance $AC$ equals the distance $BC$. Use this condition, the coordinates (9.2) and the Taylor expansion (9.1) to derive the following expession

$$x^2 + \left(\frac{1}{2}f''x^2 + R\right)^2 = R^2 \, , \qquad (9.3)$$

where $f''$ denotes the second derivative the extremum: $f'' = d^2 f/dx^2$.

By expanding the square one can obtain an expression for the radius of curvature $R$. However, when expanding the square, it is only necessary to account for terms up to order $x^2$. The reason for this is that in the Taylor expansion (9.1) the third order term $x^3$ as well as all the higher terms have been neglected. (This is because the tangent-circle in figure 9.1 approximates the function $f(x)$ well only near its maximum.) It is therefore not consistent to retain terms of third and higher order in the remainder of the calculation.

**Problem d:** Use this to derive from (9.3) the following relation between the radius of curvature and the second derivative

$$R = \frac{-1}{f''} \, . \qquad (9.4)$$

This expression relates the radius of the tangent-circle to the second derivative. The approximation of a function by its tangent-circle plays an important role in reflection seismology where it is used to derive the *15-degrees approximation* that accounts for near-vertical wave propagation in the Earth very efficiently [20] [103].

Of course, a stationary point can be either a minimum or a maximum: these two situations are shown in figure 9.2.

**Problem e:** Use (9.4) to show that for the curve in the left hand panel
(a minimum) the radius of curvature is negative and that for the
curve in the right hand panel (a maximum) the radius of curvature
is positive.

The result in the last problem of course depends critically on the following
property of a stationary point:

$$\left. \begin{array}{l} \text{when } f(x) \text{ is a minimum: } \dfrac{\partial^2 f}{\partial x^2} > 0 \text{ ;} \\[4mm] \text{when } f(x) \text{ is a maximum: } \dfrac{\partial^2 f}{\partial x^2} < 0 \text{ .} \end{array} \right\} \tag{9.5}$$

The curve in the left hand panel of figure 9.2 is called *concave*, while the
curve in the right hand panel is called *convex*. This implies that a concave
curve has a negative radius of curvature and a convex curve has positive
radius of curvature.

The distinction between these different kinds of extrema is crucial in
the stability properties of physical systems. Suppose that the function
denotes a potential $V(x)$, then according to (4.30) the force associated
with potential is given by

$$\mathbf{F}(\mathbf{x}) = -\nabla V(\mathbf{x}) \text{ .} \tag{4.30}$$

In one dimension this expression is given by

$$F(x) = -\frac{dV}{dx} \text{ .} \tag{9.6}$$

**Problem f:** Show that at the points $B$ in the panels of figure 9.2 the
force vanishes.

Suppose a particle that is influenced by the potential is at rest at one
of the points $B$ in figure 9.2. Since the force vanishes at these points the
particle will remain forever at that point. For this reason the points where
$\nabla V = 0$ are called the *equilibrium points*.

However, when we move the particle slightly away from the equilibrium
points, the force may either push it back towards the equilibrium point,
or push it further away. In the first case the equilibrium is *stable*, in the
second case it is *unstable*.

**Problem g:** Draw the direction of the force at the points $A$ and $C$ in the
two panels of figure 9.2 and deduce that the equilibrium in the left
hand panel is stable while the equilibrium in the right hand panel

is unstable. Use (9.5) to show that:

$$
\left.
\begin{array}{c}
\text{the equilibrium is stable when } \dfrac{\partial^2 V}{\partial x^2} > 0 \; ; \\[3mm]
\text{the equilibrium is unstable when } \dfrac{\partial^2 V}{\partial x^2} < 0 \; .
\end{array}
\right\}
\qquad (9.7)
$$

If you find these arguments difficult, you can think of a ball that can move along the curves in the panels of figure 9.2 in a gravitational field. At the point $B$ the ball is in an area that is flat. If it does not move it will remain at that point forever. When the ball in the left hand panel moves away from the equilibrium point, it rolls uphill and the gravitational force will send it back towards the equilibrium points $B$. However, when the ball in the right hand panel moves away from point $B$, it will roll downhill further from the equilibrium point. The equilibrium at point $B$ in the right hand panel is unstable. These properties are all related to the second derivative of a function.

## 9.2 The shortest distance between two points

In this section we give a proof that the shortest distance between two points is a straight line. This may appear to be a trivial problem, but it is shown here because it sets the stage for the Laplacian. It also gives a brief introduction to *variational calculus*.

**Problem a:** The shortest path between two points is only a straight line in a rectangular geometry. Take a globe and a rubber band. Stretch the band over the globe so that it passes across Amsterdam and San Francisco. The trajectory of the rubber band over the globe is not a straight line. (Instead it is a great circle.) Use this to explain why on a flight between North America and Europe one often flies over Greenland.

In this section we work in a rectangular geometry and expect the shortest curve between two points to be a straight line.

The problem is shown geometrically in figure 9.3. Two points $A$ and $B$ are given in the $(x, y)$-plane; these points are fixed. We are looking for the function $y = h(x)$ that describes the curve with the smallest length.

**Problem b:** We first need to determine the length of the curve given a certain shape $h(x)$. Consider an increment $dx$ of the $x$-variable. Show that this increment corresponds with an increment $dy =$

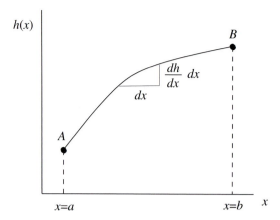

Fig. 9.3.    The relation between the derivative of a function and the arclength of
the corresponding curve.

$(dh/dx)\, dx$ of the $y$-variable. Use this result to derive that the length
$L$ of the curve is given by:

$$L(h) = \int_a^b \sqrt{1 + h_x^2}\, dx \; , \tag{9.8}$$

where $h_x = dh/dx$.

Note that the length of the curve is a function of the shape $h(x)$ of curve;
for this reason the notation $L(h)$ is used.

We want to find the function $h(x)$ that minimizes the length $L$. Unfortunately we cannot simply differentiate $L$ with respect to $h$ because
$h(x)$ is a function rather than a simple variable. However, we can use the
concept of stationarity that was introduced in section 9.1. When $L(h)$ is
minimized, the length of the curve does not change to first order when
$h(x)$ is perturbed. Consider figure 9.4 in which a perturbation $\varepsilon(x)$ is
added to the original function $h(x)$. Since the endpoints of the curve are
fixed, the perturbation is required to vanish at the endpoints:

$$\varepsilon(a) = \varepsilon(b) = 0 \; . \tag{9.9}$$

In order to solve the problem we need to find the change $\delta L$ in the length
of the curve that is caused by the perturbation $\varepsilon(x)$.

**Problem c:** Replace $h$ in (9.8) by $h + \varepsilon$, carry out a first order Taylor
expansion of the integrand with respect to $\varepsilon$ and use this to show
that the perturbation of the length of the curve is given by

$$\delta L(h) = \int_a^b \frac{h_x \varepsilon_x}{\sqrt{1 + h_x^2}}\, dx \; . \tag{9.10}$$

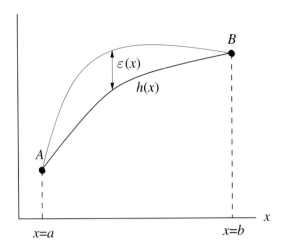

Fig. 9.4. The unperturbed function $h(x)$ and the perturbation $\varepsilon(x)$ that vanishes at the endpoint of the interval.

It order that the analysis is as transparent as possible we will make an approximation to (9.10) by restricting ourselves to the special case that the slope of curve is small. Since $h_x = dh/dx$ denotes the slope of the curve, this corresponds to the condition $h_x \ll 1$. In this case the term $h_x$ in the denominator can be ignored, and the perturbation is given by

$$\delta L(h) = \int_a^b h_x \varepsilon_x dx \ . \tag{9.11}$$

(The result that we obtain is not affected by this approximation; the only reason for making the approximation is that we don't want the principle of variational calculus to be hidden by analytical complexity.)

Expression (9.11) gives the first order perturbation of the length with respect to $\varepsilon(x)$. The condition of stationarity tells us that for the shortest curve, (9.11) must be equal to zero for *any* small perturbation $\varepsilon(x)$. However, we have not yet used the constraint (9.9) which states that the endpoints of the curve are fixed.

**Problem d:** Carry out an integration by parts of (9.11) and use the constraint (9.9) to show that the first order perturbation is given by

$$\delta L(h) = -\int_a^b \frac{d^2 h}{dx^2} \varepsilon \, dx \ . \tag{9.12}$$

For the function $h(x)$ that minimizes the length of the curve this integral must vanish for *all* perturbations $\varepsilon(x)$. This is the case when

$$\frac{d^2 h}{dx^2} = 0 \ . \tag{9.13}$$

This expression states that the curve that has the smallest length has no curvature, and it therefore is straight line.

**Problem e:** Let the $y$-coordinates of the points $A$ and $B$ be denoted by $y_A$ and $y_B$ respectively. Integrate (9.13) subject to the condition that the curve goes through the points $A$ and $B$ and convince yourself that the solution is indeed given by a straight line.

In the following section this minimization problem is generalized to two dimensions.

## 9.3 The shape of a soap film

In the last section you derived that the requirement that a curve between two points has the smallest length implies that its curvature vanishes. Suppose we consider a two-dimensional surface in three dimensions, and that we seek the surface that minimizes its surface area given the locations of the edges of the surface. Physically this problem is described by a soap film that is suspended within a wire frame. The surface tension of the soap film minimizes the surface area of the soap film subject to the constraint that the edges of the soap film are fixed. In this section, the shape of the soap film is described by giving the $z$-coordinate as a function of the $x$- and $y$-coordinates: $z = h(x, y)$.

At first sight one might guess that the soap film will be plane, because any deviations of the soap film from this plane will increase its surface area. Since the curvature of a plane vanishes in the $x$-direction as well as in the $y$-direction, the analogy with expression (9.13) suggests that the soap film satisfies the following equations: $\partial^2 h/\partial x^2 = \partial^2 h/\partial y^2 = 0$. That this is not in general true can be seen in the soap film shown in figure 9.5. In this example the wire frame that defines the edge of the soap film is not confined to a plane. This results in a soap film that is curved and hence it is not possible that both $\partial^2 h/\partial x^2$ and $\partial^2 h/\partial y^2$ are equal to zero. In this section we derive an equation for the shape of the soap film.

The position of a point on the soap film is given by the following position vector:

$$\mathbf{r} = \begin{pmatrix} x \\ y \\ h(x, y) \end{pmatrix}. \tag{9.14}$$

In problem e of section 3.4 you showed that the surface area $dS$ of a sphere that corresponds to increments $d\theta$ and $d\varphi$ of the angles on the sphere is given by

$$dS = \left| \frac{\partial \mathbf{r}}{\partial \theta} \times \frac{\partial \mathbf{r}}{\partial \varphi} \right| d\theta d\varphi . \tag{3.34}$$

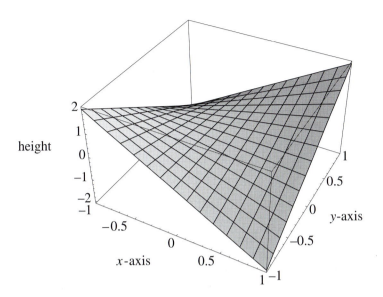

Fig. 9.5. The shape of a soap film whose edges are fixed at the outer edges of the box.

**Problem a:** Use the reasoning used in problem e of section 3.4 to show that increments of the surface area of the soap film satisfy the following expression

$$dS = \left| \frac{\partial \mathbf{r}}{\partial x} \times \frac{\partial \mathbf{r}}{\partial y} \right| dxdy \ . \tag{9.15}$$

**Problem b:** Use (9.14) and (9.15) to derive that the total surface area is given by

$$S = \iint \sqrt{1 + |\nabla h|^2} \, dxdy \ . \tag{9.16}$$

Note the analogy between this expression and (9.8).

Since the surface tension that governs the shape of the soap film tends to minimize the surface area of the soap film, the shape $h(x, y)$ follows from the requirement that $h(x, y)$ is the function that minimizes the surface area (9.16). As in section 9.2 the solution follows from the requirement that for the function $h(x, y)$ that minimizes the surface area, the surface area is stationary for perturbations of $h(x, y)$. This means that when $h(x, y)$ is replaced by $h(x, y) + \varepsilon(x, y)$ the first order change of the surface area $S$ vanishes.

**Problem c:** Make the substitution $h(x, y) \rightarrow h(x, y) + \varepsilon(x, y)$, use the identity $|\nabla h|^2 = (\nabla h \cdot \nabla h)$ in (9.16) and linearize the result in $\varepsilon(x, y)$ to derive that the perturbation of the surface area is to first order in $\varepsilon(x, y)$ given by

$$\delta S(h) = \iint \frac{(\nabla h \cdot \nabla \varepsilon)}{\sqrt{1 + |\nabla h|^2}} dx dy . \tag{9.17}$$

In order to concentrate on the essentials we assume that the deflection of the soap surface is small, which means that $|\nabla h| \ll 1$. Under this assumption the $|\nabla h|$ term in the denominator can be ignored and the perturbation is given by

$$\delta S(h) = \iint (\nabla h \cdot \nabla \varepsilon) \, dx dy . \tag{9.18}$$

In the problem we are considering here, the edge of the soap film is kept at a fixed location. This means that the perturbation $\varepsilon(x, y)$ must vanish at the edge of the soap film:

$$\varepsilon(x, y) = 0 \qquad \text{at the edge of the soap film .} \tag{9.19}$$

This constraint has not yet been taken into account; it is incorporated in the following two problems.

**Problem d:** Derive the identity $(\nabla h \cdot \nabla \varepsilon) = \nabla \cdot (\varepsilon \nabla h) - \varepsilon \nabla \cdot \nabla h$.

**Problem e:** Insert this result in (9.18), apply Gauss's theorem to the first term and use (9.19) to show that the resulting integral over the edge of the surface vanishes, so that the perturbation of the surface area is given by

$$\delta S(h) = -\iint \varepsilon \, (\nabla \cdot \nabla h) \, dx dy . \tag{9.20}$$

The second derivative $(\nabla \cdot \nabla h)$ is called the *Laplacian* of $h$. This operator is often denoted by the notation $\Delta$. However, since the term $\nabla \cdot \nabla$ is reminiscent of the square of the vector $\nabla$, the notation $\nabla^2$ is also used and in this book this latter notation is the one we will mostly use for the Laplacian. The requirement that the first order perturbation of the soap film vanishes implies that $(\nabla \cdot \nabla h)$ must be equal to zero. The soap film therefore satisfies the following differential equation:

$$\nabla^2 h = 0 . \tag{9.21}$$

In mathematical physics this equation is called the *Laplace equation*. Before analyzing this equation in more detail we first define the Laplacian more precisely.

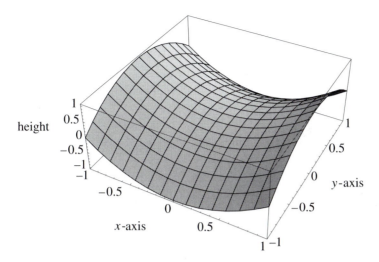

Fig. 9.6. The shape of the same soap film as in the previous figure after a rotation through 45 degrees. (New edges define the area of this soap film.) The shape of the soap film is given by the function $h(x, y) = x^2 - y^2$; the shape of the soap film in the previous figure is given by $h(x, y) = 2xy$.

**Problem f:** Use the definition $\nabla^2 = \nabla \cdot \nabla$ to show that

$$\nabla^2 = \operatorname{div} \operatorname{grad} = \frac{\partial^2}{\partial x^2} + \frac{\partial^2}{\partial y^2} \qquad \text{(in two dimensions)} . \qquad (9.22)$$

Analogously the Laplacian in three dimensions is defined as

$$\nabla^2 = \operatorname{div} \operatorname{grad} = \frac{\partial^2}{\partial x^2} + \frac{\partial^2}{\partial y^2} + \frac{\partial^2}{\partial z^2} \qquad \text{(in three dimensions)} . \qquad (9.23)$$

Let us now consider the shape of the soap film that is shown in figure 9.5. At the beginning of this section it was argued that the curvatures $\partial^2 h/\partial x^2$ and $\partial^2 h/\partial y^2$ of the soap film in the $x$- and $y$-directions could not both be equal to zero. Instead, (9.21) implies that the *sum* of the curvatures in the $x$- and $y$-directions vanishes. This can be seen by rotating the soap film through 45 degrees as shown in figure 9.6.

**Problem g:** Show that for the surface in figure 9.6 the curvature in the $x$-direction is positive and the curvature in the $y$-direction is negative.

**Problem h:** The surface in figure 9.6 is given by

$$h(x, y) = x^2 - y^2 . \qquad (9.24)$$

Show that for this surface $\nabla^2 h = 0$ and that the curvatures in the $x$-direction and in the $y$-direction cancel.

These results imply that the soap film behaves in a fundamentally different way than the rubber band between two fixed points that we treated in the previous section. The rubber band follows a straight line and the curvature is equal to zero while for the soap film the *sum* of the curvatures in orthogonal direction vanishes. However, as shown in figures 9.5 and 9.6 this does not imply that the soap film is a planar surface. The fact that the curvature terms $\partial^2 h/\partial x^2$ and $\partial^2 h/\partial y^2$ have opposite signs implies that when the function is concave in one direction it must be convex in the other direction. This means that when the function is a maximum in one direction, it must be a minimum in the other direction. The surface therefore has the shape of saddle.

Let us now consider a stationary point of the function: that is a point where $\partial h/\partial x = \partial h/\partial y = 0$. This point can be neither a minimum nor a maximum because when it is a minimum for variations in one direction, it *must* be a maximum for variations in the other direction. This means that:

**Theorem**  *A function that satisfies $\nabla^2 h = 0$ cannot have an extremum; the function can only have a maximum or minimum at the edge of the domain on which it is defined.*

We will see in section 9.5 that this has important consequences. When a function is a minimum all the second derivatives must analogously to (9.5) be positive, while for an extremum that is a maximum all the second derivatives must be negative.

**Problem i:** Show that:

$$\left. \begin{array}{l} \text{when } h(\mathbf{r}) \text{ is a minimum: } \nabla^2 h > 0 \;; \\[2mm] \text{when } h(\mathbf{r}) \text{ is a maximum: } \nabla^2 h < 0 \,. \end{array} \right\} \qquad (9.25)$$

**Problem j:** The above condition is a necessary condition but not a sufficient condition. (This means that every minimum must satisfy the requirement $\nabla^2 h > 0$, but conversely when $\nabla^2 h > 0$ is satisfied the function is not necessarily a minimum.) Show this by analyzing the stationary point of the function $h(x, y, z) = x^2 + y^2 - z^2$.

## 9.4 Sources of curvature

The main point of the previous section is that a function that satisfies the Laplace equation (9.21) can have nonzero curvature. In general there are two reasons why a soap film can be curved. The first reason is that the edges of the soap film are not confined to a common plane, as shown in figures 9.5 and 9.6. Since the edges are part of the soap film, this necessarily implies that the soap film cannot be confined to a plane.

There is another reason why a soap film can be curved. Let us consider a soap film whose edges are located in the $(x, y)$-plane:

$$h(x, y) = 0 \qquad \text{at the edge} . \qquad (9.26)$$

However, let us suppose that gravity pulls the soap film down. Such a soap film will bend down from the edges and will therefore be curved. This means that an external force can also lead to curvature of the soap film.

In this section we derive the differential equation for a soap film that is subject to a gravitational force. This is achieved by minimizing the total energy of the soap film. The gravitational energy is given by $\rho gh$, where $\rho$ is the mass-density per unit area and $g$ is the acceleration of gravity. The surface energy of the soap film is some positive constant $k$ times the surface area of the soap film.

**Problem a:** Use these results and (9.16) to show that the energy is given by:

$$E = \iint \left( k\sqrt{1 + |\nabla h|^2} + \rho gh \right) dx dy . \qquad (9.27)$$

The shape of the soap film is determined by the condition that the energy is a minimum.

**Problem b:** In the previous section we minimized the first term in the integrand. Generalize the derivation of the previous section to take the second term in (9.27) into account as well to derive that the soap film under gravity satisfies the following differential equation:

$$\nabla^2 h = \frac{\rho g}{k} . \qquad (9.28)$$

This equation is called *Poisson's equation*. The gravitational force has the effect that the total curvature of the soap film can be nonzero. This corresponds to the reasoning at the beginning of this section that gravity will make the soap film sag, so that total curvature can be nonzero despite the fact that the edges of the soap film are confined to a plane. Note that gravity acts as a source term in the differential equation (9.28).

There is an interesting analogy between the soap film in a gravitational field and the electric potential that is generated by electric charges. According to equation (5.13) the electric field generated by a charge density $\rho$ is given by

$$(\nabla \cdot \mathbf{E}) = \rho(\mathbf{r})/\varepsilon_0 . \qquad (5.13)$$

By analogy with (4.30) the potential associated with this electric field is given by

$$\mathbf{E} = -\nabla V \ . \tag{9.29}$$

**Problem c:** Show that the potential satisfies

$$\nabla^2 V = -\rho/\varepsilon_0 \ . \tag{9.30}$$

This means that the electric potential satisfies Poisson's equation (9.28) as well. The electric charge acts as the source of the electric potential just as the gravitational force acts as the source of the deflection of the soap film.

The analogy between the deflection of the soap film and the electric potential is interesting. Let us consider once more a soap film that is not subject to a gravitational force. The equation (9.28) of the soap film followed from the requirement that the squared-gradient $|\nabla h|^2$ integrated over the surface was minimized, because then the term $\sqrt{1 + |\nabla h|^2}$ in (9.16) is minimized is well. The analogy with the electric potential means that in free space ($\rho = 0$) the electric potential behaves in such a way that the squared-gradient $|\nabla V|^2$ of the potential integrated over the volume is minimized. However, according to (9.29) the gradient of the electric potential is the electric field. This means that the electric field behaves in such a way that the volume integral $\int |\mathbf{E}|^2 \, d^3x$ is minimized. However, this quantity is nothing but the energy of a static electric field [42]. This implies that the electric field is distributed in such a way that the energy of the electric field is minimized.

## 9.5 The instability of matter

The title of this section may surprise you, but the results that will be obtained here imply that according to the laws of classical physics the structure of matter and the mass distribution in the universe cannot be in a stable equilibrium. This result was derived by Earnshaw in 1842 in his work 'On the nature of the molecular forces which regulate the constitution of the luminiferous ether' [28]. Let us consider a particle that is in three dimensions subject to a potential $V(\mathbf{r})$ that accounts for the gravitational attraction by other masses and the electrostatic force due to other charges. Let us suppose that at some point in space the particle is in equilibrium. This means that the force acting on the particle vanishes at that point, or equivalently that the gradient of the potential vanishes at that point: $\nabla V = 0$.

**Problem a:** Generalize expression (9.7) to three dimensions to show that this equilibrium point is only stable when the curvature of the potential in the three coordinate directions is positive:

$$\frac{\partial^2 V}{\partial x^2} > 0 \quad \text{and} \quad \frac{\partial^2 V}{\partial y^2} > 0 \quad \text{and} \quad \frac{\partial^2 V}{\partial z^2} > 0 . \tag{9.31}$$

According to (9.30), in free space ($\rho = 0$) the potential that corresponds to the gravitational force exerted by other masses, and the electrostatic force generated by other charges, satisfies Laplace's equation $\nabla^2 V = 0$.

**Problem b:** Show that Laplace's equation implies that when the curvature of the potential is positive in one direction, the curvature must be negative in at least one other direction.

**Problem c:** Use this to deduce that when the equilibrium point is stable to perturbations in one direction, it must be unstable for perturbations in at least one other direction.

An equilibrium point is in general unstable when it is unstable for perturbations in at least one of the directions. Using problem c this means that an equilibrium point for the potential $V(\mathbf{r})$ that satisfies Poisson's equation cannot be stable.

This has far-reaching consequences. Let us consider a crystal. Within the framework of classical physics, each ion in the crystal moves in an electric potential that is generated by all the other ions in the crystal. This potential satisfies Laplace's equation at the location of the ion that we are considering because the net charge density $\rho$ of the *other* ions is zero at that point. This means that the motion of the ion at its equilibrium point is not stable. When this is the case, the crystal is not stable because small perturbations of each ion from its equilibrium point lead to unstable motions of the ions. This implies that according to the laws of classical physics, matter is not stable! This result, known as *Earnshaw's theorem* [28], states that the equilibrium points of any configuration of static electrical, magnetic and gravitational fields are unstable.

Hopefully you are convinced that matter is stable, whatever the unsettling prediction of Earnshaw's theorem may be. The only way that we can resolve this paradox is if matter does not satisfy the equations of classical electrostatics. This implies that quantum effects must play a crucial role in the stability of matter.

Earnshaw's theorem also applies to the gravitational field and it states that equilibrium points in the gravitational field are not stable. This means that a universe in equilibrium would not be stable! Let us first consider our solar system. It is essential for the solar system that the planets and their moons are in motion. This means that the solar system is not in a static equilibrium. In fact, the motion of the planets and moons is governed by the combination of the gravitational force plus the inertia force that is associated with the motion of the planets. The miracle of planetary motion is that the gravitational *motion* is stable. (You showed in section 5.5 that the motion of planetary orbits is only stable in less than four spatial dimensions.) The situation is comparable to the movement of a bicycle which derives its stability from the motion of the wheels that rotate. In the same way, it is the combination of the gravitational field and the inertial forces due to the motion of the planets that leads to stable planetary orbits. On scales much larger than the solar system relativistic effects are important and Poisson's equation is not sufficient to describe the gravitational field on a cosmological scale.

In 1842, Earnshaw [28] carried out his work not to study the structure of matter, but to study the structure of the ether which at that time was seen as the carrier of electromagnetic waves. He viewed the ether as a system of interacting particles and showed that the interaction between these particles could not be governed by Newton's law of gravitation using the analysis shown in this section. From the requirement of stability he derived conditions for the potential that governs the interaction between etheral particles. Note that in 1842 Earnshaw could not invoke quantum mechanics in his description of interacting microscopic particles because that theory had not yet been formulated.

## 9.6 Where does lightning start?

During a thunderstorm, the vertical motion of ice particles causes the separation of positive and negative electric charges in the atmosphere. This charge separation leads to an electric field within the atmosphere. When the field strength at a certain location is sufficiently strong, atoms are ionized and a current flows. This current induces further ionization which extends the path of the current. This physical process describes the events that initiate a lightning bolt. In principle there seems to be no reason why a lightning bolt cannot start in mid-air, far away from the Earth's surface and far from the electrical charges that induce the electric field. In this section we will show that lightning can only start at very specific locations.

To see this we will study the Laplacian of the square of the length of a vector field **E**.

**Problem a:** Compute the Laplacian of $E^2 = E_x^2 + E_y^2 + E_z^2$ and show that is it given by

$$\nabla^2 E^2 = 2 \left( |\nabla E_x|^2 + |\nabla E_y|^2 + |\nabla E_z|^2 \right) + 2 \left( \mathbf{E} \cdot \nabla E \right). \quad (9.32)$$

(Up to this point the Laplacian has always acted on a scalar; when it acts on a vector as in (9.32) it means that the Laplacian of every component is taken so that $\nabla^2 \mathbf{E}$ stands for a vector with components $\nabla^2 E_x$, $\nabla^2 E_y$ and $\nabla E_z$, respectively.)

Let us now consider a vector field $\mathbf{E}$ that satisfies the Laplace equation:

$$\nabla^2 \mathbf{E} = 0. \tag{9.33}$$

**Problem b:** Show that a vector field that satisfies the Laplace equation (9.33) satisfies the following inequality:

$$\nabla^2 E^2 \geq 0. \tag{9.34}$$

**Problem c:** Use this to show that the strength of the vector field $(E^2)$ cannot have a maximum in the region where (9.33) is satisfied.

Note that (9.34) does not exclude the possibility that $E^2$ has a minimum. However, remember that $E^2$ is the (squared) length of a vector, it therefore satisfies the inequality $E^2 \geq 0$ as well.

The result you derived in problem c is all we need to solve our problem. Away from the charges in the atmosphere, a static electric field satisfies (9.33), see ref. [42]. According to problem c, the electric field strength $E^2$ cannot have a maximum in the region of the atmosphere where there are no charges that are the source of the field. Lightning will initiate where the field strength is strongest. This means that lightning must initiate either in the regions where Laplace equation (9.33) does not hold (at the charges that induce the electric field), or at the boundary of the area (the Earth's surface). This implies that lighting must start either at the charges that induce the field or at the Earth's surface.

**Problem d:** The source of the magnetic field of the Earth is located in the Earth's core. Away from this source, the magnetic field satisfies Laplace equation (9.33) (at least when the field is stationary)[42]. Show that the magnetic field of the Earth cannot have its maximum strength outside the Earth.

## 9.7 The Laplacian in spherical and cylindrical coordinates

In many applications it is useful to use the Laplacian in spherical or cylindrical coordinates. In principle this result can be derived by applying the transformation rules to the second partial derivatives in the Laplacian when changing from Cartesian to spherical or cylindrical coordinates. However, this route is unnecessarily complex, especially since we have done most of the work already. The key element in the derivation was derived in problem f of section 9.3 where you showed that the Laplacian is the divergence of the gradient: $\nabla^2 = \operatorname{div}\operatorname{grad}$. In sections 4.6 and 5.4 you have already derived expressions for the gradient and the divergence in spherical and cylindrical coordinates, and all you need to do is to insert the expression for the gradient in curvilinear coordinates into the expression of the divergence in curvilinear coordinates.

**Problem a:** Use (4.47) and (5.16) to show that the Laplacian of a function $f$ in cylindrical coordinates is given by

$$\nabla^2 f = \frac{1}{r}\frac{\partial}{\partial r}\left(r\frac{\partial f}{\partial r}\right) + \frac{1}{r^2}\frac{\partial^2 f}{\partial \varphi^2} + \frac{\partial^2 f}{\partial z^2}\,. \tag{9.35}$$

**Problem b:** Find the expressions in the sections 4.6 and 5.4 that allow you to derive that the Laplacian of a function $f$ in spherical coordinates is given by

$$\nabla^2 f = \frac{1}{r^2}\frac{\partial}{\partial r}\left(r^2\frac{\partial f}{\partial r}\right) + \frac{1}{r^2\sin\theta}\frac{\partial}{\partial \theta}\left(\sin\theta\frac{\partial f}{\partial \theta}\right) + \frac{1}{r^2\sin^2\theta}\frac{\partial^2 f}{\partial \varphi^2}\,. \tag{9.36}$$

In later parts of this book extensive use is made of these expressions of the Laplacian.

## 9.8 Averaging integrals for harmonic functions

In this section we focus on functions $f$ that satisfy Laplace's equation (9.21), i.e. $\nabla^2 f = 0$. Functions whose Laplacian is zero are called *harmonic functions*. They play an important role in mathematical physics. We will show in section 15.1 that the real and imaginary parts of analytic functions in the complex plane are harmonic functions. Let us first focus on a harmonic function $f(x, y)$ in two dimensions. In this section we derive that the function value at a certain point is equal to the average of that function over a circle centered around that point with an *arbitrary* radius. To see this we use a system of cylindrical coordinates and choose the origin of the cylindrical coordinates at the point that we consider. (Remember that we are free to choose the origin of the coordinate system where we want.)

**Problem a:** Show that $f$ satisfies the following differential equation:

$$\frac{1}{r}\frac{\partial}{\partial r}\left(r\frac{\partial f}{\partial r}\right) + \frac{1}{r^2}\frac{\partial^2 f}{\partial \varphi^2} = 0\,. \tag{9.37}$$

**Problem b:** Integrate this expression over a disk of radius $R$ centered at the origin and derive that

$$\int_0^R \int_0^{2\pi} \left[ \frac{\partial}{\partial r} \left( r \frac{\partial f}{\partial r} \right) + \frac{1}{r} \frac{\partial^2 f}{\partial \varphi^2} \right] d\varphi dr = 0 .$$ (9.38)

(Note the powers of $r$ in this expression.)

**Problem c:** Carry out the $\varphi$-integration in the last term of the integrand to show that this term gives a vanishing contribution.

It is convenient to introduce at this point the average $\bar{f}(r)$ of the function $f$ over a circle with radius $r$:

$$\bar{f}(r) \equiv \frac{1}{2\pi} \int_0^{2\pi} f(r, \varphi) \, d\varphi .$$ (9.39)

**Problem d:** Use (9.38) and the result of problem c to derive that $\bar{f}$ satisfies the following equation:

$$\left[ r \frac{\partial \bar{f}}{\partial r} \right]_{r=0}^{r=R} = 0 .$$ (9.40)

**Problem e:** This expression holds for any radius $R$, hence $r \partial \bar{f} / \partial r$ is independent of $r$, so that it is a constant: $r \partial \bar{f} / \partial r = C$. Integrate this expression to derive that

$$\bar{f}(r) = C \, \ln r + A ,$$ (9.41)

where $A$ is an unknown integration constant.

**Problem f:** The integration constant $C$ must be equal to zero because $\bar{f}(r)$ is finite as $r \to 0$. Evaluate (9.41) at the origin, determine the constant $A$ and show that $f(r)$ satisfies

$$f(r = 0) = \frac{1}{2\pi} \int_0^{2\pi} f(r, \varphi) \, d\varphi .$$ (9.42)

This expression states that the value of a harmonic function at the origin is the average of the function over a circle with arbitrary radius. The amazing property is that this holds for any value of the radius, provided $f$ is harmonic everywhere within the circle with radius $r$. Note that we are free to choose the origin of the coordinate system so that the averaging integral (9.42) holds for any point.

The averaging integral can also be used to prove that a harmonic function cannot have a minimum or a maximum in the region where it is defined. Suppose the function had a maximum at a certain location. There then exists a circle around this point where the function has smaller values than at the maximum (otherwise it would not be a maximum). Equation (9.42) cannot hold for this circle because the right hand side would be smaller than the left hand side. Therefore it is impossible for $f$ to have a maximum.

**Problem g:** Generalize the derivation in this section to spherical coordinates and derive the following expression

$$f(r = 0) = \frac{1}{4\pi} \int_0^\pi \int_0^{2\pi} f(r, \theta, \varphi) \sin \theta \, d\varphi d\theta \, . \tag{9.43}$$

**Problem h:** Show that the right hand side of this expression is the average of that function over a sphere with arbitrary radius $r$ centered around that point $r = 0$.

This means that, in three dimensions, the value of a harmonic function at a certain point is equal to the average of the function over a sphere with arbitrary radius $r$ that is centered on that point.

# 10

---

# Conservation laws

In physics one frequently handles the change of a property with time by considering properties that do *not* change with time. For example, when two particles collide elastically, the momentum and the energy of each particle may change. However, this change can be found from the consideration that the total momentum and energy of the system are conserved. Often in physics, such conservation laws are the main ingredients for describing a system. In this chapter we deal with conservation laws for continuous systems. These are systems in which the physical properties are a continuous function of the space coordinates. Examples are the motion in a fluid or solid, and the temperature distribution in a body. The introduced conservation laws are not only of great importance in physics, they also provide worthwhile exercises in the use of vector calculus introduced in the previous chapters.

## 10.1 The general form of conservation laws

In this section a general derivation of conservation laws is given. Suppose we consider a physical quantity $Q$. This quantity could denote the mass density of a fluid, the heat content within a solid or any other type of physical variable. In fact, there is no reason why $Q$ should be a scalar, it could also be a vector (such as the momentum density) or a higher order tensor. Let us consider a volume $V$ in space that does not change with time. This volume is bounded by a surface $\partial V$. The total amount of $Q$ within this volume is given by the integral $\int_V QdV$. The rate of change of this quantity with time is given by $(\partial/\partial t)\int_V QdV$.

In general, there are two reasons for the quantity $\int_V QdV$ to change with time. First, the field $Q$ may have sources or sinks within the volume $V$. The net source of the field $Q$ per unit volume is denoted by the symbol $S$. The total source of $Q$ within the volume is simply the volume integral

$\int_V S dV$ of the source density. Second, it may be that the quantity $Q$ is transported in the medium. With this transport process is associated, a current $\mathbf{J}$ of the quantity $Q$. As an example one can think of $Q$ as being the mass-density of a fluid. In that case $\int_V Q dV$ is the total mass of the fluid in the volume. This total mass can change because there is a source of fluid within the volume (i.e. a tap or a bathroom sink), or the total mass may change because of the flow through the boundary of the volume.

The rate of change of $\int_V Q dV$ by the current is given by the *inward* flux of the current $\mathbf{J}$ through the surface $\partial V$. If we retain the convention that the surface element $d\mathbf{S}$ points out off the volume, the inward flux is given by $-\oint_{\partial V} \mathbf{J} \cdot d\mathbf{S}$. Together with the rate of change due to the source density $S$ within the volume this implies that the rate of change of the total amount of $Q$ within the volume satisfies:

$$\frac{\partial}{\partial t} \int_V Q dV = -\oint_{\partial V} \mathbf{J} \cdot d\mathbf{S} + \int_V S dV . \tag{10.1}$$

Using Gauss's law (7.1), the surface integral on the right hand side can be written as $-\int_V (\nabla \cdot \mathbf{J}) dV$, so that the expression above is equivalent with

$$\frac{\partial}{\partial t} \int_V Q dV + \int_V (\nabla \cdot \mathbf{J}) dV = \int_V S dV . \tag{10.2}$$

Since the volume $V$ is assumed to be fixed with time, the time-derivative of the volume integral is the volume integral of the time-derivative: $(\partial/\partial t) \int_V Q dV = \int_V (\partial Q/\partial t) dV$. It should be noted that (10.2) holds for *any* volume $V$. If the volume is an infinitesimal volume, the volume integrals in (10.2) can be replaced by the integrand multiplied by the infinitesimal volume. This means that (10.2) is equivalent to:

$$\frac{\partial Q}{\partial t} + (\nabla \cdot \mathbf{J}) = S . \tag{10.3}$$

This is the general form of a conservation law in physics; it simply states that the rate of change of a quantity is due to the sources (or sinks) of that quantity and due to the divergence of the current of that quantity. Of course, the general conservation law (10.3) is not very meaningful as long as we don't provide expressions for the current $\mathbf{J}$ and the source $S$. In this section we will see some examples in which the current and the source follow from physical theory, but we will also encounter examples in which they follow from an 'educated' guess.

Equation (10.3) will not be completely new to you. In section 7.4 the probability-density current for a quantum-mechanical system was derived.

**Problem a:** Use the derivation in this section to show that expression

(7.16) can be written as

$$\frac{\partial}{\partial t}|\psi|^2 + (\nabla \cdot \mathbf{J}) = 0 , \tag{10.4}$$

with $\mathbf{J}$ given by (7.17).

This equation constitutes a conservation law for the probability density of a particle. Note that (7.16) could be derived rigorously from the Schrödinger equation (7.14), so that the conservation law (10.4) and the expression for the current $\mathbf{J}$ follow from the basic equation of the system.

**Problem b:** Why is the source term on the right hand side of (10.4) equal to zero?

## 10.2 The continuity equation

In this section we consider the conservation of mass in a continuous medium such as a fluid or a solid. In that case, the quantity $Q$ is the mass-density $\rho$. If we assume that mass is neither created nor destroyed, the source term vanishes: $S = 0$. The vector $\mathbf{J}$ is the mass current and denotes the flow of mass per unit volume. Consider a small volume $\delta V$. The mass within this volume is equal to $\rho \delta V$. If the velocity of the medium is denoted by $\mathbf{v}$, the mass flow is given by $\rho \delta V \mathbf{v}$. Dividing this by the volume $\delta V$ one obtains the mass flow per unit volume; this quantity is called the mass-density current:

$$\mathbf{J} = \rho \mathbf{v} . \tag{10.5}$$

Using these results, the principle of the conservation of mass can be expressed as

$$\frac{\partial \rho}{\partial t} + \nabla \cdot (\rho \mathbf{v}) = 0 . \tag{10.6}$$

This expression plays a very important role in continuum mechanics and is called the *continuity equation*.

Up to this point the reasoning has been based on a volume $V$ that did not change with time. This means that our treatment was strictly Eulerian: we considered the change of physical properties at a fixed location. As an alternative, a Lagrangian description of the same process can be given. In such an approach one specifies how physical properties change as they are moved along by the flow. In this approach one seeks an expression for the total time-derivative $d/dt$ of physical properties rather than expressions for the partial derivative $\partial/\partial t$. It follows from (4.39) that these two derivatives are related in the following way:

$$\frac{d}{dt} = \frac{\partial}{\partial t} + (\mathbf{v} \cdot \nabla) . \tag{10.7}$$

This distinction between the total time-derivative and the partial time-derivative is treated in detail in section 4.5.

**Problem a:** Show that the total derivative of the mass-density is given by:

$$\frac{d\rho}{dt} + \rho(\nabla \cdot \mathbf{v}) = 0 \ . \tag{10.8}$$

**Problem b:** This expression gives the change in the density when one follows the flow. Let us consider a infinitesimal volume $\delta V$ that is carried around with the flow. The mass of this volume is given by $\delta m = \rho \delta V$. The mass within that volume is conserved (why?), so that $\delta \dot{m} = 0$. (The dot denotes the total time-derivative, *not* the partial time-derivative.) Use this expression and (10.8) to show that $(\nabla \cdot \mathbf{v})$ is the rate of change of the volume normalized by size of the volume:

$$\frac{\delta \dot{V}}{\delta V} = (\nabla \cdot \mathbf{v}) \ . \tag{10.9}$$

We have learned a new meaning of the divergence of the velocity field: it equals the relative change in volume per unit time.

## 10.3 Conservation of momentum and energy

In the description of a point mass in classical mechanics, the conservation of momentum and energy can be derived from Newton's third law. The same is true for a continuous medium such as a fluid or a solid. In order to formulate Newton's law for a continuous medium we start with a Lagrangian point of view and consider a volume $\delta V$ *that moves with the flow*. The mass of this volume is given by $\delta m = \rho \delta V$. This mass is constant because the volume is defined to move with the flow, hence mass cannot flow in or out of the volume. Let the force per unit volume be denoted by $\mathbf{F}$, so that the total force acting on the volume is $\mathbf{F}\delta V$. The force $\mathbf{F}$ contains both forces generated by external agents (such as gravity) and internal agents such as the pressure force $-\nabla p$ or the effect of internal stresses $(\nabla \cdot \boldsymbol{\sigma})$. The stress tensor $\boldsymbol{\sigma}$ is treated in section 21.10. Newton's law applied to the volume $\delta V$ takes the form:

$$\frac{d}{dt}(\rho \delta V \mathbf{v}) = \mathbf{F}\delta V \ . \tag{10.10}$$

Since the mass $\delta m = \rho \delta V$ is constant with time it can be taken outside the derivative in (10.10). Dividing the resulting expression by $\delta V$ leads to the Lagrangian form of the equation of motion:

$$\rho \frac{d\mathbf{v}}{dt} = \mathbf{F} \ . \tag{10.11}$$

Note that the density appears *outside* the time-derivative, despite the fact that the density may vary with time. Using the prescription (10.7) one obtains the Eulerian form of Newton's law for a continuous medium:

$$\rho \frac{\partial \mathbf{v}}{\partial t} + \rho \mathbf{v} \cdot \nabla \mathbf{v} = \mathbf{F} \ . \tag{10.12}$$

This equation is not yet in the general form (10.3) of conservation laws because in the first term on the left hand side we have the density *times* a time-derivative, and because the second term on the left hand side is not the divergence of some current.

**Problem a:** Use (10.12) and the continuity equation (10.6) to show that:

$$\frac{\partial(\rho \mathbf{v})}{\partial t} + \nabla \cdot (\rho \mathbf{v} \mathbf{v}) = \mathbf{F} \ . \tag{10.13}$$

This expression does take the form of a conservation law; it expresses that the momentum (density) $\rho \mathbf{v}$ is conserved. (For brevity we will often not include the affix 'density' in the description of the different quantities, but remember that all quantities are given per unit volume.) The source of momentum is given by the force $\mathbf{F}$, and this reflects that forces are the cause of changes in momentum. In addition there is a momentum current $\mathbf{J} = \rho \mathbf{v} \mathbf{v}$ that describes the transport of momentum by the flow. This momentum current is not a simple vector, it is a dyad and hence is represented by a $3 \times 3$ matrix. This is not surprising since the momentum is a vector with three components and each component can be transported in three spatial directions.

Sometimes there is a certain arbitrariness in what we call the current and what we call the source. As an example let us consider (10.13) again. According to (4.11) the pressure force is given by $\mathbf{F} = -\nabla p$. Suppose that there are no other forces acting on the fluid, the right hand side of (10.13) is given by $-\nabla p$. This term does not have the form of the divergence, but we can rewrite it by using that:

$$\frac{\partial p}{\partial x_i} = \sum_j \frac{\partial}{\partial x_j} (p \delta_{ij}) \ , \tag{10.14}$$

where $\delta_{ij}$ is the Kronecker delta. This quantity is defined as follows:

$$\delta_{ij} = \begin{cases} 1 & \text{when} & i = j \\ 0 & \text{when} & i \neq j \end{cases} \ . \tag{10.15}$$

The matrix elements of the identity matrix $\mathbf{I}$ are given by the Kronecker delta: $I_{ij} = \delta_{ij}$.

**Problem b:** Show that expression (10.14) may be written in vector form
as $\nabla p = \nabla \cdot (p\mathbf{I})$. Show that the law of conservation of momentum
is given by

$$\frac{\partial(\rho\mathbf{v})}{\partial t} + \nabla \cdot \mathbf{S} = \mathbf{0} , \qquad (10.16)$$

where $\mathbf{S}$ is defined by

$$\mathbf{S} = \rho\mathbf{v}\mathbf{v} + p\mathbf{I} . \qquad (10.17)$$

This quantity describes the *radiation stress*[9] which accounts for the internal stress in an acoustic medium that is generated by the waves that
propagate through the medium. Of course, when other external forces are
present, they will lead to a right hand side of (10.16) that is nonzero. This
example shows that it can be arbitrary whether a certain physical effect
is accounted for by a source term or by a current. This arbitrariness is
caused by the fact that it is not clear what is internal to the system and
what is external. In (10.13) the pressure force is treated as an external
force whereas in (10.17) the pressure contributes to a current within the
system. There is no objective reason to prefer one or the other of the two
formulations.

You may find the inner products of vectors and the $\nabla$-operator in expressions such (10.12) confusing, and indeed a notation such as $\rho\mathbf{v} \cdot \nabla\mathbf{v}$
can be a source of error and confusion. When working with quantities like
this it is clearer to explicitly write out the components of all vectors or
tensors. In component form an equation such as (10.12) is written as:

$$\rho\frac{\partial v_i}{\partial t} + \sum_j \rho v_j \partial_j v_i = F_i . \qquad (10.18)$$

**Problem c:** Rewrite the continuity equation (10.6) in component form
and redo the derivation of problem a with all equations in component form to arrive at the conservation law of momentum in component form:

$$\frac{\partial(\rho v_i)}{\partial t} + \sum_j \partial_j(\rho v_j v_i) = F_i . \qquad (10.19)$$

In order to derive the law of energy conservation we start by deriving
the conservation law for the kinetic energy (density)

$$E_K = \frac{1}{2}\rho v^2 = \sum_i \frac{1}{2}\rho v_i v_i . \qquad (10.20)$$

**Problem d:** Express the partial time-derivative $\partial(\rho v^2)/\partial t$ in the time-
derivatives $\partial(\rho v_i)/\partial t$ and $\partial v_i/\partial t$, use (10.18) and (10.19) to elimi-

nate these time-derivatives and write the final results as:

$$\sum_i \frac{\partial(\frac{1}{2}\rho v_i v_i)}{\partial t} = -\sum_{i,j} \partial_j \left(\frac{1}{2}\rho v_i v_i v_j\right) + \sum_j v_j F_j .$$ (10.21)

**Problem e:** Use definition (10.20) to rewrite the expression above as the conservation law of kinetic energy:

$$\frac{\partial E_K}{\partial t} + \nabla \cdot (\mathbf{v} E_K) = (\mathbf{v} \cdot \mathbf{F}) .$$ (10.22)

This equation states that the kinetic energy current is given by $\mathbf{J} = \mathbf{v} E_K$, and this term describes how kinetic energy is transported by the flow. The term $(\mathbf{v} \cdot \mathbf{F})$ on the right hand side denotes the source of kinetic energy. It was shown in section 4.4 that $(\mathbf{v} \cdot \mathbf{F})$ is the power delivered by the force $\mathbf{F}$. This means that (10.22) states that the power produced by the force $\mathbf{F}$ is the source of kinetic energy.

In order to invoke the potential energy as well we assume for the moment that the force $\mathbf{F}$ is the gravitational force. Suppose there is a gravitational potential $V(\mathbf{r})$, then the gravitational force is given by

$$\mathbf{F} = -\rho \nabla V ,$$ (10.23)

and the potential energy $E_P$ is given by

$$E_P = \rho V .$$ (10.24)

**Problem f:** Take the (partial) time-derivative of (10.24), use the continuity equation (10.6) to eliminate $\partial \rho / \partial t$, use that the potential $V$ does not depend explicitly on time and employ (10.23) and (10.24) to derive the conservation law of potential energy:

$$\frac{\partial E_P}{\partial t} + \nabla \cdot (\mathbf{v} E_P) = -(\mathbf{v} \cdot \mathbf{F}) .$$ (10.25)

Note that this conservation law is very similar to the conservation law (10.22) for kinetic energy. The meaning of the second term on the left hand side will be clear to you by now; it denotes the divergence of the current $\mathbf{v} E_P$ of potential energy. Note that the right hand side of (10.24) has the opposite sign to the right hand side of (10.22). This reflects the fact that when the force $\mathbf{F}$ acts as a source of kinetic energy, it acts as a sink of potential energy; the opposite signs imply that kinetic and potential energy are converted into each other. However, the total energy $E = E_K + E_P$ should have no source or sink.

**Problem g:** Show that the total energy is source-free:

$$\frac{\partial E}{\partial t} + \nabla \cdot (\mathbf{v} E) = 0 .$$ (10.26)

## 10.4 The heat equation

In the previous section we saw that the momentum and energy current
could be derived from Newton's law. Such a rigorous derivation is not
always possible. In this section the transport of heat is treated, and we will
see that the law for heat transport cannot be derived rigorously. Consider
the general conservation equation (10.3), where $T$ is the temperature.
(Strictly speaking we should derive the heat equation using a conservation
law for the thermal energy rather than the temperature. The thermal
energy is given by $CT$, with $C$ the heat capacity. When the specific heat is
constant, the distinction between thermal energy and temperature implies
multiplication by a constant, but for simplicity this multiplication is left
out here.)

The source term in the conservation equation is simply the amount
of heat (normalized by the heat capacity) supplied to the medium. An
example of such a source is the decay of radioactive isotopes that forms
a major source in the heat budget of the Earth. The transport of heat is
affected by the heat current $\mathbf{J}$. In the Earth, heat can be transported by
two mechanisms: heat conduction and heat advection. The first process is
similar to the process of diffusion; it accounts for the fact that heat flows
from warm regions to colder regions. The second process accounts for the
heat that is transported by the flow field $\mathbf{v}$ in the medium. Therefore, the
current $\mathbf{J}$ can be written as a sum of two components:

$$\mathbf{J} = \mathbf{J}^{conduction} + \mathbf{J}^{advection} \ . \tag{10.27}$$

The heat advection is given by

$$\mathbf{J}^{advection} = \mathbf{v}T \ , \tag{10.28}$$

which reflects that heat is simply carried around by the flow. This view-
point of the process of heat transport is in fact too simplistic in many
situations. Fletcher [33] describes how the human body loses heat during
outdoor activities through four processes: conduction, advection, evaporation and radiation. He describes in detail the conditions under which
each of these processes dominate, and how the associated heat loss can be
reduced. In the physics of the atmosphere, energy transport by radiation
and by evaporation (or condensation) also plays a crucial role.

For the moment we focus on heat conduction. In general, heat flows
from warm regions to colder regions. The vector $\nabla T$ points from cold
regions to warmer regions. It is therefore logical that the heat conduction
points in the opposite direction from the temperature gradient:

$$\mathbf{J}^{conduction} = -\kappa \nabla T \ , \tag{10.29}$$

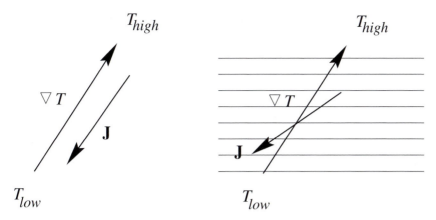

Fig. 10.1. Heat flow and temperature gradient in an isotropic medium (left hand panel) and in a medium consisting of alternating layers of copper and styrofoam (right hand panel).

see the left hand panel of figure 10.1. The constant $\kappa$ is the heat conductivity. (For a given value of $\nabla T$ the heat conduction increases when $\kappa$ increases, hence it indeed measures the conductivity.) However, the simple law (10.29) does not hold for every medium. Consider a medium consisting of alternating layers of a good heat conductor (such as copper) and a poor heat conductor (such as styrofoam). In such a medium the heat will be preferentially transported along the planes of the good heat conductor and the conductive heat flow $\mathbf{J}^{conduction}$ and the temperature gradient are not antiparallel, see the right hand panel in figure 10.1. In that case there is a matrix operator that relates $\mathbf{J}^{conduction}$ and $\nabla T$: $J_i^{conduction} = -\sum_j \kappa_{ij}\partial_j T$, with $\kappa_{ij}$ the heat conductivity tensor. In this section we restrict ourselves to the simple conduction law (10.29). Combining this law with the expressions (10.27), (10.28) and the conservation law (10.3) for heat gives:

$$\frac{\partial T}{\partial t} + \nabla \cdot (\mathbf{v}T - \kappa\nabla T) = S . \tag{10.30}$$

As a first example we consider a solid in which there is no flow ($\mathbf{v} = 0$). For a constant heat conductivity $\kappa$, (10.30) reduces to:

$$\frac{\partial T}{\partial t} = \kappa\nabla^2 T + S . \tag{10.31}$$

The expression is called the 'heat equation', despite the fact that it holds only under special conditions. This expression is identical to Fick's law which accounts for diffusion processes. This is not surprising since heat is transported by a diffusive process in the absence of advection.

We now consider heat transport in a one-dimensional medium (such as a bar) when there is no source of heat. In that case the heat equation reduces to

$$\frac{\partial T}{\partial t} = \kappa \frac{\partial^2 T}{\partial x^2} \ .$$
(10.32)

If we know the temperature throughout the medium at some initial time (i.e. $T(x, t = 0)$ is known), then (10.32) can be used to compute the temperature at later times. As a special case we consider a Gaussian-shaped temperature distribution at $t = 0$:

$$T(x, t = 0) = T_0 \exp\left(-\frac{x^2}{L^2}\right) \ .$$
(10.33)

**Problem a:** Sketch this temperature distribution and indicate the role of the constants $T_0$ and $L$.

We assume that the temperature profile maintains a Gaussian shape at later times but that the peak value and the width may change, i.e. we consider a solution of the following form:

$$T(x, t) = F(t)e^{-H(t)x^2} .$$
(10.34)

At this point the functions $F(t)$ and $H(t)$ are not yet known.

**Problem b:** Show that these functions satisfy the initial conditions:

$$F(0) = T_0, \qquad H(0) = 1/L^2 \ .$$
(10.35)

**Problem c:** Show that for the special solution (10.34) the heat equation reduces to:

$$\frac{\partial F}{\partial t} - x^2 F \frac{\partial H}{\partial t} = \kappa \left(4FH^2x^2 - 2FH\right) \ .$$
(10.36)

It is possible to derive equations for the time evolution of $F$ and $H$ by recognizing that (10.36) can only be satisfied for *all values of* $x$ when all terms proportional to $x^2$ balance and when the terms independent of $x$ balance.

**Problem d:** Use this to show that $F(t)$ and $H(t)$ satisfy the following differential equations:

$$\frac{\partial F}{\partial t} = -2\kappa FH \ ,$$
(10.37)

$$\frac{\partial H}{\partial t} = -4\kappa H^2 \ .$$
(10.38)

It is easiest to solve the last equation first because it contains only $H(t)$ whereas (10.37) contains both $F(t)$ and $H(t)$.

**Problem e:** Solve (10.38) with the initial condition (10.35) and show that:

$$H(t) = \frac{1}{4\kappa t + L^2} \; . \tag{10.39}$$

**Problem f:** Solve (10.37) with the initial condition (10.35) and show that:

$$F(t) = T_0 \frac{L}{\sqrt{4\kappa t + L^2}} \; . \tag{10.40}$$

Inserting these solutions into (10.34) gives the temperature field at all times $t \geq 0$:

$$T(x, t) = T_0 \frac{L}{\sqrt{4\kappa t + L^2}} \exp \left( -\frac{x^2}{4\kappa t + L^2} \right) . \tag{10.41}$$

**Problem g:** Sketch the temperature for several later times and describe using the solution (10.41) how the temperature profile changes as time progresses.

The total heat $Q^{total}(t)$ at time $t$ is given by $Q^{total}(t) = C \int_{-\infty}^{\infty} T(x, t) dx$, where $C$ is the heat capacity.

**Problem h:** Show that the total heat does not change with time for the solution (10.41). Hint: reduce any integral of the form $\int_{-\infty}^{\infty} e^{-\alpha x^2} dx$ to the integral $\int_{-\infty}^{\infty} e^{-u^2} du$ with a suitable change of variables. You don't even have to use that $\int_{-\infty}^{\infty} e^{-u^2} du = \sqrt{\pi}$.

**Problem i:** Show that for *any* solution of the heat equation (10.32), where the heat flux vanishes at the endpoints ($\kappa \partial_x T(x = \pm\infty, t) = 0$), the total heat $Q^{total}(t)$ is constant in time.

**Problem j:** What happens to the special solution (10.41) when the temperature field evolves backward in time? Consider in particular times earlier than $t = -L^2/4\kappa$.

**Problem k:** The peak value of the temperature field (10.41) decays as $1/\sqrt{4\kappa t + L^2}$ with time. Do you expect that in more dimensions this decay will be more rapid or slower with time? Don't do any calculations but use your common sense only!

Up to this point, we have considered the conduction of heat in a medium without flow ($\mathbf{v} = 0$). In many applications the flow in the medium plays

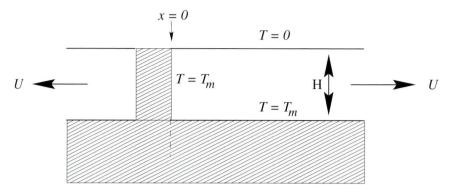

Fig. 10.2.   Sketch of the cooling model of the oceanic lithosphere.

a crucial role in redistributing heat. This is particularly the case when heat is the source of convective motion, as for example in the Earth's mantle, the atmosphere and the central heating systems in buildings. As an example of the role of advection we consider the cooling model of the oceanic lithosphere proposed by Parsons and Sclater [66].

At the mid-oceanic ridges, lithospheric material with thickness $H$ is produced. At a ridge the temperature of this material is essentially the temperature $T_m$ of mantle material. As shown in figure 10.2, this implies that at $x = 0$ and at depth $z = H$ the temperature is given by the mantle temperature: $T(x = 0, z) = T(x, z = H) = T_m$. We assume that the velocity with which the plate moves away from the ridge is constant:

$$\mathbf{v} = U\hat{\mathbf{x}} . \tag{10.42}$$

We consider the situation in which the temperature is stationary. This does not imply that the flow vanishes; it means that the partial time-derivatives vanish: $\partial T/\partial t = 0$, $\partial \mathbf{v}/\partial t = 0$.

**Problem l:** Show that in the absence of heat sources ($S = 0$) the conservation equation (10.30) reduces to:

$$U\frac{\partial T}{\partial x} = \kappa\left(\frac{\partial^2 T}{\partial x^2} + \frac{\partial^2 T}{\partial z^2}\right) . \tag{10.43}$$

In general the thickness of the oceanic lithosphere is less than 100 km, whereas the width of ocean basins is several thousand kilometers.

**Problem m:** Use this fact to explain that the following expression is a reasonable approximation to (10.43):

$$U\frac{\partial T}{\partial x} = \kappa\frac{\partial^2 T}{\partial z^2} . \tag{10.44}$$

**Problem n:** Show that with the replacement $\tau = x/U$ this expression is identical to the heat equation (10.32).

Note that $\tau$ is the time it has taken the oceanic plate to move from its point of creation ($x = 0$) to the point under consideration ($x$), hence the time $\tau$ is simply the *age* of the oceanic lithosphere. This implies that solutions of the one-dimensional heat equation can be used to describe the cooling of oceanic lithosphere with the age of the lithosphere taken as the time variable. Accounting for cooling with such a model leads to a prediction of the depth of the ocean that increases as $\sqrt{t}$ with the age of the lithosphere. For ages less than about 100 Myear this is in very good agreement with the observed ocean depth [66].

## 10.5 The explosion of a nuclear bomb

As an example of the use of conservation equations we study the condition under which a ball of uranium or plutonium can explode through a nuclear chain reaction. The starting point is once again the general conservation law (10.3), where $Q$ is the concentration $N(\mathbf{r}, t)$ of neutrons per unit volume. We assume that the material is solid and that there is no flow: $\mathbf{v} = 0$. The neutron concentration is affected by two processes. First, the neutrons experience normal diffusion. For simplicity we assume that the neutron current is given by (10.29): $\mathbf{J} = -\kappa\nabla N$, with $\kappa$ a constant. Second, neutrons are produced in the nuclear chain reaction. For example, when an atom of $U_{235}$ absorbs one neutron, it may undergo fission and emit three free neutrons. This effectively constitutes a source of neutrons. The intensity of this source depends on the neutrons that are around to produce the fission of atoms. This implies that the source term is proportional to the neutron concentration: $S = \lambda N$, where $\lambda$ is a positive constant that depends on the details of the nuclear reactions.

**Problem a:** Show that the neutron concentration satisfies:

$$\frac{\partial N}{\partial t} = \kappa\nabla^2 N + \lambda N \ . \tag{10.45}$$

This equation needs to be supplemented with boundary conditions. We assume that the radioactive material is in the shape of a sphere of radius $R$. At the edge of the sphere the neutron concentration vanishes while at the center of the sphere the neutron concentration must remain finite for finite times:

$$N(r = R, t) = 0 \quad \text{and} \quad N(r = 0, t) \text{ is finite} \ . \tag{10.46}$$

We restrict our attention to solutions that are spherically symmetric: $N = N(r, t)$.

**Problem b:** Apply separation of variables by writing the neutron concentration as $N(r,t) = F(r)H(t)$ and show that $F(r)$ and $H(t)$ satisfy the following equations:

$$\frac{\partial H(t)}{\partial t} = \mu H(t) \; , \tag{10.47}$$

$$\nabla^2 F(r) + \frac{(\lambda - \mu)}{\kappa} F(r) = 0 \; , \tag{10.48}$$

where $\mu$ is a separation constant that is not yet known.

**Problem c:** Show that for positive $\mu$ there is an exponential growth of the neutron concentration with characteristic growth time $\tau = 1/\mu$.

**Problem d:** Use the Laplacian in spherical coordinates to rewrite (10.48). Make the substitution $F(r) = f(r)/r$ and show that $f(r)$ satisfies:

$$\frac{\partial^2 f}{\partial r^2} + \frac{(\lambda - \mu)}{\kappa} f = 0 \; . \tag{10.49}$$

**Problem e:** Derive the boundary conditions at $r = 0$ and $r = R$ for $f(r)$.

**Problem f:** Show that (10.49) with the boundary conditions derived in problem e can only be satisfied when

$$\mu = \lambda - \left(\frac{n\pi}{R}\right)^2 \kappa \qquad \text{for integer } n \; . \tag{10.50}$$

**Problem g:** Show that for $n = 0$ the neutron concentration vanishes so that we only need to consider values $n \geq 1$.

Equation (10.50) gives the growth rate of the neutron concentration. It can be seen that the effects of unstable nuclear reactions and of neutron diffusion oppose each other. The $\lambda$ term accounts for the growth of the neutron concentration through fission reactions: this term makes the inverse growth rate $\mu$ more positive. Conversely, the $\kappa$ term accounts for diffusion: this term gives a negative contribution to $\mu$.

**Problem h:** What value of $n$ gives the largest growth rate? Show that exponential growth of the neutron concentration (i.e. a nuclear explosion) can only occur when

$$R > \pi \sqrt{\frac{\kappa}{\lambda}} \; . \tag{10.51}$$

This implies that a nuclear explosion can only occur when the ball of fissionable material is larger than a certain critical size. If the ball is smaller than the critical size, more neutrons diffuse out of it than are created by fission, hence the nuclear reaction stops. In some of the earliest nuclear devices an explosion was created by bringing two half spheres that each were stable together to form one whole sphere that was unstable.

**Problem g:** Suppose you have a ball of fissionable material that is just unstable and that you shape this material into a cube rather than a ball. Do you expect this cube to be stable or unstable? Don't use any equations!

## 10.6 Viscosity and the Navier–Stokes equation

Many fluids exhibit a certain degree of viscosity. In this section it will be shown that viscosity can be seen as an *ad-hoc* description of the momentum current in a fluid due to small-scale movements in the fluid. The starting point of the analysis is the equation of momentum conservation in a fluid:

$$\frac{\partial(\rho\mathbf{v})}{\partial t} + \nabla\cdot(\rho\mathbf{vv}) = \mathbf{F} \tag{10.13}$$

In a real fluid, motion takes place at a large range of length scales from microscopic eddies to organized motions with a size comparable to the size of the fluid body. Whenever we describe a fluid, it is impossible to account for the motion at the very small length scales. This is not only so in analytical descriptions, but it is in particular the case in numerical simulations of fluid flow. For example, in current weather prediction schemes the motion of the air is computed on a grid with a distance of about 100 km between the grid points. When you look at the weather it is obvious that there is considerable motion at smaller length scales (e.g. cumulus clouds indicating convection, fronts, etc.). In general one cannot simply ignore the motion at these short length scales because these small-scale fluid motions transport significant amounts of momentum, heat and other quantities such as moisture [64].

One way to account for the effect of the small-scale motion is to express the small-scale motion in the large-scale motion. It is not obvious that this is consistent with reality, but it appears to be the only way to avoid a complete description of the small-scale motion of the fluid (which would be impossible).

In order to do this, we assume that there is some length scale which separates the small-scale flow from the large-scale flow, and we decompose the velocity into a long-wavelength component $\mathbf{v}^L$ and a short-wavelength

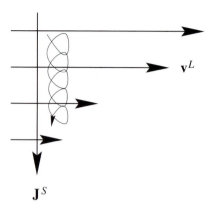

Fig. 10.3.   The direction of momentum transport within a large-scale flow by small-scale motions.

component $\mathbf{v}^S$:

$$\mathbf{v} = \mathbf{v}^L + \mathbf{v}^S \ . \tag{10.52}$$

In addition, we take spatial averages over a length scale that corresponds to the length scale that distinguishes the large-scale flow from the small-scale flow. This average is indicated by angle brackets: $\langle \cdots \rangle$. The average of the small-scale flow is zero, $\langle \mathbf{v}^S \rangle = 0$, while the average of the large-scale flow is equal to the large-scale flow, $\langle \mathbf{v}^L \rangle = \mathbf{v}^L$, because the large-scale flow by definition does not vary over the averaging length. For simplicity we assume that the density does not vary.

**Problem a:** Show that the momentum equation for the large-scale flow is given by:

$$\frac{\partial (\rho \mathbf{v}^L)}{\partial t} + \nabla \cdot (\rho \mathbf{v}^L \mathbf{v}^L) + \nabla \cdot (\langle \rho \mathbf{v}^S \mathbf{v}^S \rangle) = \mathbf{F} \ . \tag{10.53}$$

Show in particular why this expression contains a contribution that is quadratic in the small-scale flow, but that the terms that are linear in $\mathbf{v}^S$ do not contribute.

All the terms in (10.53) are familiar, except the last term on the left hand side. This term exemplifies the effect of the small-scale flow on the large-scale flow since it accounts for the transport of momentum by the small-scale flow. It seems that at this point further progress is impossible without knowing the small-scale flow $\mathbf{v}^S$. One way to make further progress is to express the small-scale momentum current $\langle \rho \mathbf{v}^S \mathbf{v}^S \rangle$ in the large scale flow.

Consider the large-scale flow shown in figure 10.3. Whatever the small-scale motions are, in general they will have the character of mixing. In

the example in the figure, the momentum is large at the top of the figure
and smaller at the bottom. As a first approximation one may assume that
the small-scale motions transport momentum in the direction opposite to
the momentum gradient of the large-scale flow. By analogy with (10.29)
we can approximate the momentum transport by the small-scale flow by:

$$\mathbf{J}^S \equiv \left\langle \rho \mathbf{v}^S \mathbf{v}^S \right\rangle \approx -\mu \nabla \mathbf{v}^L \,, \tag{10.54}$$

where $\mu$ plays the role of a diffusion constant.

**Problem b:** Insert this relation into (10.53), and drop the superscript $L$
of $\mathbf{v}^L$ to show that large-scale flow satisfies:

$$\frac{\partial(\rho \mathbf{v})}{\partial t} + \nabla \cdot (\rho \mathbf{v} \mathbf{v}) = \mu \nabla^2 \mathbf{v} + \mathbf{F} \,. \tag{10.55}$$

This equation is called the Navier–Stokes equation. The first term on
the right hand side accounts for the momentum transport by small-scale
motions. Effectively this leads to the viscosity of the fluid.

**Problem c:** Viscosity tends to damp motion at smaller length scales
more than motion at larger length scales. Show that the term $\mu \nabla^2 \mathbf{v}$
indeed affects shorter length scales more than larger length scales.

**Problem d:** Do you think this treatment of the momentum flux due to
small-scale motion is realistic? Can you think of an alternative?

Despite reservations that you may (or may not) have against the treatment of viscosity in this section, you should realize that the Navier–Stokes
equation (10.55) is widely used in fluid mechanics.

## 10.7 Quantum mechanics and hydrodynamics

As we saw in section 7.4, the behavior of atomic particles is described by
Schrödinger's equation:

$$i\hbar \frac{\partial \psi(\mathbf{r}, t)}{\partial t} = -\frac{\hbar^2}{2m} \nabla^2 \psi(\mathbf{r}, t) + V(\mathbf{r}) \psi(\mathbf{r}, t) \,, \tag{7.14}$$

rather than Newton's law. In this section we reformulate the linear wave
equation (7.14) as the laws of conservation of mass and momentum for a
normal fluid. In order to do this we write the wavefunction $\psi$ as

$$\psi = \sqrt{\rho} \times e^{(i/\hbar)\varphi}. \tag{10.56}$$

This equation is simply the decomposition of a complex function into its
absolute value and its phase, hence $\rho$ and $\varphi$ are real functions. The factor
$\hbar$ is a constant added here for notational convenience. This constant is
characteristic of quantum mechanics and is called Planck's constant.

**Problem a:** Insert the decomposition (10.56) in Schrödinger's equation (7.14), divide by $\sqrt{\rho}e^{(i/\hbar)\varphi}$ and separate the result into real and imaginary parts to show that $\rho$ and $\varphi$ satisfy the following differential equations:

$$\partial_t\rho + \nabla \cdot \left(\rho\frac{1}{m}\nabla\varphi\right) = 0 , \tag{10.57}$$

$$\partial_t\varphi + \frac{1}{2m}|\nabla\varphi|^2 + \frac{\hbar^2}{8m}\left(\frac{1}{\rho^2}|\nabla\rho|^2 - \frac{2}{\rho}\nabla^2\rho\right) = -V . \tag{10.58}$$

The problem is that at this point we do not yet have a velocity. Let us define the following velocity vector:

$$\mathbf{v} \equiv \frac{1}{m}\nabla\varphi . \tag{10.59}$$

**Problem b:** Show that this definition of the velocity is identical to the velocity obtained in (7.20).

**Problem c:** Show that with this definition of the velocity, (10.57) is identical to the continuity equation:

$$\frac{\partial\rho}{\partial t} + \nabla \cdot (\rho\mathbf{v}) = 0 . \tag{10.6}$$

**Problem d:** In order to reformulate (10.58) as an equation of conservation of momentum, differentiate (10.58) with respect to $x_i$. Then use the definition (10.59) and the relation between force and potential $(\mathbf{F} = -\nabla V)$ to write the result as:

$$\partial_t v_i + \frac{1}{2}\sum_j \partial_i(v_j v_j) + \frac{\hbar^2}{8m}\left[\partial_i\left(\frac{1}{\rho^2}|\nabla\rho|^2\right) - 2\partial_i\left(\frac{1}{\rho}\nabla^2\rho\right)\right] = \frac{1}{m}F_i . \tag{10.60}$$

The second term on the left hand side does not look very much like the term $\sum_j \partial_j(\rho v_j v_i)$ in the left hand side of (10.13). To make progress we need to rewrite the term $\sum_j \partial_i(v_j v_j)$ as a term of the form $\sum_j \partial_j(v_j v_i)$. In general these terms are different.

**Problem e:** Show that for the special case that the velocity is the gradient of a scalar function (as in expression (10.59)):

$$\sum_j \frac{1}{2}\partial_i(v_j v_j) = \sum_j \partial_j(v_j v_i) . \tag{10.61}$$

With this step we can rewrite the second term on the left hand side of (10.60). Part of the third term in (10.60) we will designate as $Q_i$:

$$Q_i \equiv -\frac{1}{8}\left[\partial_i\left(\frac{1}{\rho^2}|\nabla\rho|^2\right) - 2\partial_i\left(\frac{1}{\rho}\nabla^2\rho\right)\right]. \tag{10.62}$$

**Problem f:** Use (10.6) and (10.60)–(10.62) to derive that:

$$\partial_t(\rho\mathbf{v}) + \nabla\cdot(\rho\mathbf{vv}) = \frac{\rho}{m}\left(\mathbf{F}+\hbar^2\mathbf{Q}\right). \tag{10.63}$$

Note that this equation is identical to the momentum equation (10.13). This implies that the Schrödinger equation is equivalent to the continuity equation (10.6) and to the momentum equation (10.13) for a classical fluid. In section 7.4 we saw that atomic particles behave as waves rather than as point-like particles. In this section we have discovered that these particles also behave like a fluid! This has led to hydrodynamic formulations of quantum mechanics [52][37]. In general, quantum-mechanical phenomena depend critically on Planck's constant. Quantum mechanics reduces to classical mechanics in the limit $\hbar \rightarrow 0$. The only place where Planck's constant occurs in (10.63) is in the additional force: $\mathbf{Q}$ multiplied by the square of Planck's constant. This implies that the action of the force term $\mathbf{Q}$ is fundamentally quantum-mechanical; it has no analogue in classical mechanics.

**Problem g:** Suppose we consider a particle in one dimension that is represented by the following wave function:

$$\psi(x,t) = e^{-x^2/L^2}e^{i(kx-\omega t)}. \tag{10.64}$$

Sketch the corresponding probability-density $\rho$ and use (10.62) to deduce that the quantum force $Q$ acts to broaden the wavefunction with time.

This example shows that (at least for this case) the quantum force $\mathbf{Q}$ makes the wavefunction 'spread out' with time. This reflects the fact that if a particle propagates with time, its position becomes more and more uncertain.

# 11

## Scale analysis

In most situations, the equations that we would like to solve in mathematical physics are too complicated to solve analytically. One of the reasons for this is often that an equation contains many different terms which make the problem simply too complex to be manageable. However, many of these terms may in practice be very small. Ignoring these small terms can simplify the problem to such an extent that it can be solved in closed form. Moreover, by deleting terms that are small one is able to focus on the terms that are significant and that contain the relevant physics. In this sense, ignoring small terms can actually give a better physical insight into the processes that really do matter.

Scale analysis is a technique in which one estimates the different terms in an equation by considering the scale over which the relevant parameters vary. This is an extremely powerful tool for simplifying problems. A comprehensive overview of this technique with many applications is given by Kline [45] and in chapter 6 of Lin *et al.* [49]. Interesting examples of the application of scaling arguments to biology are given by Vogel [94].

With the application of scale analysis one caveat must be made. One of the major surprises of classical physics of the twentieth century was the discovery of chaos in dynamical systems [86]. In a chaotic system small changes in the initial conditions lead to a change in the time evolution of the system that grows exponentially with time. Deleting small terms from the equation of motion of such a system can have a similar effect; the effects of omitting small terms can lead to changes in the system that may grow exponentially with time. This means that for chaotic systems one must be careful in omitting small terms from the equations.

The principle of scale analysis is first applied to the problem of determining the sense of rotation of a vortex in an emptying bathtub. Many of the equations that are used in physics are differential equations. For this reason it is crucial in scale analysis to be able to estimate the order of magnitude of derivatives. The estimation of derivatives is therefore treated in section 11.2. In subsequent sections this is then applied to a variety of different problems.

## 11.1 The vortex in a bathtub

Often when you empty a bathtub you can observe a beautiful vortex above the drain in the bathtub. When you ask people about the sense of rotation of this vortex many people will reply that the sense of rotation is in the counterclockwise direction on the northern hemisphere and in the clockwise direction in the southern hemisphere (or vice versa). This statement is wrong and the easiest way to verify this is to take a bath, empty the bathtub and observe that it is as easy to obtain a vortex that rotates in the clockwise direction as it is to create a vortex that rotates in the counterclockwise direction. In fact, if you have ever had the opportunity to take a bath in Nairobi, you would have had the chance to observe that on the equator (where the effective rotation of the Earth vanishes) a bathtub drains in exactly the same way as at higher latitudes.

Yet when you ask your friends you may find that it is very difficult to convince them of these facts. In that case you may be able to convince them with a calculation. One approach would be take the equations of fluid flow on a rotating Earth and compute numerically in all details the flow in your bathtub. However, this approach is not only terribly impractical but it is also very difficult to carry out because you know neither the detailed shape of your bathtub nor the initial conditions of the water in the bathtub before you drain it. An even more serious drawback of this approach is that this numerical simulation does not give much physical insight. At best it gives a perfect simulation of the real bathtub, but in that case it would be better to do the experiment in the real bathtub.

As an alternative we can estimate the forces that are acting on the fluid in the vortex. In figure 11.1 the forces are shown that act on a fluid parcel in a vortex on the Earth that rotates with angular velocity $\Omega$. The dominant forces are the pressure force $\mathbf{F}_{pres}$ that is associated with the pressure gradient in the fluid, the Coriolis force $\mathbf{F}_{Cor}$ due to the rotation of the Earth and the centrifugal force $\mathbf{F}_{cent}$ that is due to the circular motion of the fluid in the vortex. The rotational velocity of the vortex is denoted by $\omega$. The only force that depends on the position on the Earth is the Coriolis force $\mathbf{F}_{Cor}$. In figure 11.1 this force points to the right hand side of the velocity vector; this is the case in the northern hemisphere whereas in the southern hemisphere this force would point to the left hand side of the velocity vector. The centrifugal force always points outward from the vortex while the pressure force always points

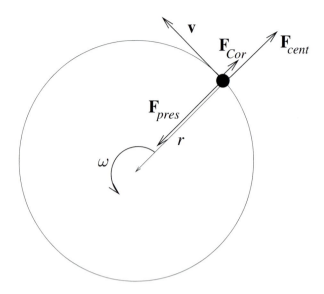

Fig. 11.1.   The forces that act in a fluid vortex on the northern hemisphere of a rotating Earth.

into the vortex. This means that any asymmetry between the behavior of a vortex in the northern hemisphere and one in the southern hemisphere must be due to the Coriolis force.

   This means that in order to test our hypothesis we need to estimate the strength of the Coriolis force compared to the other forces that are operative. In the balance of forces, the pressure force will be balanced by the Coriolis force, or by the centrifugal force or by both of them. For this reason, it is sufficient to estimate the strength of the Coriolis force compared to the centrifugal force. We will derive in section 12.3 that the Coriolis force is given by $\mathbf{F}_{Cor} = -2\mathbf{\Omega} \times \mathbf{v}$. Since we are only estimating orders of magnitude this means that the Coriolis force is of the order

$$F_{Cor} \sim \Omega v \ . \tag{11.1}$$

The strength of the centrifugal force is given by

$$F_{cent} = \frac{v^2}{r} \ . \tag{11.2}$$

**Problem a:** Use figure 11.1 to deduce that $v/r = \omega$ and then use this result to show that the ratio of the Coriolis force to the centrifugal force is approximately given by

$$\frac{F_{Cor}}{F_{cent}} \sim \frac{\Omega}{\omega} \ . \tag{11.3}$$

The ratio of the Coriolis force to the centrifugal force is thus of the order of the ratio of the rotation rate of the Earth to the rotation rate of the vortex in the bathtub.

**Problem b:** Use this result to show that

$$\frac{F_{Cor}}{F_{cent}} \sim 10^{-5} . \tag{11.4}$$

This means that the Coriolis force is much smaller than all other forces that are operating on the fluid parcel. The asymmetry in the balance of forces created by the Coriolis force is in practice also much smaller than the asymmetry in the shape of the bathtub and the asymmetry in the initial conditions of the fluid before you drain the bathtub. We can conclude from this that the Earth's rotation is negligible in the dynamics of the vortex; hence the sense of rotation of a vortex in a bathtub does not depend on the geographic location of the bathtub. This example shows that one can sometimes learn more from simple physical arguments and estimates of orders of magnitude than from very complex numerical calculations.

It follows from (11.3) that the centrifugal force and the Coriolis force are of the same order of magnitude when the rotation rate of the fluid in the bathtub is of the same order of magnitude as the Earth's rotation.

**Problem c:** Estimate the ratio of the Coriolis force to the centrifugal force for the atmospheric motion around a depression.

You will have found that for the atmosphere the Coriolis force *is* one of the dominant forces. This is the reason that in the northern hemisphere the air moves in the counterclockwise direction around a low-pressure area and in the clockwise direction around a high-pressure center (and vice-versa on the southern hemisphere). A low-pressure area is indeed very similar to the vortex in a bathtub. This analogy is in fact the cause of the misconception that the vortex in a bathtub always turns in one direction depending on geographic location. The major difference is, however, that the relative strengths of the forces that are operative in the atmosphere and in a bathtub are completely different. This result has been derived here with a simple scale analysis.

## 11.2 Three ways to estimate a derivative

In this section three different ways to estimate the derivative of a function $f(x)$ are derived. The *first* way to estimate the derivative is to realize that the derivative is nothing but the slope of the function $f(x)$. Consider figure 11.2 in which the function $f(x)$ is assumed to be known at neighboring points $x$ and $x + h$.

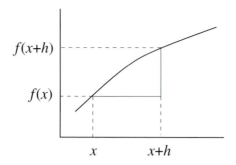

Fig. 11.2.   The slope of a function $f(x)$ that is known at positions $x$ and $x + h$.

**Problem a:** Deduce from the geometry of this figure that the slope of the function at $x$ is approximately given by $[f(x + h) - f(x)]/h$.

Since the slope is the derivative this means that the derivative of the function is approximately given by

$$\frac{df}{dx} \approx \frac{f(x + h) - f(x)}{h} \, . \tag{11.5}$$

The *second* way to derive the same result is to realize that the derivative is defined by the following limit:

$$\frac{df}{dx} \equiv \lim_{h \to 0} \frac{f(x + h) - f(x)}{h} \, . \tag{11.6}$$

If we consider the right hand side of this expression without taking the limit, we do not quite obtain the derivative, but as long as $h$ is sufficiently small we obtain the approximation (11.5).

The problem with estimating the derivative of $f(x)$ in the previous ways is that although we obtain an estimate of the derivative, but we do not know how good these estimates are. We know that if $f(x)$ were a straight line, which has a constant slope, the estimate (11.5) would be exact. Hence it is the deviation of $f(x)$ from a straight line that makes (11.5) only an approximation. This means that it is the curvature of $f(x)$ that accounts for the error in the approximation (11.5). The *third* way of estimating the derivative provides this error estimate as well.

**Problem b:** Consider the Taylor series (2.17). Truncate this series after the second order term and solve the resulting expression for $df/dx$ to derive that

$$\frac{df}{dx} = \frac{f(x + h) - f(x)}{h} - \frac{1}{2} \frac{d^2 f}{dx^2} h + \cdots \, , \tag{11.7}$$

where the dots indicate terms of order $h^2$ and higher.

In the limit $h \to 0$ the last term vanishes and (11.6) is obtained. When one ignores the last term in (11.7) for finite $h$ one once again obtains the approximation (11.5).

**Problem c:** Use (11.7) to show that the error made in the approximation (11.5) indeed depends on the *curvature* of the function $f(x)$.

The approximation (11.5) has a variety of applications. The first is the numerical solution of differential equations. Suppose one has a differential equation that cannot be solved in closed form. To fix our minds, consider the differential equation

$$\frac{df}{dx} = G(f(x), x) , \qquad (11.8)$$

with initial value

$$f(0) = f_0 . \qquad (11.9)$$

If this equation cannot be solved in closed form, one can solve it numerically by evaluating the function $f(x)$ not for every value of $x$, but only at a finite number of $x$-values that are separated by a distance $h$. These points $x_n$ are given by $x_n = nh$, and the function $f(x)$ at location $x_n$ is denoted by $f_n$:

$$f_n \equiv f(x_n) . \qquad (11.10)$$

**Problem d:** Show that the derivative $df/dx$ at location $x_n$ can be approximated by:

$$\frac{df}{dx}(x_n) = \frac{1}{h}(f_{n+1} - f_n) . \qquad (11.11)$$

**Problem e:** Insert this result into the differential equation (11.8) and solve the resulting expression for $f_{n+1}$ to show that:

$$f_{n+1} = f_n + hG(f_n, x_n) . \qquad (11.12)$$

This is all we need to numerically solve the differential equation (11.8) with the boundary condition (11.9). Once $f_n$ is known, (11.12) can be used to compute $f_{n+1}$. This means that the function can be computed at all values of the grid points $x_n$ recursively. To start this process, one uses the boundary condition (11.9) that gives the value of the function at location $x_0 = 0$. This technique for estimating the derivative of a function can be extended to higher order derivatives as well so that second order differential equations can also be solved numerically. In practice, one has to pay serious attention to the *stability* of the numerical solution. The requirements of stability and numerical efficiency have led to many refinements of the numerical methods for solving differential equations.

The interested reader can consult Press *et al.* [69] for an introduction and many practical algorithms.

The estimate (11.5) has a second important application because it allows us to estimate the order of magnitude of a derivative. Suppose a function $f(x)$ varies over a characteristic range of values $F$ and that this variation takes place over a characteristic distance $L$. It follows from (11.5) that the derivative of $f(x)$ is of the order of the ratio of the variation of the function $f(x)$ divided by the length scale over which the function varies. In other words:

$$\left| \frac{df}{dx} \right| \approx \frac{\text{variation of the function } f(x)}{\text{length scale of the variation}} \sim \frac{F}{L} \; . \tag{11.13}$$

In this expression the term $\sim F/L$ indicates that the derivative is of the order $F/L$. Note that this is not in general an accurate estimate of the precise value of the function $f(x)$, it only provides us with an estimate of the order of magnitude of a derivative. However, this is all we need to carry out scale analysis.

**Problem f:** Suppose $f(x)$ is a sinusoidal wave with amplitude $A$ and wavelength $\lambda$:

$$f(x) = A \sin\left(\frac{2\pi x}{\lambda}\right) \; . \tag{11.14}$$

Show that (11.13) implies that the order of magnitude of the derivative of this function is given by $|df/dx| \sim O\left(A/\lambda\right)$. Compare this estimate of the order of magnitude with the true value of the derivative and pay attention both to the numerical value as well as to the spatial variation.

From the previous estimate we can learn two things. First, the estimate (11.13) is only a rough estimate that *locally* can be very poor. One should always be aware that (11.13) may break down at certain points and that this can cause errors in the subsequent scale analysis. Second, (11.13) differs by a factor $2\pi$ from the true derivative. However, $2\pi \approx 6.28$, which is not a small number. Therefore you must be aware that hidden numerical factors may enter scaling arguments.

## 11.3 The advective terms in the equation of motion

As a first example of scale analysis we consider the role of advective terms in the equation of motion. As shown in (10.12) the equation of motion for a continuous medium is given by

$$\frac{\partial \mathbf{v}}{\partial t} + \mathbf{v} \cdot \nabla \mathbf{v} = \frac{1}{\rho} \mathbf{F} \; . \tag{11.15}$$

Note that we have divided by the density compared to the original expression (10.12). This equation can describe the propagation of acoustic waves when **F** is the pressure force, and it accounts for elastic waves when **F** is given by the elastic forces in the medium. We consider the situation in which waves with a wavelength $\lambda$ and a period $T$ propagate through the medium.

The advective terms **v**·∇**v** often pose a problem in solving this equation. This is because the partial time-derivative $\partial \mathbf{v}/\partial t$ is *linear* in the velocity **v** but the advective terms $\mathbf{v} \cdot \nabla \mathbf{v}$ are *nonlinear* in the velocity **v**. Since linear equations are in general much easier to solve than nonlinear equations it is very useful to know under which conditions the advective terms $\mathbf{v} \cdot \nabla \mathbf{v}$ can be ignored compared with the partial derivative $\partial \mathbf{v}/\partial t$.

**Problem a:** Let the velocity of the continuous medium have a characteristic value $V$. Show that $|\partial \mathbf{v}/\partial t| \sim V/T$ and that $|\mathbf{v} \cdot \nabla \mathbf{v}| \sim V^2/\lambda$.

**Problem b:** Show that this means that the ratio of the advective terms to the partial time-derivative is given by

$$\frac{|\mathbf{v} \cdot \nabla \mathbf{v}|}{|\partial \mathbf{v}/\partial t|} \sim \frac{V}{c} , \tag{11.16}$$

where $c = \lambda/T$ is the phase velocity with which the waves propagate through the medium.

This result implies that the advective terms can be ignored when the velocity of the medium itself is much less than the velocity of the waves propagating through the medium:

$$V \ll c . \tag{11.17}$$

In other words, when the amplitude of the wave motion is so small that the velocity of the particle motion in the medium is much less than the phase velocity of the waves, one can ignore the advective terms in the equation of motion.

**Problem c:** Suppose an earthquake causes a ground displacement of 1 mm at a frequency of 1 Hz at a large distance. The wave velocity of seismic $P$-waves is of the order of 5 km/s near the surface. Show that in that case $V/c \sim 10^{-6}$.

The small value of $V/c$ implies that for the propagation of elastic waves due to earthquakes one can ignore advective terms in the equation of motion. Note, however, that this is not necessarily true near the earthquake where the motion is much more violent and where the associated velocity of the rocks is not necessarily much smaller than the wave velocity.

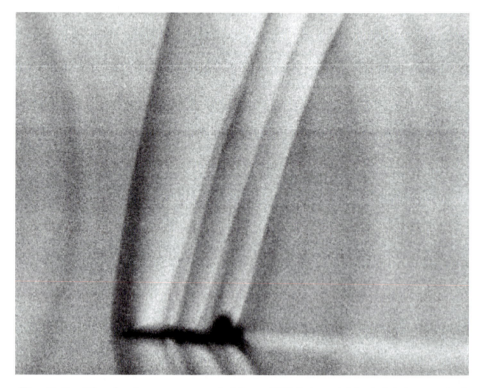

Fig. 11.3. The shock waves generated by a T38 flying at Mach 1.1 (a speed of 1.1 times the speed of sound) as made visible with the schlieren method.

There are a number of physical phenomena that are intimately related to the presence of the advective terms in the equation of motion. One important phenomenon is the occurrence of shock waves when the motion of the medium exceeds the wave velocity. A prime example of shock waves is the sonic boom made by an aircraft that moves at a velocity greater than the speed of sound [44][88]. A spectacular example can be seen in figure 11.3 where the shock waves generated by a T38 flying at a speed of Mach 1.1 at an altitude of 13.700 ft can be seen. These shock waves are visualized using the schlieren method [48] which is an optical technique to convert phase differences of light waves into amplitude differences.

Another example of shock waves is the formation of the *hydraulic jump*. You may not know what a hydraulic jump is, but you have surely seen one! Consider water flowing down a channel such as a mountain stream as shown in figure 11.4. The flow velocity is denoted by $v$. At the bottom of the channel a rock disrupts the flow. This rock generates water waves that propagate with a velocity $c$ compared to the moving water. When the flow velocity is less than the wave velocity ($v < c$, see the left hand panel of figure 11.4) the waves propagate upstream with an absolute velocity

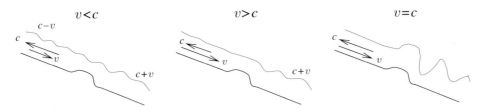

Fig. 11.4. Waves on a water flowing over a rock when $v < c$ (left hand panel), $v > c$ (middle panel) and $v = c$ (right hand panel).

$c - v$ and propagate downstream with an absolute velocity $c + v$. When the flow velocity is larger than the wave velocity ($v > c$, see the middle panel of figure 11.4) the waves move only downstream because the wave velocity is not sufficiently large to move the waves against the current. The most interesting case is when the flow velocity equals the wave velocity ($v = c$, see the right hand panel of figure 11.4). In that case the waves that move upstream have an absolute velocity given by $c - v = 0$. In other words, these waves do not move with respect to the rock that generates the waves. This wave is continuously excited by the rock, and through a process similar to an oscillator that is driven at its resonance frequency the wave grows and grows until it ultimately breaks and becomes turbulent. This is the reason why one sees strong turbulent waves over boulders and other irregularities in streams. For further details on channel flow and hydraulic jumps the reader should consult chapter 9 of Whitaker [98]. In general, the advective terms play a crucial role steepening and breaking of waves and in the formation of shock waves. This is described in much detail by Whitham [99].

## 11.4 Geometric ray theory

Geometric ray theory is an approximation that accounts for the propagation of waves along lines through space. The theory finds its conceptual roots in optics, where for a long time one has observed that a light beam propagates along a well-defined trajectory through lenses and many other optical devices. Mathematically, this behavior of waves is accounted for in geometric ray theory, or more briefly 'ray theory.'

Ray theory is derived here for the acoustic wave equation rather than for the propagation of light because pressure waves are described by a scalar equation rather than the vector equation that governs the propagation of electromagnetic waves. The starting point is the acoustic wave equation (7.7):

$$\rho \nabla \cdot \left( \frac{1}{\rho} \nabla p \right) + \frac{\omega^2}{c^2} p = 0. \tag{11.18}$$

For simplicity the source term on the right hand side has been set to zero. In addition, the relation $c^2 = \kappa/\rho$ has been used to eliminate the bulk modulus $\kappa$ in favor of the wave velocity $c$. Both the density $\rho$ and the wave velocity are arbitrary functions of space.

In general it is not possible to solve this differential equation in closed form. Instead we seek an approximation by writing the pressure as:

$$p(\mathbf{r}, \omega) = A(\mathbf{r}, \omega)e^{i\psi(\mathbf{r}, \omega)} , \qquad (11.19)$$

with $A$ and $\psi$ real functions. Any function $p(\mathbf{r}, \omega)$ can be written in this way.

**Problem a:** Insert the solution (11.19) in the acoustic wave equation (11.18), separate the real and imaginary parts of the resulting equation to deduce that (11.18) is equivalent to the following equations:

$$\underbrace{\nabla^2 A}_{(1)} - \underbrace{A|\nabla\psi|^2}_{(2)} - \underbrace{\frac{1}{\rho}(\nabla\rho \cdot \nabla A)}_{(3)} + \underbrace{\frac{\omega^2}{c^2}A}_{(4)} = 0 , \qquad (11.20)$$

and

$$2(\nabla A \cdot \nabla\psi) + A\nabla^2\psi - \frac{1}{\rho}(\nabla\rho \cdot \nabla\psi)A = 0 . \qquad (11.21)$$

These equations are even harder to solve than the acoustic wave equation because they are nonlinear in the unknown functions $A$ and $\psi$ whereas the acoustic wave equation is linear in the pressure $p$. However, (11.20) and (11.21) form a good starting point for making the ray-geometric approximation. First we analyze (11.20).

Assume that the density varies on a length scale $L_\rho$, and that the amplitude $A$ of the wave field varies on a characteristic length scale $L_A$. Furthermore the wavelength of the waves is denoted by $\lambda$.

**Problem b:** Explain that the wavelength is the length scale over which the phase $\psi$ of the waves varies.

**Problem c:** Use the results of section 11.2 to obtain the following estimates of the order of magnitude of the terms (1)–(4) in equation (11.20):

$$\left.\begin{array}{cc} \left|\nabla^2 A\right| \sim \dfrac{A}{L_A^2}, & A|\nabla\psi|^2 \sim \dfrac{A}{\lambda^2} , \\[4mm] \left|\dfrac{1}{\rho}(\nabla\rho \cdot \nabla A)\right| \sim \dfrac{A}{L_A L_\rho}, & \dfrac{\omega^2}{c^2}A \sim \dfrac{A}{\lambda^2} . \end{array}\right\} \qquad (11.22)$$

To make further progress we assume that the length scales of both the density variations and the amplitude variations are much longer than a wavelength: $\lambda \ll L_A$ and $\lambda \ll L_\rho$.

**Problem d:** Show that under this assumption terms (1) and (3) in equation (11.20) are much smaller than terms (2) and (4).

**Problem e:** Convince yourself that ignoring terms (1) and (3) in (11.20) gives the following (approximate) expression:

$$|\nabla\psi|^2 = \frac{\omega^2}{c^2} . \tag{11.23}$$

**Problem f:** The approximation (11.23) was obtained under the premise that $|\nabla\psi| \sim 1/\lambda$. Show that this assumption is satisfied by the function $\psi$ in (11.23).

Whenever one makes approximations by deleting terms that scale analysis predicts to be small, one has to check that the final solution is consistent with the scale analysis that is used to derive the approximation.

Note that the original equation (11.20) contains both the amplitude $A$ and the phase $\psi$ but that (11.23) contains only the phase. The approximation that we have made has thus decoupled the phase from the amplitude; this simplifies the problem considerably. The frequency enters the right hand side of this equation only through a simple multiplication with $\omega^2$. The frequency dependence of $\psi$ can be found by substituting

$$\psi(\mathbf{r}, \omega) = \omega\tau(\mathbf{r}) . \tag{11.24}$$

**Problem g:** Show that after this substitution (11.23) and (11.21) are given by

$$|\nabla\tau(\mathbf{r})|^2 = \frac{1}{c^2} , \tag{11.25}$$

and

$$2(\nabla A \cdot \nabla\tau) + A\nabla^2\tau - \frac{1}{\rho}(\nabla\rho \cdot \nabla\tau) A = 0 . \tag{11.26}$$

According to (11.25) the function $\tau(\mathbf{r})$ does not depend on frequency. Note that (11.26) for the amplitude does not contain any frequency dependence either. This means that the amplitude also does not depend on frequency: $A = A(\mathbf{r})$. This has important consequences for the shape of the wave field in the ray-geometric approximation. Suppose that the wave field is excited by a source function $s(t)$ in the time domain that is represented in the frequency domain by a complex function $S(\omega)$. (The forward and backward Fourier transform is defined by (14.42) and (14.43).) In the frequency domain the response is given by (11.19) multiplied with the source function $S(\omega)$. Using that $A$ and $\tau$ do not depend on frequency, the pressure in the time domain can be written as:

$$p(\mathbf{r}, t) = \int_{-\infty}^{\infty} A(\mathbf{r})e^{i\omega\tau(\mathbf{r})}e^{-i\omega t}S(\omega)d\omega \ . \qquad (11.27)$$

**Problem h:** Use this expression to show that the pressure in the time domain can be written as:

$$p(\mathbf{r}, t) = A(\mathbf{r})s(t - \tau(\mathbf{r})) \ . \qquad (11.28)$$

This is an important result because it implies that the time dependence of the wave field is everywhere given by the same source-time function $s(t)$. In a ray-geometric approximation the shape of the waveforms is everywhere the same. There are no frequency-dependent effects in a ray-geometric approximation.

**Problem i:** Explain why this implies that geometric ray theory cannot be used to explain why the sky is blue.

The absence of any frequency-dependent wave propagation effects is both the strength and the weakness of ray theory. It is a strength because the wave fields can be computed in a simple way once $\tau(\mathbf{r})$ and $A(\mathbf{r})$ are known. The theory also tells us that this is an adequate description of the wave field as long as the frequency is sufficiently high that $\lambda \ll L_A$ and $\lambda \ll L_\rho$. However, many wave propagation phenomena are in practice frequency-dependent, and it is the weakness of ray theory that it cannot account for these phenomena.

According to (11.28) the function $\tau(\mathbf{r})$ accounts for the time delay of the waves to travel to the point $\mathbf{r}$. Therefore, $\tau(\mathbf{r})$ is the *travel time* of the wave field. The travel time is described by the differential equation (11.25); this equation is called the *eikonal equation*.

**Problem j:** Show that it follows from the eikonal equation that $\nabla\tau$ can be written as:

$$\nabla\tau = \hat{\mathbf{n}}/c \ , \qquad (11.29)$$

where $\hat{\mathbf{n}}$ is a unit vector. Show also that $\hat{\mathbf{n}}$ is perpendicular to the surface $\tau = const$.

The vector $\hat{\mathbf{n}}$ defines the direction of the *rays* along which the wave energy propagates through the medium. Taking suitable derivatives of (11.29) one can derive the *equation of kinematic ray-tracing*. This is a second order differential equation for the position of the rays; details are given by Virieux [93] or Aki and Richards [3].

Once $\tau(\mathbf{r})$ is known, one can compute the amplitude $A(\mathbf{r})$ from (11.26). We have not yet applied any scale analysis to this expression. We will not do so, because it can be solved exactly. Let us first simplify this differential equation by considering the dependence on the density $\rho$ in more detail.

**Problem k:** Write $A=\rho^\alpha B$, where the constant $\alpha$ is not yet determined. Show that the transport equation results in the following differential equation for $B(\mathbf{r})$:

$$(2\alpha - 1)(\nabla\rho \cdot \nabla\tau)B + 2\rho(\nabla B \cdot \nabla\tau) + \rho B\nabla^2\tau = 0 . \qquad (11.30)$$

Choose the constant $\alpha$ in such a way that the gradient of the density disappears from the equation and show that the remaining terms can be written as $\nabla \cdot (B^2\nabla\tau) = 0$. Finally show using (11.29) that this implies the following differential equation for the amplitude:

$$\nabla \cdot \left(\frac{1}{\rho c}A^2\hat{\mathbf{n}}\right) = 0 . \qquad (11.31)$$

Equation (11.31) states that the divergence of the vector $(A^2/\rho c)\,\hat{\mathbf{n}}$ vanishes, hence the flux of this vector through any closed surface that does not contain the source of the wave-field vanishes, see section 7.1. This is not surprising, because the vector $(A^2/\rho c)\,\hat{\mathbf{n}}$ accounts for the energy flux of acoustic waves. Expression (11.31) implies that the net flux of this vector through any closed surface is equal to zero. This means that all the energy that flows into the surface must also flow out through the surface again. The transport equation in the form (11.31) is therefore a statement of energy conservation. Virieux [93] or Aki and Richards [3] show how one can compute this amplitude once the location of the rays is known.

An interesting complication arises when the energy is focussed at a point or on a surface in space. Such an area of focussing is called a *caustic*. A familiar example of a caustic is the rainbow. One can show that at a caustic, the ray-geometric approximation leads to an infinite amplitude of the wave field [93].

**Problem l:** Show that when the amplitude becomes infinite in a finite region of space the condition $\lambda \ll L_A$ must be violated.

This means that ray theory is not valid in or near a caustic. A clear account of the physics of caustics can be found in refs. [13] and [46]. The former reference contains many beautiful images of caustics.

## 11.5 Is the Earth's mantle convecting?

The Earth is a body that continuously loses heat to outer space. This heat is partly a remnant of the heat that has been converted from the

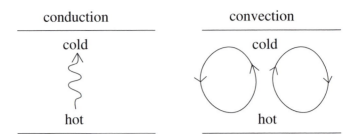

Fig. 11.5.   Two alternatives for the heat transport in the Earth. In the left hand panel the material does not move and heat is transported by conduction. In the right hand panel the material flows and heat is tranported by convection.

gravitational energy during the Earth's formation and partly, and more importantly, this heat is generated by the decay of unstable isotopes in the Earth. This heat is transported to the Earth's surface, and the question we aim to address here is: is the heat transported by conduction or by convection?

   If the material in the Earth does not flow, heat can only be transported by conduction as shown in the left hand panel of figure 11.5. This means that it is the average transfer of the molecular motion from warm regions to cold regions that is responsible for the transport of heat. On the other hand, if the material in the Earth does flow, heat can be carried by the flow as shown in the right hand panel of figure 11.5. This process is called convection.

   The starting point of the analysis is the heat equation (10.30). In the absence of source terms, this equation, for the special case of a constant heat conduction coefficient $\kappa$, may be written as:

$$\frac{\partial T}{\partial t} + \nabla \cdot (\mathbf{v}T) = \kappa \nabla^2 T . \tag{11.32}$$

The term $\nabla \cdot (\mathbf{v}T)$ describes the convective heat transport while the term $\kappa \nabla^2 T$ accounts for the conductive heat transport.

**Problem a:** Let the characteristic velocity be denoted by $V$, the characteristic length scale by $L$, and the characteristic temperature perturbation by $T$. Show that the ratio of the convective heat transport to the conductive heat transport is of the following order:

$$\frac{\text{convective heat transport}}{\text{conductive heat transport}} \sim \frac{VL}{\kappa} . \tag{11.33}$$

   This estimate gives the ratio of the two modes of heat transport, but it does not help us too much yet because we do not know the order of magnitude $V$ of the flow velocity. This quantity can be obtained from the Navier–Stokes equation of section 10.6:

$$\frac{\partial(\rho \mathbf{v})}{\partial t} + \nabla \cdot (\rho \mathbf{v} \mathbf{v}) = \mu \nabla^2 \mathbf{v} + \mathbf{F}. \tag{10.55}$$

The force $\mathbf{F}$ on the right hand side is the buoyancy force that is associated with the flow, while the term $\mu \nabla^2 \mathbf{v}$ accounts for the viscosity of the flow with viscosity coefficient $\mu$. The mantle of the Earth is extremely viscous and mantle convection (if it exists at all) is a very slow process. We therefore assume that the inertia term $\partial(\rho \mathbf{v})/\partial t$ and the advection term $\nabla \cdot (\rho \mathbf{v} \mathbf{v})$ are small compared to the viscous term $\mu \nabla^2 \mathbf{v}$. (This assumption would have to be supported by a proper scale analysis.) Under this assumption, the mantle flow is predominantly governed by a balance between the viscous force and the buoyancy force:

$$\mu \nabla^2 \mathbf{v} = -\mathbf{F} . \tag{11.34}$$

The next step is to relate the buoyancy to the temperature perturbation $T$. A temperature perturbation $T$ from a reference temperature $T_0$ leads to a density perturbation $\rho$ from the reference density $\rho_0$ given by:

$$\rho = -\alpha T . \tag{11.35}$$

In this expression $\alpha$ is the thermal expansion coefficient which accounts for the expansion or contraction of material due to temperature changes.

**Problem b:** Explain why for most materials $\alpha > 0$. A notable exception is water at temperatures below $4\,^\circ\mathrm{C}$.

**Problem c:** Write $\rho(T_0 + T) = \rho_0 + \rho$ and use the Taylor expansion (2.11) truncated after the first order term to show that the expansion coefficient is given by $\alpha = -\partial \rho / \partial T$.

**Problem d:** The buoyancy force is given by Archimedes's law which states that this force equals the weight of the displaced fluid. Use this result, (11.34) and (11.35) in a scale analysis to show that the velocity is of the following order:

$$V \sim \frac{g \alpha T L^2}{\mu} , \tag{11.36}$$

where $g$ is the acceleration of gravity.

**Problem e:** Use this to derive that the ratio of the convective heat transport to the conductive heat transport is given by

$$\frac{\text{convective heat transport}}{\text{conductive heat transport}} \sim \frac{g \alpha T L^3}{\mu \kappa} . \tag{11.37}$$

The right hand side of this expression is dimensionless, and is called the *Rayleigh number* which is denoted by $Ra$:

$$Ra \equiv \frac{g\alpha T L^3}{\mu\kappa} . \tag{11.38}$$

The Rayleigh number is an indicator for the mode of heat transport. When $Ra \gg 1$, heat is predominantly transported by convection. When the thermal expansion coefficient $\alpha$ is large and when the viscosity $\mu$ and the heat conduction coefficient $\kappa$ are small the Rayleigh number is large and heat is transported by convection.

**Problem f:** Explain physically why a large value of $\alpha$ and small values of $\mu$ and $\kappa$ lead to convective heat transport rather than conductive heat transport.

Dimensionless numbers play a crucial role in fluid mechanics. A discussion of the Rayleigh number and other dimensionless diagnostics such as the Prandtl number and the Grashof number can be found in section 14.2 of Tritton [89]. The implications of the different values of the Rayleigh number on the character of convection in the Earth's mantle is discussed in refs. [63] and [91]. Of course, if one wants to use a scale analysis one must know the values of the physical properties involved. For the Earth's mantle, the thermal expansion coefficient $\alpha$ is not very well known because of the complications involved in laboratory measurements of the thermal expansion under the extremely high ambient pressure of the Earth's mantle [21].

## 11.6 Making an equation dimensionless

Usually the terms in equations that one wants to analyze have a physical dimension such as temperature, velocity, etc. It can sometimes be useful to rescale all the variables in the equation in such a way that the rescaled variables are dimensionless. This is convenient when setting up numerical solutions of the equations, but in general it also introduces dimensionless numbers that govern the physics of the problem in a natural way. As an example we will apply this technique to the heat equation (11.32).

Any variable can be made dimensionless by dividing out a constant that has the dimension of the variable. As an example, let the characteristic temperature variation be denoted by $T_0$, then the dimensional temperature perturbation can be written as:

$$T = T_0 T' . \tag{11.39}$$

The quantity $T'$ is dimensionless. Note that in this section $T$ is the absolute temperature whereas in the previous section $T$ denoted the temperature perturbation.

In this section, dimensionless variables are denoted by a prime. For example, let the characteristic time used to scale the time variable be denoted by $\tau$:

$$t = \tau t'. \tag{11.40}$$

We can still leave $\tau$ open and later choose a value that simplifies the equations as much as possible. Of course when we want to express the heat equation (11.32) in the new time variable we need to specify how the dimensional time-derivative $\partial/\partial t$ is related to the dimensionless time-derivative $\partial/\partial t'$.

**Problem a:** Use the chain rule for differentiation to show that

$$\frac{\partial}{\partial t} = \frac{1}{\tau}\frac{\partial}{\partial t'}. \tag{11.41}$$

**Problem b:** Let the velocity be scaled with the characteristic velocity (11.36):

$$\mathbf{v} = \frac{g\alpha T_0 L^2}{\mu}\mathbf{v}', \tag{11.42}$$

and let the position vector be scaled with the characteristic length $L$ of the system: $\mathbf{r} = L\mathbf{r}'$. Use a result similar to (11.41) to convert the spatial derivatives to the new space coordinate and rescale all terms in the heat equation (11.32) to derive the following dimensionless form of this equation

$$\underbrace{\frac{1}{\tau}\frac{\partial T'}{\partial t'}}_{(1)} + \underbrace{\frac{g\alpha T_0 L}{\mu}\nabla'\cdot\left(\mathbf{v}'T'\right)}_{(2)} = \underbrace{\frac{\kappa}{L^2}\nabla'^2 T}_{(3)}, \tag{11.43}$$

where $\nabla'$ is the gradient operator with respect to the dimensionless coordinates $\mathbf{r}'$.

At this point we have not yet specified the time scale $\tau$ for the scaling of the time variable. There are three possible situations that determine $\tau$. First, when term (2) is much larger than term (3), time derivative (1) balances term (2). Second, when term (3) is much larger than term (2), the time derivative balances term (3). The third possibility is that terms (2) and (3) are of the same order of magnitude, in which case the balance of terms is more complex.

**Problem c:** Show that the ratio of term (2) to term (3) is given by the Rayleigh number which is defined in (11.38).

In this expression all the primed terms are (by definition) of order one. At this point we assume that the convective heat transport dominates the conductive heat transport, i.e. that $Ra \gg 1$. This means that term (2) is much larger than term (3), hence the time derivative in term (1) must be of the same order as the convective term (2). Terms (1) and (2) can therefore only balance when

$$\frac{1}{\tau} = \frac{g\alpha T_0 L}{\mu} \,. \tag{11.44}$$

This condition determines the time scale of the evolution of the convecting system.

**Problem d:** Show that with this choice of $\tau$ the dimensionless heat equation is given by:

$$\frac{\partial T'}{\partial t'} + \nabla' \cdot \left( \mathbf{v}' T' \right) = \frac{1}{Ra} \nabla'^2 T' \,, \tag{11.45}$$

where $Ra$ is the Rayleigh number.

The advantage of this dimensionless equation over the original heat equation is that (11.45) contains only a single constant $Ra$, whereas the dimensional heat equation (11.32) depends on a large number of constants. In addition, the scaling of the heat equation has led in a natural way to the key role of the Rayleigh number in the mode of heat transport in a fluid. Since the Rayleigh number is assumed to be large, $1/Ra \ll 1$ and the last term can be seen as a small perturbation. One can either delete this term, or treat it with perturbation theory as described in chapter 22. The last term in (11.45) contains the highest spatial derivatives of that equation because it contains second order space derivatives whereas the other terms either have no space derivatives (term (1)) or only first order space derivatives (term (2)). Deleting this term turns (11.45) from a second order differential equation (in the space variables) into a first order differential equation. This entails a change in the number of boundary conditions that need to be imposed. As shown in section 22.7 the last term in (11.45) constitutes a *singular perturbation*, and one should be careful in deleting this (small) term altogether.

Transforming dimensional equations into dimensionless equations is often used to derive the relevant dimensionless physical constants of the system as well as to set up algorithms for solving systems numerically. The basic rationale behind this approach is that the physical units that are used are completely arbitrary. It is immaterial whether we express length in meters or in inches, but of course the numerical value of a given length changes when we change from meters to inches. Making the system dimensionless removes all physical units from the system because all the resulting terms in the equation are dimensionless.

# 12

## Linear algebra

In this chapter several elements of linear algebra are treated that have important applications in (geo)physics or that serve to illustrate methodologies used in other areas of mathematical physics.

### 12.1 Projections and the completeness relation

In mathematical physics, projections play an extremely important role. This is true not only in linear algebra, but also in the analysis of linear systems such as linear filters in data processing (see section 14.10) and the analysis of vibrating systems such as the normal modes of the Earth which is treated in section 19.7. Let us consider a vector $\mathbf{v}$ that we want to project along a unit vector $\hat{\mathbf{n}}$, see figure 12.1. In the examples in this section we work in a three-dimensional space, but the arguments presented here can be generalized to any number of dimensions.

We denote the projection of $\mathbf{v}$ along $\hat{\mathbf{n}}$ as $\mathbf{Pv}$, where $\mathbf{P}$ stands for the projection operator. In a three-dimensional space this operator can be represented by a $3 \times 3$ matrix. It is our goal to find the operator $\mathbf{P}$ in terms of the unit vector $\hat{\mathbf{n}}$ as well as the matrix form of this operator. By definition the projection of $\mathbf{v}$ is directed along $\hat{\mathbf{n}}$, hence:

$$\mathbf{Pv} = C\hat{\mathbf{n}} \ . \tag{12.1}$$

This means that we know the projection operator once the constant $C$ is known.

**Problem a:** Express the length of the vector $\mathbf{Pv}$ in terms of the length of the vector $\mathbf{v}$ and the angle $\varphi$ of figure 12.1 and express the angle $\varphi$ in terms of the inner product of the vectors $\mathbf{v}$ and $\hat{\mathbf{n}}$ to show that:
$C = (\hat{\mathbf{n}} \cdot \mathbf{v})$.

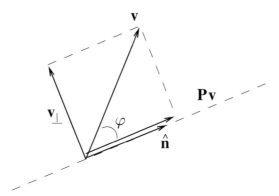

Fig. 12.1.   Definition of the geometric variables for the projection of a vector.

Inserting this expression for the constant $C$ in (12.1) leads to an expression for the projection $\mathbf{Pv}$:

$$\mathbf{Pv} = \hat{\mathbf{n}}\,(\hat{\mathbf{n}} \cdot \mathbf{v})\ . \tag{12.2}$$

**Problem b:** Show that the component $\mathbf{v}_\perp$ perpendicular to $\hat{\mathbf{n}}$ as defined in figure 12.1 is given by:

$$\mathbf{v}_\perp = \mathbf{v} - \hat{\mathbf{n}}\,(\hat{\mathbf{n}} \cdot \mathbf{v})\ . \tag{12.3}$$

**Problem c:** As an example, consider the projection along the unit vector along the $x$-axis: $\hat{\mathbf{n}} = \hat{\mathbf{x}}$. Show using (12.2) and (12.3) that in that case:

$$\mathbf{Pv} = \begin{pmatrix} v_x \\ 0 \\ 0 \end{pmatrix} \quad \text{and} \quad \mathbf{v}_\perp = \begin{pmatrix} 0 \\ v_y \\ v_z \end{pmatrix}\ .$$

**Problem d:** When we project the vector $\mathbf{Pv}$ once more along the same unit vector $\hat{\mathbf{n}}$ the vector will not change. We therefore expect that $\mathbf{P}(\mathbf{Pv}) = \mathbf{Pv}$. Show using (12.2) that this is indeed the case. Since this property holds for any vector $\mathbf{v}$ we can also write it as:

$$\mathbf{P}^2 = \mathbf{P}\ . \tag{12.4}$$

**Problem e:** If $\mathbf{P}$ were a scalar the expression above would imply that $\mathbf{P}$ is the identity operator $\mathbf{I}$ or that $\mathbf{P} = 0$. Can you explain why (12.4) does *not* imply that $\mathbf{P}$ is either the identity operator or that is equal to zero?

In (12.2) we derived the action of the projection operator on a vector $\mathbf{v}$. Since this expression holds for any vector $\mathbf{v}$ it can be used to derive an explicit form of the projection operator:

$$\mathbf{P} = \hat{\mathbf{n}}\hat{\mathbf{n}}^T\ . \tag{12.5}$$

This expression should not be confused with the inner product $(\hat{\mathbf{n}} \cdot \hat{\mathbf{n}})$; instead it denotes the dyad of the vector $\hat{\mathbf{n}}$ and itself. The superscript $T$ denotes the transpose of a vector or matrix. The transpose of a vector (or matrix) is found by interchanging its rows and columns. For example, the transpose $\mathbf{A}^T$ of a matrix $\mathbf{A}$ is defined by:

$$A_{ij}^T = A_{ji} , \tag{12.6}$$

and the transpose of the vector $\mathbf{u}$ is defined by:

$$\mathbf{u}^T = (u_x, u_y, u_z) \qquad \text{when} \qquad \mathbf{u} = \begin{pmatrix} u_x \\ u_y \\ u_z \end{pmatrix} , \tag{12.7}$$

i.e. taking the transpose converts a column vector into a row vector. The projection operator $\mathbf{P}$ is written in (12.5) as a dyad. In general the dyad $\mathbf{T}$ of two vectors $\mathbf{u}$ and $\mathbf{v}$ is defined as

$$\mathbf{T} = \mathbf{u}\mathbf{v}^T . \tag{12.8}$$

This expression simply means that the components $T_{ij}$ of the dyad are defined by

$$T_{ij} = u_i v_j , \tag{12.9}$$

where $u_i$ is the $i$-component of $\mathbf{u}$ and $v_j$ is the $j$-component of $\mathbf{v}$.

In the literature you will find different notations for the inner product of two vectors. The inner product of the vectors $\mathbf{u}$ and $\mathbf{v}$ is sometimes written as

$$(\mathbf{u} \cdot \mathbf{v}) = \mathbf{u}^T \mathbf{v} . \tag{12.10}$$

**Problem f:** Considering the vector $\mathbf{v}$ as a $3 \times 1$ matrix and the vector $\mathbf{v}^T$ as a $1 \times 3$ matrix, show that the notation used on the right hand sides of (12.10) and (12.8) is consistent with the normal rules for matrix multiplication.

Equation (12.5) relates the projection operator $\mathbf{P}$ to the unit vector $\hat{\mathbf{n}}$. From this the representation of the projection operator as a $3 \times 3$-matrix can be found by computing the dyad $\hat{\mathbf{n}}\hat{\mathbf{n}}^T$.

**Problem g:** Show that the operator for the projection along the unit vector

$$\hat{\mathbf{n}} = \frac{1}{\sqrt{14}} \begin{pmatrix} 1 \\ 2 \\ 3 \end{pmatrix}$$

is given by

$$\mathbf{P} = \frac{1}{14} \begin{pmatrix} 1 & 2 & 3 \\ 2 & 4 & 6 \\ 3 & 6 & 9 \end{pmatrix} .$$

Verify explicitly that for this example $\mathbf{P}\hat{\mathbf{n}} = \hat{\mathbf{n}}$, and explain this result.

Up to this point we have projected the vector $\mathbf{v}$ along a single unit vector $\hat{\mathbf{n}}$. Suppose we have a set of mutually orthogonal unit vectors $\hat{\mathbf{n}}_i$. The fact that these unit vectors are mutually orthogonal means that the different unit vectors are perpendicular to each other: $(\hat{\mathbf{n}}_i \cdot \hat{\mathbf{n}}_j) = 0$ when $i \neq j$. We can project $\mathbf{v}$ on each of these unit vectors and add these projections. This gives us the projection of $\mathbf{v}$ on the subspace spanned by the unit vectors $\hat{\mathbf{n}}_i$:

$$\mathbf{P}\mathbf{v} = \sum_i \hat{\mathbf{n}}_i \left( \hat{\mathbf{n}}_i \cdot \mathbf{v} \right) . \tag{12.11}$$

When the unit vectors $\hat{\mathbf{n}}_i$ span the full space we are working in, the projected vector is identical to the original vector. To see this, consider for example a three-dimensional space. Any vector can be decomposed into its components along the $x$-, $y$- and $z$-axes, and this can be written as:

$$\mathbf{v} = v_x \hat{\mathbf{x}} + v_y \hat{\mathbf{y}} + v_z \hat{\mathbf{z}} = \hat{\mathbf{x}} \left( \hat{\mathbf{x}} \cdot \mathbf{v} \right) + \hat{\mathbf{y}} \left( \hat{\mathbf{y}} \cdot \mathbf{v} \right) + \hat{\mathbf{z}} \left( \hat{\mathbf{z}} \cdot \mathbf{v} \right) ; \tag{12.12}$$

note that this expression has the same form as (12.11). This implies that when in (12.11) we sum over a set of unit vectors that completely spans the space we are working in, the right hand side of (12.11) is identical to the original vector $\mathbf{v}$, i.e. $\sum_i \hat{\mathbf{n}}_i \left( \hat{\mathbf{n}}_i \cdot \mathbf{v} \right) = \mathbf{v}$. The operator of the left hand side of this equality is therefore identical to the identity operator $\mathbf{I}$:

$$\sum_{i=1}^{N} \hat{\mathbf{n}}_i \hat{\mathbf{n}}_i^T = \mathbf{I} . \tag{12.13}$$

Keep in mind that $N$ is the dimension of the space we are working in; if we sum over a smaller number of unit vectors we project on a subspace of the $N$-dimensional space. Expression (12.13) expresses that the vectors $\hat{\mathbf{n}}_i$ (with $i = 1, \dots, N$) can be used to give a complete representation of any vector. Such a set of vectors is called a *complete set*, and expression (12.13) is called the *closure relation*.

**Problem h:** Verify explicitly that when the unit vectors $\hat{\mathbf{n}}_i$ are chosen to be the unit vectors $\hat{\mathbf{x}}$, $\hat{\mathbf{y}}$ and $\hat{\mathbf{z}}$ along the $x$-, $y$- and $z$-axes that the right hand side of (12.13) is given by the $3 \times 3$ identity matrix.

There are, of course, many different ways of choosing a set of three orthogonal unit vectors in three dimensions. Expression (12.13) should hold for every choice of a complete set of unit vectors.

**Problem i:** Verify explicitly that when the unit vectors $\hat{\mathbf{n}}_i$ are chosen to be the unit vectors $\hat{\mathbf{r}}$, $\hat{\boldsymbol{\theta}}$ and $\hat{\boldsymbol{\varphi}}$ defined in (3.6) for a system of

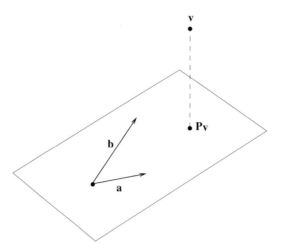

Fig. 12.2.   Definition of the geometric variables for the projection on a plane.

spherical coordinates the right hand side of (12.13) is given by the $3 \times 3$ identity matrix.

## 12.2 A projection on vectors that are not orthogonal

In the previous section we considered a projection on a set of orthogonal unit vectors. In this section we consider an example of a projection on a set of vectors that are not necessarily orthogonal. Consider two vectors **a** and **b** in a three-dimensional space. These two vectors span a two-dimensional plane. In this section we determine the projection of a vector **v** on the plane spanned by the vectors **a** and **b**, see figure 12.2 for the geometry of the problem. The projection of **v** on the plane is denoted by **Pv**.

By definition the projected vector **Pv** lies in the plane spanned by **a** and **b**, and this vector can thus be written as:

$$\mathbf{Pv} = \alpha\mathbf{a} + \beta\mathbf{b} \ . \tag{12.14}$$

The task of finding the projection can therefore be reduced to finding the two coefficients $\alpha$ and $\beta$. These constants follow from the requirement that the vector joining **v** with its projection **Pv** is perpendicular to both **a** and **b**, see figure 12.2.

**Problem a:** Show that this requirement is equivalent to the following system of equations for $\alpha$ and $\beta$:

$$\left.\begin{array}{l} \alpha\left(\mathbf{a}\cdot\mathbf{a}\right) + \beta\left(\mathbf{a}\cdot\mathbf{b}\right) = \left(\mathbf{a}\cdot\mathbf{v}\right), \\ \alpha\left(\mathbf{a}\cdot\mathbf{b}\right) + \beta\left(\mathbf{b}\cdot\mathbf{b}\right) = \left(\mathbf{b}\cdot\mathbf{v}\right). \end{array}\right\} \tag{12.15}$$

**Problem b:** Show that the solution of this system is given by

$$\left.\begin{array}{l} \alpha = \dfrac{\left[b^2\mathbf{a} - (\mathbf{a}\cdot\mathbf{b})\,\mathbf{b}\right]\cdot\mathbf{v}}{a^2b^2 - (\mathbf{a}\cdot\mathbf{b})^2}\,, \\[4mm] \beta = \dfrac{\left[a^2\mathbf{b} - (\mathbf{a}\cdot\mathbf{b})\,\mathbf{a}\right]\cdot\mathbf{v}}{a^2b^2 - (\mathbf{a}\cdot\mathbf{b})^2}\,, \end{array}\right\} \qquad (12.16)$$

where $a$ denotes the length of the vector $\mathbf{a}$: $a \equiv |\mathbf{a}|$, and a similar notation is used for the vector $\mathbf{b}$.

**Problem c:** Show using (12.14) and (12.16) that the projection operator $\mathbf{P}$ for the projection on the plane is given by

$$\mathbf{P} = \frac{1}{a^2b^2 - (\mathbf{a}\cdot\mathbf{b})^2}\left[b^2\mathbf{a}\mathbf{a}^T + a^2\mathbf{b}\mathbf{b}^T - (\mathbf{a}\cdot\mathbf{b})\left(\mathbf{a}\mathbf{b}^T + \mathbf{b}\mathbf{a}^T\right)\right].$$

$$(12.17)$$

This example shows that a projection on a set of nonorthogonal basis vectors is much more complex than the projection on a set of orthonormal basis vectors. A different way of finding the projection operator of expression (12.17) is by first finding two *orthogonal* unit vectors in the plane spanned by $\mathbf{a}$ and $\mathbf{b}$ and then using (12.11). One unit vector can be found by dividing $\mathbf{a}$ by its length to give the unit vector $\hat{\mathbf{a}} = \mathbf{a}/|\mathbf{a}|$. The second unit vector can be found by considering the component $\mathbf{b}_\perp$ of $\mathbf{b}$ perpendicular to $\hat{\mathbf{a}}$ and by normalizing the resulting vector to form the unit vector $\hat{\mathbf{b}}_\perp$ that is perpendicular to $\hat{\mathbf{a}}$, see figure 12.3.

**Problem d:** Use (12.3) to find $\hat{\mathbf{b}}_\perp$ and show that the projection operator $\mathbf{P}$ of (12.17) can also be written as

$$\mathbf{P} = \hat{\mathbf{a}}\hat{\mathbf{a}}^T + \hat{\mathbf{b}}_\perp\hat{\mathbf{b}}_\perp^T\,. \qquad (12.18)$$

Note that this expression is consistent with (12.11).

Up to this point the plane has been defined by the vectors $\mathbf{a}$ and $\mathbf{b}$ (or equivalently by the orthonormal unit vectors $\hat{\mathbf{a}}$ and $\hat{\mathbf{b}}_\perp$). However, a plane can also be defined by the unit vector $\hat{\mathbf{n}}$ that is perpendicular to the plane, see figure 12.3. In fact, the unit vectors $\hat{\mathbf{a}}$, $\hat{\mathbf{b}}_\perp$ and $\hat{\mathbf{n}}$ form a complete orthonormal basis of the three-dimensional space. According to (12.13) this implies that $\hat{\mathbf{a}}\hat{\mathbf{a}}^T + \hat{\mathbf{b}}_\perp\hat{\mathbf{b}}_\perp^T + \hat{\mathbf{n}}\hat{\mathbf{n}}^T = \mathbf{I}$. With (12.18) this implies that the projection operator $\mathbf{P}$ can also be written as

$$\mathbf{P} = \mathbf{I} - \hat{\mathbf{n}}\hat{\mathbf{n}}^T\,. \qquad (12.19)$$

**Problem e:** Give an alternative derivation of this result. Hint: let the operator in (12.19) act on an arbitrary vector $\mathbf{v}$.

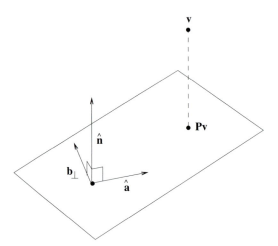

Fig. 12.3.   Definition of the normal vector to a plane.

### 12.3 The Coriolis force and centrifugal force

As an example of working with the cross-product of vectors we consider the inertia forces that occur in the mechanics of rotating coordinate systems. This is of great importance in the Earth sciences, because the rotation of the Earth plays a crucial role in the motion of wind in the atmosphere and currents in the ocean. In addition, the Earth's rotation is essential for the generation of the magnetic field of the Earth in the outer core.

In order to describe the motion of a particle in a rotating coordinate system we need to characterize the rotation somehow. This can be achieved by introducing a vector $\boldsymbol{\Omega}$ that is aligned with the rotation axis and whose length is given by the *rate* of rotation expressed in radians per seconds.

**Problem a:** Compute the direction of $\boldsymbol{\Omega}$ and the length $\Omega = |\boldsymbol{\Omega}|$ for the Earth's rotation.

Let us assume we are considering a vector $\mathbf{q}$ that is constant in the rotating coordinate system. In a nonrotating system this vector changes with time because it co-rotates with the rotating system. The vector $\mathbf{q}$ can be decomposed into a component $\mathbf{q}_{//}$ along the rotation vector and a component $\mathbf{q}_{\perp}$ perpendicular to the rotation vector. In addition, a vector $\mathbf{b}$ is defined in figure 12.4 that is perpendicular to both $\mathbf{q}_{\perp}$ and $\boldsymbol{\Omega}$ in such a way that $\boldsymbol{\Omega}$, $\mathbf{q}_{\perp}$ and $\mathbf{b}$ form a right handed orthogonal system.

**Problem b:** Show that:

$$\left.\begin{aligned}
\mathbf{q}_{//} &= \hat{\boldsymbol{\Omega}}\left(\hat{\boldsymbol{\Omega}} \cdot \mathbf{q}\right), \\
\mathbf{q}_{\perp} &= \mathbf{q} - \hat{\boldsymbol{\Omega}}\left(\hat{\boldsymbol{\Omega}} \cdot \mathbf{q}\right), \\
\mathbf{b} &= \hat{\boldsymbol{\Omega}} \times \mathbf{q}.
\end{aligned}\right\} \qquad (12.20)$$

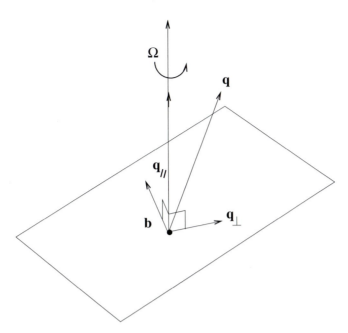

Fig. 12.4. Decomposition of a vector in a rotating coordinate system.

**Problem c:** In a fixed nonrotating coordinate system, the vector $\mathbf{q}$ rotates, hence its position is time-dependent: $\mathbf{q} = \mathbf{q}(t)$. Let us consider how the vector changes over a time interval $\Delta t$. Since the component $\mathbf{q}_{//}$ is at all times directed along the rotation vector $\mathbf{\Omega}$, it is constant in time. Over a time interval $\Delta t$ the coordinate system rotates over an angle $\Omega \Delta t$. Use this to show that the component of $\mathbf{q}$ perpendicular to the rotation vector satisfies:

$$\mathbf{q}_{\perp}(t + \Delta t) = \cos\left(\Omega \Delta t\right) \mathbf{q}_{\perp}(t) + \sin\left(\Omega \Delta t\right) \mathbf{b} , \qquad (12.21)$$

and that time evolution of $\mathbf{q}$ is therefore given by

$$\mathbf{q}(t + \Delta t) = \mathbf{q}(t) + \left[\cos\left(\Omega \Delta t\right) - 1\right] \mathbf{q}_{\perp}(t) + \sin\left(\Omega \Delta t\right) \mathbf{b} . \quad (12.22)$$

**Problem d:** The goal is to obtain the time derivative of the vector $\mathbf{q}$. This quantity can be computed using the rule $d\mathbf{q}/dt = \lim_{\Delta t \to 0} [\mathbf{q}(t + \Delta t) - \mathbf{q}(t)]/\Delta t$. Use this and (12.22) to show that

$$\dot{\mathbf{q}} = \Omega \mathbf{b} , \qquad (12.23)$$

where the dot denotes the time derivative. Use (12.20) to show that the time derivative of the vector $\mathbf{q}$ is given by

$$\dot{\mathbf{q}} = \mathbf{\Omega} \times \mathbf{q} . \qquad (12.24)$$

At this point the vector $\mathbf{q}$ can be any vector that corotates with the rotating coordinate system. In this rotating coordinate system, three Cartesian basis vectors $\hat{\mathbf{x}}$, $\hat{\mathbf{y}}$ and $\hat{\mathbf{z}}$ can be used as a basis to decompose the position vector:

$$\mathbf{r}_{rot} = x\hat{\mathbf{x}} + y\hat{\mathbf{y}} + z\hat{\mathbf{z}} \ . \tag{12.25}$$

Since these basis vectors are constant in the rotating coordinate system, they satisfy (12.24) so that:

$$\left.\begin{array}{l} d\hat{\mathbf{x}}/dt = \boldsymbol{\Omega} \times \hat{\mathbf{x}} \ , \\ d\hat{\mathbf{y}}/dt = \boldsymbol{\Omega} \times \hat{\mathbf{y}} \ , \\ d\hat{\mathbf{z}}/dt = \boldsymbol{\Omega} \times \hat{\mathbf{z}} \ . \end{array}\right\} \tag{12.26}$$

It should be noted that we have *not* assumed that the position vector $\mathbf{r}_{rot}$ in (12.25) rotates with the coordinate system, we have only assumed that the unit vectors $\hat{\mathbf{x}}$, $\hat{\mathbf{y}}$ and $\hat{\mathbf{z}}$ rotate with the coordinate system. Of course, this will leave an imprint on the velocity and the acceleration. In general the velocity and the acceleration follow by differentiating (12.25) with time. If the unit vectors $\hat{\mathbf{x}}$, $\hat{\mathbf{y}}$ and $\hat{\mathbf{z}}$ were fixed, they would not contribute to the time derivative. However, the unit vectors $\hat{\mathbf{x}}$, $\hat{\mathbf{y}}$ and $\hat{\mathbf{z}}$ rotate with the coordinate system and the associated time derivative is given by (12.26).

**Problem e:** Differentiate the position vector in (12.25) with respect to time and show that the velocity vector $\mathbf{v}$ is given by:

$$\mathbf{v} = \dot{x}\hat{\mathbf{x}} + \dot{y}\hat{\mathbf{y}} + \dot{z}\hat{\mathbf{z}} + \boldsymbol{\Omega} \times \mathbf{r} \ . \tag{12.27}$$

The terms $\dot{x}\hat{\mathbf{x}} + \dot{y}\hat{\mathbf{y}} + \dot{z}\hat{\mathbf{z}}$ denote the velocity as seen in the rotating coordinate system; this velocity is denoted by $\mathbf{v}_{rot}$. The velocity vector can therefore be written as:

$$\mathbf{v} = \mathbf{v}_{rot} + \boldsymbol{\Omega} \times \mathbf{r} \ . \tag{12.28}$$

**Problem f:** Give an interpretation of the last term in this expression.

**Problem g:** The acceleration follows by differentiating (12.27) for the velocity once more with respect to time. Show that the acceleration is given by

$$\mathbf{a} = \ddot{x}\hat{\mathbf{x}} + \ddot{y}\hat{\mathbf{y}} + \ddot{z}\hat{\mathbf{z}} + 2\boldsymbol{\Omega} \times (\dot{x}\hat{\mathbf{x}} + \dot{y}\hat{\mathbf{y}} + \dot{z}\hat{\mathbf{z}}) + \boldsymbol{\Omega} \times (\boldsymbol{\Omega} \times \mathbf{r}) \ . \tag{12.29}$$

The terms $\ddot{x}\hat{\mathbf{x}} + \ddot{y}\hat{\mathbf{y}} + \ddot{z}\hat{\mathbf{z}}$ on the right hand side denote the acceleration as seen in the rotating coordinate system; this quantity will be denoted by $\mathbf{a}_{rot}$. The terms $\dot{x}\hat{\mathbf{x}} + \dot{y}\hat{\mathbf{y}} + \dot{z}\hat{\mathbf{z}}$ again denote the velocity $\mathbf{v}_{rot}$ as seen in the rotating coordinate system. The left hand side is by Newton's law equal to $\mathbf{F}/m$, where $\mathbf{F}$ is the force acting on the particle.

**Problem h:** Use this to show that in the rotating coordinate system Newton's law is given by:

$$m\mathbf{a}_{rot} = \mathbf{F} - 2m\mathbf{\Omega} \times \mathbf{v}_{rot} - m\mathbf{\Omega} \times (\mathbf{\Omega} \times \mathbf{r}) \ . \qquad (12.30)$$

The rotation manifests itself through two additional forces. The term $-2m\mathbf{\Omega} \times \mathbf{v}_{rot}$ describes the *Coriolis force* and the term $-m\mathbf{\Omega} \times (\mathbf{\Omega} \times \mathbf{r})$ describes the centrifugal force.

**Problem i:** Show that the centrifugal force is perpendicular to the rotation axis and is directed from the rotation axis towards the particle.

**Problem j:** Air flows from high-pressure areas to low-pressure areas. As air flows in the northern hemisphere from a high-pressure area to a low-pressure area, is it deflected towards the right or towards the left when seen from above?

**Problem k:** Compute the magnitude of the centrifugal force and the Coriolis force you experience due to the Earth's rotation when you ride your bicycle. Compare this with the force $m\mathbf{g}$ you experience due to the gravitational attraction of the Earth. It suffices to compute the orders of magnitude of the different terms. In the northern hemisphere does the Coriolis force deflect you to the left or to the right? Have you ever noticed a tilt due to the Coriolis force while riding your bicycle?

In meteorology and oceanography it is often convenient to describe the motion of air or water along the Earth's surface using a Cartesian coordinate system that rotates with the Earth with unit vectors pointing eastwards ($\hat{\mathbf{e}}_1$), northwards ($\hat{\mathbf{e}}_2$) and upwards ($\hat{\mathbf{e}}_3$), see figure 12.5. These unit vectors can be related to the unit vectors $\hat{\mathbf{r}}$, $\hat{\boldsymbol{\varphi}}$ and $\hat{\boldsymbol{\theta}}$ that are defined in (3.7). Let the velocity in the eastward direction be denoted by $u$, the velocity in the northward direction by $v$ and the vertical velocity by $w$.

**Problem l:** Show that:

$$\hat{\mathbf{e}}_1 = \hat{\boldsymbol{\varphi}} \ , \quad \hat{\mathbf{e}}_2 = -\hat{\boldsymbol{\theta}} \ , \quad \hat{\mathbf{e}}_3 = \hat{\mathbf{r}} \ , \qquad (12.31)$$

and that the velocity in this rotating coordinate system is given by

$$\mathbf{v} = u\hat{\mathbf{e}}_1 + v\hat{\mathbf{e}}_2 + w\hat{\mathbf{e}}_3 \ . \qquad (12.32)$$

**Problem m:** We assume that the axes of the spherical coordinate system are chosen in such a way that the direction $\theta = 0$ is aligned with the rotation axis. This is a different way of saying that the rotation

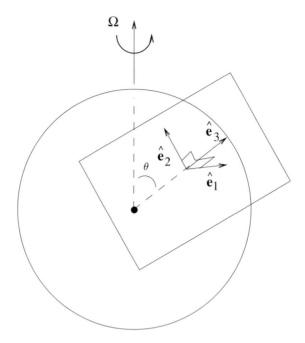

Fig. 12.5.   Definition of a local Cartesian coordinate system that is aligned with the Earth's surface.

vector is parallel to the $z$-axis: $\boldsymbol{\Omega} = \Omega \hat{\mathbf{z}}$. Use the first two expressions of (3.13) to show that the rotation vector has the following expansion in terms of the unit vectors $\hat{\mathbf{r}}$ and $\hat{\boldsymbol{\theta}}$:

$$\boldsymbol{\Omega} = \Omega \left( \cos \theta \, \hat{\mathbf{r}} - \sin \theta \, \hat{\boldsymbol{\theta}} \right) . \tag{12.33}$$

**Problem n:** In the rotating coordinate system, the Coriolis force is given by $\mathbf{F}_{cor} = -2m\boldsymbol{\Omega} \times \mathbf{v}$. Use expressions (12.31)–(12.33) and (3.11) for the cross-product of the unit vectors to show that the Coriolis force is given by

$$\mathbf{F}_{cor} = 2m\Omega \sin \theta \, u \, \hat{\mathbf{r}} + 2m\Omega \cos \theta \, u \, \hat{\boldsymbol{\theta}} + 2m\Omega \left( v \cos \theta \, - w \sin \theta \right) \hat{\boldsymbol{\varphi}} . \tag{12.34}$$

**Problem o:** Both the ocean and the atmosphere are shallow in the sense that the vertical length scale (a few kilometers for the ocean and around 10 kilometers for the atmosphere) is much less than the horizontal length scale. This causes the vertical velocity to be much smaller than the horizontal velocity. For this reason the vertical velocity $w$ can be neglected in (12.34). Use this approximation and definition (12.31) to show that the horizontal component $\mathbf{a}_{cor}^{H}$ of the

Coriolis acceleration is given in this approach by:

$$\mathbf{a}_{cor}^{H} = -f\hat{\mathbf{e}}_3 \times \mathbf{v} , \tag{12.35}$$

with

$$f = 2\Omega \cos\theta . \tag{12.36}$$

This result is widely used in meteorology and oceanography, because (12.35) states that, in the Cartesian coordinate system aligned with the Earth's surface, the Coriolis force generated by the rotation around the true Earth's axis of rotation is identical to the Coriolis force generated by the rotation around a vertical axis with a rotation rate given by $\Omega \cos\theta$. This rotation rate is largest at the poles, where $\cos\theta = \pm 1$, and vanishes at the equator, where $\cos\theta = 0$. The parameter $f$ in (12.35) acts as a coupling parameter; it is called the *Coriolis parameter*. (In the literature on geophysical fluid dynamics one often uses latitude rather than the co-latitude $\theta$ that is used here, and for this reason one often sees a sin term rather than a cos term in the definition of the Coriolis parameter.) In many applications one disregards the dependence of $f$ on the colatitude $\theta$; in that approach $f$ is a constant and one speaks of the *f-plane approxima-tion*. However, the dependence of the Coriolis parameter on $\theta$ is crucial in explaining a number of atmospheric and oceanographic phenomena such as the propagation of Rossby waves and the formation of the Gulfstream. In a further refinement one linearizes the dependence of the Coriolis pa-rameter with colatitude. This leads to the *$\beta$-plane approximation*. Details can be found in the books of Holton [39] and Pedlosky [67].

## 12.4 The eigenvalue decomposition of a square matrix

In this section we consider the way in which a square $N \times N$ matrix $\mathbf{A}$ operates on a vector. Since a matrix describes a linear transformation from a vector to a new vector, the action of the matrix $\mathbf{A}$ can be quite complicated. However, suppose the matrix has a set of eigenvectors $\hat{\mathbf{v}}^{(n)}$. We assume these eigenvectors are normalized, hence a caret is used in the notation $\hat{\mathbf{v}}^{(n)}$. These eigenvectors are extremely useful because the action of $\mathbf{A}$ on an eigenvector $\hat{\mathbf{v}}^{(n)}$ is very simple:

$$\mathbf{A}\hat{\mathbf{v}}^{(n)} = \lambda_n \hat{\mathbf{v}}^{(n)} , \tag{12.37}$$

where $\lambda_n$ is the eigenvalue of the eigenvector $\hat{\mathbf{v}}^{(n)}$. When $\mathbf{A}$ acts on an eigenvector, the resulting vector is parallel to the original vector, and the only effect of $\mathbf{A}$ on this vector is to elongate the vector (when $\lambda_n \geq 1$), compress the vector (when $0 \leq \lambda_n < 1$) or reverse the vector (when $\lambda_n < 0$). We restrict ourselves here to matrices that are real and symmetric.

**Problem a:** Show that for such a matrix the eigenvalues are real and the eigenvectors are orthogonal.

The fact that the eigenvectors $\hat{\mathbf{v}}^{(n)}$ are normalized and mutually orthogonal can be expressed as

$$\left(\hat{\mathbf{v}}^{(n)} \cdot \hat{\mathbf{v}}^{(m)}\right) = \delta_{nm} , \tag{12.38}$$

where $\delta_{nm}$ is the Kronecker delta which is defined as

$$\delta_{nm} = \begin{cases} 1 & \text{when} & n = m \\ 0 & \text{when} & n \neq m \end{cases} . \tag{10.15}$$

The eigenvectors $\hat{\mathbf{v}}^{(n)}$ can be used to define the columns of a matrix $\mathbf{V}$:

$$\mathbf{V} = \begin{pmatrix} \vdots & \vdots & & \vdots \\ \hat{\mathbf{v}}^{(1)} & \hat{\mathbf{v}}^{(2)} & \cdots & \hat{\mathbf{v}}^{(N)} \\ \vdots & \vdots & & \vdots \end{pmatrix} ; \tag{12.39}$$

this definition implies that

$$V_{ij} \equiv v_i^{(j)} . \tag{12.40}$$

**Problem b:** Use the orthogonality of the eigenvectors $\hat{\mathbf{v}}^{(n)}$ (expression (12.38)) to show that the matrix $\mathbf{V}$ is unitary, i.e. to show that

$$\mathbf{V}^T \mathbf{V} = \mathbf{I} , \tag{12.41}$$

where $\mathbf{I}$ is the identity matrix with elements $I_{kl} = \delta_{kl}$. The superscript $T$ denotes the transpose.

There are $N$ eigenvectors which are mutually orthonormal in an $N$-dimensional space. These eigenvectors therefore form a complete set and analogously to (12.13) the completeness relation can be expressed as

$$\mathbf{I} = \sum_{n=1}^{N} \hat{\mathbf{v}}^{(n)} \hat{\mathbf{v}}^{(n)T} . \tag{12.42}$$

When the terms in this expression operate on a arbitrary vector $\mathbf{p}$, an expansion of $\mathbf{p}$ in the eigenvectors that is completely analogous to (12.11) is obtained:

$$\mathbf{p} = \sum_{n=1}^{N} \hat{\mathbf{v}}^{(n)} \hat{\mathbf{v}}^{(n)T} \mathbf{p} = \sum_{n=1}^{N} \hat{\mathbf{v}}^{(n)} \left(\hat{\mathbf{v}}^{(n)} \cdot \mathbf{p}\right) . \tag{12.43}$$

This is a useful expression because it can be used to simplify the effect of the matrix $\mathbf{A}$ on an arbitrary vector $\mathbf{p}$.

**Problem c:** Let $\mathbf{A}$ act on (12.43) and show that:

$$\mathbf{A}\mathbf{p} = \sum_{n=1}^{N} \lambda_n \hat{\mathbf{v}}^{(n)} \left( \hat{\mathbf{v}}^{(n)} \cdot \mathbf{p} \right) . \qquad (12.44)$$

This expression has an interesting geometric interpretation. When $\mathbf{A}$ acts on $\mathbf{p}$, the vector $\mathbf{p}$ is projected on each of the eigenvectors; this is described by the term $(\hat{\mathbf{v}}^{(n)} \cdot \mathbf{p})$. The corresponding eigenvector $\hat{\mathbf{v}}^{(n)}$ is multiplied by the eigenvalue $\hat{\mathbf{v}}^{(n)} \to \lambda_n \hat{\mathbf{v}}^{(n)}$ and the result is summed over all the eigenvectors. The action of $\mathbf{A}$ can thus be reduced to a projection on eigenvectors, a multiplication with the corresponding eigenvalue and a summation over all eigenvectors. The eigenvalue $\lambda_n$ can be seen as the sensitivity of the eigenvector $\hat{\mathbf{v}}^{(n)}$ to the matrix $\mathbf{A}$.

**Problem d:** Expression (12.44) holds for every vector $\mathbf{p}$. Use this to show that $\mathbf{A}$ can be written as:

$$\mathbf{A} = \sum_{n=1}^{N} \lambda_n \hat{\mathbf{v}}^{(n)} \hat{\mathbf{v}}^{(n)T} . \qquad (12.45)$$

**Problem e:** Show that with the definition (12.39) this result can also be written as:

$$\mathbf{A} = \mathbf{V}\mathbf{\Sigma}\mathbf{V}^T , \qquad (12.46)$$

where $\mathbf{\Sigma}$ is a matrix that has the eigenvalues on its diagonal and whose other elements are equal to zero:

$$\mathbf{\Sigma} = \begin{pmatrix} \lambda_1 & 0 & \cdots & 0 \\ 0 & \lambda_2 & \cdots & 0 \\ \vdots & \vdots & \ddots & \vdots \\ 0 & 0 & & \lambda_N \end{pmatrix} . \qquad (12.47)$$

Hint: let (12.46) act on an arbitrary vector, use definition (12.40) and see what happens.

## 12.5 Computing a function of a matrix

Expansion (12.45) (or equivalently (12.46)) is very useful because it provides a way to compute the inverse of a matrix and to complete complex functions of a matrix such as the exponential of a matrix. Let us first use (12.45) to compute the inverse $\mathbf{A}^{-1}$ of the matrix. In order to do this we must know the effect of $\mathbf{A}^{-1}$ on the eigenvectors $\hat{\mathbf{v}}^{(n)}$.

**Problem a:** Use the relation $\hat{\mathbf{v}}^{(n)} = \mathbf{I}\hat{\mathbf{v}}^{(\mathbf{n})} = \mathbf{A}^{-1}\mathbf{A}\hat{\mathbf{v}}^{(n)}$ to show that $\hat{\mathbf{v}}^{(n)}$ is also an eigenvector of the inverse $\mathbf{A}^{-1}$ with eigenvalue $1/\lambda_n$:

$$\mathbf{A}^{-1}\hat{\mathbf{v}}^{(n)} = \frac{1}{\lambda_n}\hat{\mathbf{v}}^{(n)} \; . \tag{12.48}$$

**Problem b:** Use this result and the eigenvector decomposition (12.43) to show that the effect of $\mathbf{A}^{-1}$ on a vector $\mathbf{p}$ can be written as

$$\mathbf{A}^{-1}\mathbf{p} = \sum_{n=1}^{N} \frac{1}{\lambda_n}\hat{\mathbf{v}}^{(n)}\left(\hat{\mathbf{v}}^{(n)}\cdot\mathbf{p}\right) \; . \tag{12.49}$$

Also show that this implies that $\mathbf{A}^{-1}$ can also be written as:

$$\mathbf{A}^{-1} = \mathbf{V}\mathbf{\Sigma}^{-1}\mathbf{V}^T \; , \tag{12.50}$$

with

$$\mathbf{\Sigma}^{-1} = \begin{pmatrix} 1/\lambda_1 & 0 & \cdots & 0 \\ 0 & 1/\lambda_2 & \cdots & 0 \\ \vdots & \vdots & \ddots & \vdots \\ 0 & 0 & & 1/\lambda_N \end{pmatrix} \; . \tag{12.51}$$

This is an important result: it means that once we have computed the eigenvectors and eigenvalues of a matrix, we can compute the inverse matrix very efficiently. Note that this procedure gives problems when one of the eigenvalues is equal to zero because for such an eigenvalue $1/\lambda_n$ is not defined. However, this makes sense: when one (or more) of the eigenvalues vanishes the matrix is singular and the inverse does not exist. Also when one of the eigenvalues is nonzero but close to zero, the corresponding term $1/\lambda_n$ is very large, and in practice this gives rise to numerical instabilities. In this situation the inverse of the matrix exists, but the result is very sensitive to computational (and other) errors. Such a matrix is called *poorly conditioned*.

In general, a function of a matrix, such as the exponent of a matrix, is not defined. However, suppose we have a function $f(z)$ that operates on a scalar $z$ and that this function can be written as a power series:

$$f(z) = \sum_p a_p z^p \; . \tag{12.52}$$

For example, when $f(z) = \exp(z)$, then $f(z) = \sum_{p=0}^{\infty}(1/p!)z^p$. On replacing the scalar $z$ by the matrix $\mathbf{A}$ the power series expansion can be used to *define* the effect of the function $f$ when it operates on the matrix $\mathbf{A}$:

$$f(\mathbf{A}) \equiv \sum_p a_p \mathbf{A}^p \; . \tag{12.53}$$

Although this may seem to be a simple rule to compute $f(\mathbf{A})$, it is actually not so useful because in many applications the summation (12.53) consists of infinitely many terms and the computation of $\mathbf{A}^p$ can be computationally very demanding. Again, the eigenvalue decomposition (12.45) or (12.46) allows us to simplify the evaluation of $f(\mathbf{A})$.

**Problem c:** Show that $\hat{\mathbf{v}}^{(n)}$ is also an eigenvector of $\mathbf{A}^p$ with eigenvalue $(\lambda_n)^p$, i.e. show that

$$\mathbf{A}^p\hat{\mathbf{v}}^{(n)} = (\lambda_n)^p\,\hat{\mathbf{v}}^{(n)}\;. \tag{12.54}$$

Hint: first compute $\mathbf{A}^2\hat{\mathbf{v}}^{(n)} = \mathbf{A}\left(\mathbf{A}\hat{\mathbf{v}}^{(n)}\right)$, then $\mathbf{A}^3\hat{\mathbf{v}}^{(n)}$, etc.

**Problem d:** Use this result to show that (12.46) can be generalized to:

$$\mathbf{A}^p = \mathbf{V}\mathbf{\Sigma}^p\mathbf{V}^T\;, \tag{12.55}$$

with $\mathbf{\Sigma}^p$ given by

$$\mathbf{\Sigma}^p =
\begin{pmatrix}
\lambda_1^p & 0 & \cdots & 0 \\
0 & \lambda_2^p & \cdots & 0 \\
\vdots & \vdots & \ddots & \vdots \\
0 & 0 & & \lambda_N^p
\end{pmatrix}. \tag{12.56}$$

**Problem e:** Finally use (12.53) to show that $f(\mathbf{A})$ can be written as:

$$f(\mathbf{A}) = \mathbf{V}f\left(\mathbf{\Sigma}\right)\mathbf{V}^T\;, \tag{12.57}$$

with $f\left(\mathbf{\Sigma}\right)$ given by

$$f\left(\mathbf{\Sigma}\right) =
\begin{pmatrix}
f\left(\lambda_1\right) & 0 & \cdots & 0 \\
0 & f\left(\lambda_2\right) & \cdots & 0 \\
\vdots & \vdots & \ddots & \vdots \\
0 & 0 & & f\left(\lambda_N\right)
\end{pmatrix}. \tag{12.58}$$

**Problem f:** In order to revert to an explicit eigenvector expansion, show that (12.57) can be written as:

$$f(\mathbf{A}) = \sum_{n=1}^{N} f\left(\lambda_n\right)\hat{\mathbf{v}}^{(n)}\hat{\mathbf{v}}^{(n)T}\;. \tag{12.59}$$

With this expression (or the equivalent expression (12.57)) the evaluation of $f(\mathbf{A})$ is simple once the eigenvectors and eigenvalues of $\mathbf{A}$ are known, because in (12.59) the function $f$ only acts on the eigenvalues, but not

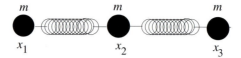

Fig. 12.6.   Definition of variables for a simple vibrating system.

on the matrix. Since the function $f$ normally acts on a scalar (such as the eigenvalues), the eigenvector decomposition has obviated the need for computing higher powers of the matrix $\mathbf{A}$. However, from a numerical point of view computing functions of matrices can be a tricky issue. For example, Moler and van Loan [56] give nineteen dubious ways to compute the exponential of a matrix.

## 12.6 The normal modes of a vibrating system

An eigenvector decomposition is not only useful for computing the inverse of a matrix and other functions of a matrix, it also provides a way to analyze characteristics of dynamical systems. As an example, a simple model for the oscillations of a vibrating molecule is discussed here. This system is the prototype of a vibrating system that has different modes of vibration. The natural modes of vibration are usually called the *normal modes* of that system. Consider the mechanical system shown in figure 12.6. Three particles with mass $m$ are coupled by two springs with spring constant $k$. It is assumed that the three masses are constrained to move along a line. The displacements of the masses from their equilibrium positions are denoted with $x_1$, $x_2$ and $x_3$. This mechanical model can considered to be a (grossly oversimplified) model of a tri-atomic molecule such as $CO_2$ or $H_2O$.

Each of the masses can experience an external force $F_i$, where the subscript $i$ denotes the mass under consideration. The equations of motion for the three masses are given by:

$$\left.\begin{aligned}
m\ddot{x}_1 &= k(x_2 - x_1) + F_1 , \\
m\ddot{x}_2 &= -k(x_2 - x_1) + k(x_3 - x_2) + F_2 , \\
m\ddot{x}_3 &= -k(x_3 - x_2) + F_3 .
\end{aligned}\right\} \qquad (12.60)$$

For the moment we consider harmonic oscillations, i.e. we assume that both the driving forces $F_i$ and the displacements $x_i$ vary with time as $e^{-i\omega t}$. The displacements $x_1$, $x_2$ and $x_3$ can be used to form a vector $\mathbf{x}$, and summarily a vector $\mathbf{F}$ can be formed from the three forces $F_1$, $F_2$ and $F_3$ that act on the three masses.

**Problem a:** Show that for an harmonic motion with frequency $\omega$ the equations of motion can be written in vector form as:

$$\left(\mathbf{A} - \frac{m\omega^2}{k}\mathbf{I}\right)\mathbf{x} = \frac{1}{k}\mathbf{F} \ , \tag{12.61}$$

with the matrix $\mathbf{A}$ given by

$$\mathbf{A} = \begin{pmatrix} 1 & -1 & 0 \\ -1 & 2 & -1 \\ 0 & -1 & 1 \end{pmatrix} \ . \tag{12.62}$$

The normal modes of the system are given by the patterns of oscillations of the system when there is no driving force. For this reason, we set the driving force $\mathbf{F}$ on the right hand side of (12.61) momentarily to zero. Equation (12.61) then reduces to a homogeneous system of linear equations; such a system of equations can only have nonzero solutions when the determinant of the matrix vanishes. Since the matrix $\mathbf{A}$ has three eigenvalues, the system can only oscillate freely at three discrete eigenfrequencies. The system can only oscillate at other frequencies when it is driven by the force $\mathbf{F}$ at such a frequency.

**Problem b:** Show that the eigenfrequencies $\omega_i$ of the vibrating system are given by

$$\omega_i = \sqrt{\frac{k\lambda_i}{m}} \ , \tag{12.63}$$

where $\lambda_i$ are the eigenvalues of the matrix $\mathbf{A}$.

**Problem c:** Show that the eigenfrequencies of the system are given by:

$$\omega_1 = 0 \ , \qquad \omega_2 = \sqrt{\frac{k}{m}} \ , \qquad \omega_3 = \sqrt{\frac{3k}{m}} \ . \tag{12.64}$$

**Problem d:** The frequencies do not give the vibrations of each of the three particles respectively. Instead these frequencies give the eigenfrequencies of the three modes of oscillation of the system. The eigenvector that corresponds to each eigenvalue gives the displacement of each particle for that mode of oscillation. Show that these eigenvectors are given by:

$$\hat{\mathbf{v}}^{(1)} = \frac{1}{\sqrt{3}}\begin{pmatrix} 1 \\ 1 \\ 1 \end{pmatrix} \ , \quad \hat{\mathbf{v}}^{(2)} = \frac{1}{\sqrt{2}}\begin{pmatrix} 1 \\ 0 \\ -1 \end{pmatrix} \ , \quad \hat{\mathbf{v}}^{(3)} = \frac{1}{\sqrt{6}}\begin{pmatrix} 1 \\ -2 \\ 1 \end{pmatrix} \ . \tag{12.65}$$

Remember that the eigenvectors can be multiplied by an arbitrary constant, and this constant is chosen in such a way that each eigenvector has length 1.

**Problem e:** Show that these eigenvectors satisfy the requirement (12.38).

**Problem f:** Sketch the motion of the three masses for each normal mode. Explain physically why the third mode with frequency $w_3$ has a higher eigenfrequency than the second mode $w_2$.

**Problem g:** Explain physically why the second mode has an eigenfrequency $w_2 = \sqrt{k/m}$ that is identical to the frequency of a single mass $m$ that is suspended by a spring with spring constant $k$.

**Problem h:** What type of motion does the first mode with eigenfrequency $w_1$ describe? Explain physically why this frequency is independent of the spring constant $k$ and the mass $m$.

Now that we know the normal modes of the system, we consider the case in which the system is driven by a force $\mathbf{F}$ that varies in time as $e^{-iwt}$. For simplicity it is assumed that the frequency $w$ of the driving force differs from the eigenfrequencies of the system: $w \neq w_i$. The eigenvectors $\hat{\mathbf{v}}^{(n)}$ defined in (12.65) form a complete orthonormal set, hence both the driving force $\mathbf{F}$ and the displacement $\mathbf{x}$ can be expanded in this set. Using (12.43) the driving force can be expanded as

$$\mathbf{F} = \sum_{n=1}^{3} \hat{\mathbf{v}}^{(n)} (\hat{\mathbf{v}}^{(n)} \cdot \mathbf{F}) \,. \tag{12.66}$$

**Problem i:** Write the displacement vector as a superposition of the normal mode displacements: $\mathbf{x} = \sum_{n=1}^{3} c_n \hat{\mathbf{v}}^{(n)}$, use expansion (12.66) for the driving force and insert these equations in the equation of motion (12.61) to solve for the unknown coefficients $c_n$. Eliminate the eigenvalues with (12.63) and show that the displacement is given by:

$$\mathbf{x} = \frac{1}{m} \sum_{n=1}^{3} \frac{\hat{\mathbf{v}}^{(n)} (\hat{\mathbf{v}}^{(n)} \cdot \mathbf{F})}{(w_n^2 - w^2)} \,. \tag{12.67}$$

This expression has a nice physical interpretation. Expression (12.67) states that the total response of the system can be written as a superposition of the different normal modes (the $\sum_{n=1}^{3} \hat{\mathbf{v}}^{(n)}$ terms). The effect that the force has on each normal mode is given by the inner product $(\hat{\mathbf{v}}^{(n)} \cdot \mathbf{F})$. This is nothing but the component of the force $\mathbf{F}$ along the eigenvector $\hat{\mathbf{v}}^{(n)}$, see (12.2). The term $1/(w_n^2 - w^2)$ gives the sensitivity of each mode to a driving force with frequency $w$; this term can be called a sensitivity term. When the driving force is close to one of the eigenfrequencies of the $n$th mode, $1/(w_n^2 - w^2)$ is very large. In that case the system is close to resonance and the resulting displacement will be very large. On the other hand, when the frequency of the driving force is very far from the eigenfrequencies of the system, $1/\left(\omega_n^2 - \omega^2\right)$ is small and the system gives a very small response. The total response can be seen as a combination of three basic operations: eigenvector expansion, projection and multiplication with a response function. Note that the same operations were used in the explanation of the action of a matrix $\mathbf{A}$ given below equation (12.44).

## 12.7 Singular value decomposition

In section 12.4 the decomposition of a square matrix in terms of eigenvectors was treated. In many practical applications, such as inverse problems, one encounters a system of equations that is not square:

$$\underbrace{\mathbf{A}}_{\substack{M \times N \\ matrix}} \quad \underbrace{\mathbf{x}}_{\substack{N \\ rows}} \quad = \quad \underbrace{\mathbf{y}}_{\substack{M \\ rows}} \tag{12.68}$$

Consider the example in which the vector $\mathbf{x}$ has $N$ components and there are $M$ equations. In that case the vector $\mathbf{y}$ has $M$ components and the matrix $\mathbf{A}$ has $M$ rows and $N$ columns, i.e. it is an $M \times N$ matrix. A relation such as (12.37) which states that $\mathbf{A}\hat{\mathbf{v}}^{(n)} = \lambda_n \hat{\mathbf{v}}^{(n)}$ cannot possibly hold because when the matrix $\mathbf{A}$ acts on an $N$-vector it produces an $M$-vector, whereas in (12.37) the vector on the right hand side has the same number of components as the vector on the left hand side. It is clear that the theory of section 12.4 cannot be applied when the matrix is not square. However, it is possible to generalize the theory of section 12.4 for when $\mathbf{A}$ is not square. For simplicity it is assumed that $\mathbf{A}$ is a real matrix.

In section 12.4 a single set of orthonormal eigenvectors $\hat{\mathbf{v}}^{(n)}$ was used to analyze the problem. Since the vectors $\mathbf{x}$ and $\mathbf{y}$ in (12.68) have different dimensions, it is necessary to expand the vector $\mathbf{x}$ in a set of $N$ orthogonal vectors $\hat{\mathbf{v}}^{(n)}$ that each have $N$ components and to expand $\mathbf{y}$ in a different set of $M$ orthogonal vectors $\hat{\mathbf{u}}^{(m)}$ that each have $M$ components. Suppose we have chosen a set $\hat{\mathbf{v}}^{(n)}$, let us define vectors $\hat{\mathbf{u}}^{(n)}$ by the following relation:

$$\mathbf{A}\hat{\mathbf{v}}^{(n)} = \lambda_n \hat{\mathbf{u}}^{(n)} . \tag{12.69}$$

The constant $\lambda_n$ should not be confused with an eigenvalue, this constant follows from the requirement that $\hat{\mathbf{v}}^{(n)}$ and $\hat{\mathbf{u}}^{(n)}$ are both vectors of unit length. At this point, the choice of $\hat{\mathbf{v}}^{(n)}$ is still open. The vectors $\hat{\mathbf{v}}^{(n)}$ will now be required to satisfy in addition to (12.69) the following condition:

$$\mathbf{A}^T \hat{\mathbf{u}}^{(n)} = \mu_n \hat{\mathbf{v}}^{(n)} , \tag{12.70}$$

where $\mathbf{A}^T$ is the transpose of $\mathbf{A}$.

**Problem a:** In order to find the vectors $\hat{\mathbf{v}}^{(n)}$ and $\hat{\mathbf{u}}^{(n)}$ that satisfy both (12.69) and (12.70), multiply (12.69) by $\mathbf{A}^T$ and use (12.70) to eliminate $\hat{\mathbf{u}}^{(n)}$. Do this to show that $\hat{\mathbf{v}}^{(n)}$ satisfies:

$$\left(\mathbf{A}^T\mathbf{A}\right)\hat{\mathbf{v}}^{(n)} = \lambda_n\mu_n\hat{\mathbf{v}}^{(n)} \ . \tag{12.71}$$

Use similar steps to show that $\hat{\mathbf{u}}^{(n)}$ satisfies

$$\left(\mathbf{A}\mathbf{A}^T\right)\hat{\mathbf{u}}^{(n)} = \lambda_n\mu_n\hat{\mathbf{u}}^{(n)} \ . \tag{12.72}$$

These equations state that the $\hat{\mathbf{v}}^{(n)}$ are the eigenvectors of $\mathbf{A}^T\mathbf{A}$ and that the $\hat{\mathbf{u}}^{(n)}$ are the eigenvectors of $\mathbf{A}\mathbf{A}^T$.

**Problem b:** Show that both $\mathbf{A}^T\mathbf{A}$ and $\mathbf{A}\mathbf{A}^T$ are real symmetric matrices and that this implies that the basis vectors $\hat{\mathbf{v}}^{(n)}$ $(n = 1, \ldots, N)$ and $\hat{\mathbf{u}}^{(m)}$ $(m = 1, \ldots, M)$ are both orthonormal:

$$\left(\hat{\mathbf{v}}^{(n)} \cdot \hat{\mathbf{v}}^{(m)}\right) = \left(\hat{\mathbf{u}}^{(n)} \cdot \hat{\mathbf{u}}^{(m)}\right) = \delta_{nm} \ . \tag{12.73}$$

Although (12.71) and (12.72) can be used to find the basis vectors $\hat{\mathbf{v}}^{(n)}$ and $\hat{\mathbf{u}}^{(n)}$, these expressions cannot be used to find the constants $\lambda_n$ and $\mu_n$, because these expressions state that the *product* $\lambda_n\mu_n$ is equal to the eigenvalues of $\mathbf{A}^T\mathbf{A}$ and $\mathbf{A}\mathbf{A}^T$. This implies that only the product of $\lambda_n$ and $\mu_n$ is defined.

**Problem c:** In order to find the relation between $\lambda_n$ and $\mu_n$, take the inner product of (12.69) with $\hat{\mathbf{u}}^{(n)}$ and use the orthogonality relation (12.73) to show that:

$$\lambda_n = \left(\hat{\mathbf{u}}^{(n)} \cdot \mathbf{A}\hat{\mathbf{v}}^{(n)}\right) \ . \tag{12.74}$$

**Problem d:** Show that for arbitrary vectors $\mathbf{p}$ and $\mathbf{q}$

$$(\mathbf{p} \cdot \mathbf{A}\mathbf{q}) = \left(\mathbf{A}^T\mathbf{p} \cdot \mathbf{q}\right) \ . \tag{12.75}$$

**Problem e:** Apply this relation to (12.74) and use (12.70) to show that

$$\lambda_n = \mu_n \ . \tag{12.76}$$

This is all the information we need to find both $\lambda_n$ and $\mu_n$. Since these quantities are equal, and since by virtue of (12.71) these eigenvectors are equal to the eigenvectors of $\mathbf{A}^T\mathbf{A}$, it follows that both $\lambda_n$ and $\mu_n$ are given by the square-root of the eigenvalues of $\mathbf{A}^T\mathbf{A}$. Note that it follows from (12.72) that the product $\lambda_n \mu_n$ also equals the eigenvalues of $\mathbf{AA}^T$. This can only be the case when $\mathbf{A}^T\mathbf{A}$ and $\mathbf{AA}^T$ have the same eigenvalues. Before we proceed, let us show that this is indeed the case. Let the eigenvalues of $\mathbf{A}^T\mathbf{A}$ be denoted by $\Lambda_n$ and the eigenvalues of $\mathbf{AA}^T$ by $\Upsilon_n$, i.e.

$$\mathbf{A}^T\mathbf{A}\hat{\mathbf{v}}^{(n)} = \Lambda_n\hat{\mathbf{v}}^{(n)} \tag{12.77}$$

and

$$\mathbf{AA}^T\hat{\mathbf{u}}^{(n)} = \Upsilon_n\hat{\mathbf{u}}^{(n)} . \tag{12.78}$$

**Problem f:** Take the inner product of (12.77) with $\hat{\mathbf{v}}^{(n)}$ to show that $\Lambda_n = (\hat{\mathbf{v}}^{(n)} \cdot \mathbf{A}^T\mathbf{A}\hat{\mathbf{v}}^{(n)})$, use the properties (12.75) and $\mathbf{A}^{TT} = \mathbf{A}$ and (12.69) to show that $\lambda_n^2 = \Lambda_n$. Use similar steps to show that $\mu_n^2 = \Upsilon_n$. With (12.76) this implies that $\mathbf{AA}^T$ and $\mathbf{A}^T\mathbf{A}$ have the same eigenvalues.

The proof that $\mathbf{AA}^T$ and $\mathbf{A}^T\mathbf{A}$ have the same eigenvalues was not only given as a check of the consistency of the theory, the fact that $\mathbf{AA}^T$ and $\mathbf{A}^T\mathbf{A}$ have the same eigenvalues has important implications. Since $\mathbf{AA}^T$ is an $M \times M$ matrix, it has $M$ eigenvalues, and since $\mathbf{A}^T\mathbf{A}$ is an $N \times N$ matrix it has $N$ eigenvalues. The only way for these matrices to have the same eigenvalues, but to have a different number of eigenvalues is for the number of nonzero eigenvalues to be given by the minimum of $N$ and $M$. In practice, some of the eigenvalues of $\mathbf{AA}^T$ may be zero, hence the number of nonzero eigenvalues of $\mathbf{AA}^T$ can be less than $M$. By the same token, the number of nonzero eigenvalues of $\mathbf{A}^T\mathbf{A}$ can be less than $N$. The number of nonzero eigenvalues will be denoted by $P$. It is not known *a priori* how many nonzero eigenvalues there are, but it follows from the arguments above that $P$ is smaller than or equal to $M$ and $N$. This implies that

$$P \leq \min(N, M) , \tag{12.79}$$

where $\min(N, M)$ denotes the minimum of $N$ and $M$. Therefore, whenever a summation over eigenvalues occurs, we need to take only $P$ eigenvalues into account. Since the ordering of the eigenvalues is arbitrary, it is assumed in the following that the eigenvectors are ordered in decreasing size: $\lambda_1 \geq \lambda_2 \geq \cdots \geq \lambda_N$. In this ordering the eigenvalues for $n > P$ are equal to zero so that the summation over eigenvalues runs from 1 to $P$.

**Problem g:** The matrices $\mathbf{AA}^T$ and $\mathbf{A}^T\mathbf{A}$ have the same eigenvalues. When you need the eigenvalues and eigenvectors, from the point of view of computational efficiency would it be more efficient to compute the eigenvalues and eigenvectors of $\mathbf{A}^T\mathbf{A}$ or of $\mathbf{AA}^T$? Consider the situations $M > N$ and $M < N$ separately.

Let us now return to the task of making an eigenvalue decomposition of the matrix $\mathbf{A}$. The vectors $\hat{\mathbf{v}}^{(n)}$ form a basis in $N$-dimensional space. Since the vector $\mathbf{x}$ is $N$-dimensional, every vector $\mathbf{x}$ can be decomposed according to (12.43): $\mathbf{x} = \sum_{n=1}^{N} \hat{\mathbf{v}}^{(n)} (\hat{\mathbf{v}}^{(n)} \cdot \mathbf{x})$.

**Problem h:** Let the matrix $\mathbf{A}$ act on this expression and use (12.69) to show that:

$$\mathbf{A}\mathbf{x} = \sum_{n=1}^{P} \lambda_n \hat{\mathbf{u}}^{(n)} \left( \hat{\mathbf{v}}^{(n)} \cdot \mathbf{x} \right) . \tag{12.80}$$

**Problem i:** This expression must hold for any vector $\mathbf{x}$. Use this property to deduce that:

$$\mathbf{A} = \sum_{n=1}^{P} \lambda_n \hat{\mathbf{u}}^{(n)} \hat{\mathbf{v}}^{(n)T} . \tag{12.81}$$

**Problem j:** The eigenvectors $\hat{\mathbf{v}}^{(n)}$ can be arranged in an $N \times N$ matrix $\mathbf{V}$, defined in (12.39). Similarly the eigenvectors $\hat{\mathbf{u}}^{(n)}$ can be used to form the columns of an $M \times M$ matrix $\mathbf{U}$:

$$\mathbf{U} = \begin{pmatrix} \vdots & \vdots & & \vdots \\ \hat{\mathbf{u}}^{(1)} & \hat{\mathbf{u}}^{(2)} & \cdots & \hat{\mathbf{u}}^{(M)} \\ \vdots & \vdots & & \vdots \end{pmatrix} . \tag{12.82}$$

Show that $\mathbf{A}$ can also be written as:

$$\mathbf{A} = \mathbf{U}\boldsymbol{\Sigma}\mathbf{V}^{T} , \tag{12.83}$$

with the diagonal matrix $\boldsymbol{\Sigma}$ defined in (12.47).

This decomposition of $\mathbf{A}$ in terms of eigenvectors is called the *singular value decomposition* of the matrix. This is frequently abbreviated to SVD.

**Problem k:** You may have noticed the similarity between (12.81) and (12.45) for a square matrix and (12.83) and (12.46). Show that for the special case $M = N$ the theory of this section is identical to the eigenvalue decomposition for a square matrix presented in section 12.4. Hint: what are the vectors $\hat{\mathbf{u}}^{(n)}$ when $M = N$?

Let us now solve the original system of linear equations (12.68) for the unknown vector $\mathbf{x}$. In order to do this, expand the vector $\mathbf{y}$ in the vectors $\hat{\mathbf{u}}^{(n)}$ that span the $M$-dimensional space: $\mathbf{y} = \sum_{m=1}^{M} \hat{\mathbf{u}}^{(m)} (\hat{\mathbf{u}}^{(m)} \cdot \mathbf{y})$, and expand the vector $\mathbf{x}$ in the vectors $\hat{\mathbf{v}}^{(n)}$ that span the $N$-dimensional space:

$$\mathbf{x} = \sum_{n=1}^{N} c_n \hat{\mathbf{v}}^{(n)} . \tag{12.84}$$

**Problem l:** At this point the coefficients $c_n$ are unknown. Insert the expansions for $\mathbf{y}$ and $\mathbf{x}$ and the expansion (12.81) for the matrix $\mathbf{A}$ in the linear system (12.68) and use the orthogonality properties of the eigenvectors to show that $c_n = (\hat{\mathbf{u}}^{(n)} \cdot \mathbf{y})/\lambda_n$, so that

$$\mathbf{x} = \sum_{n=1}^{P} \frac{1}{\lambda_n} \left( \hat{\mathbf{u}}^{(n)} \cdot \mathbf{y} \right) \hat{\mathbf{v}}^{(n)} . \tag{12.85}$$

Note that although in the original expansion (12.84) a summation is carried out over all $N$ basis vectors, in solution (12.85) a summation is carried out over the first $P$ basis vectors only. The reason for this is that the remaining eigenvectors have eigenvalues that are equal to zero so that they can be left out of the expansion (12.81) of the matrix $\mathbf{A}$. Indeed, these eigenvalues would give rise to problems because if they were retained they would lead to infinite contributions $1/\lambda \to \infty$ in solution (12.85). In practice, some eigenvalues may be nonzero, but close to zero, so that the term $1/\lambda$ gives rise to numerical instabilities. Therefore, one also often leaves out nonzero but small eigenvalues in summation (12.85).

This may appear to be an objective procedure for defining solutions for linear problems that are undetermined or for problems that are otherwise ill-conditioned, but there is a price one pays for leaving out basis vectors in the construction of the solution. The vector $\mathbf{x}$ is $N$-dimensional, hence one needs $N$ basis vectors to construct an arbitrary vector $\mathbf{x}$, see (12.84). The solution vector given in (12.85) is built by superposing only $P$ basis vectors. This implies that the solution vector is constrained to be within the $P$-dimensional subspace spanned by the first $P$ eigenvectors. Therefore, it is not clear that the solution vector in (12.85) is identical to the true vector $\mathbf{x}$. However, the point of using the singular value decomposition is that the solution is only constrained by the linear system of (12.68) within the subspace spanned by the first $P$ basis vectors $\hat{\mathbf{v}}^{(n)}$. Solution (12.85) ensures that only the components of $\mathbf{x}$ within that subspace are affected by the right hand side vector $\mathbf{y}$. This technique is extremely important in the analysis of linear inverse problems.

## 12.8 The Householder transformation

Linear systems of equations can be solved in a systematic way by sweeping columns of the matrix that defines the system of equations. As an example consider the system

$$\left. \begin{array}{rcl} x + y + 3z &=& 5 , \\ -x + 2z &=& 1 , \\ 2x + y + 2z &=& 5 . \end{array} \right\} \tag{12.86}$$

This system of equations will be written here also as:

$$\begin{pmatrix} 1 & 1 & 3 & | & 5 \\ -1 & 0 & 2 & | & 1 \\ 2 & 1 & 2 & | & 5 \end{pmatrix}, \tag{12.87}$$

which is nothing but a compressed notation of (12.86). The matrix shown in (12.87) is called the *augmented matrix* because the matrix defining the left hand side of (12.86) is augmented with the right hand side of (12.86). The linear equations can be solved by adding the first row to the second row and subtracting the first row twice from the third row, the resulting system of equations is then represented by the following augmented matrix:

$$\begin{pmatrix} 1 & 1 & 3 & | & 5 \\ 0 & 1 & 5 & | & 6 \\ 0 & -1 & -4 & | & -5 \end{pmatrix}. \tag{12.88}$$

Note that in the first column all elements below the first elements are equal to zero. By adding the second row to the third row we can also make all elements below the second element in the second column equal to zero:

$$\begin{pmatrix} 1 & 1 & 3 & | & 5 \\ 0 & 1 & 5 & | & 6 \\ 0 & 0 & 1 & | & 1 \end{pmatrix}. \tag{12.89}$$

The system is now in *upper-triangular form*; this is a different way of saying that all matrix elements below the diagonal vanish. This is convenient because the system can now be solved by *backsubstitution*. To see how this works note that the augmented matrix (12.89) is a shorthand notation for the following system of equations:

$$\left. \begin{array}{rcl} x + y + 3z & = & 5 , \\ y + 5z & = & 6 , \\ z & = & 1 . \end{array} \right\} \tag{12.90}$$

The value of $z$ follows from the last equation, given this value of $z$ the value of $y$ follows from the middle equations, and given $y$ and $z$ the value of $x$ follows from the top equation.

**Problem a:** Show that the solution of the linear equations is given by $x = y = z = 1$.

For small systems of linear equations this process for solving linear equations can be carried out by hand. For large systems of equations this process must be carried out on a computer. This is only possible when one has a systematic and efficient way of carrying out this sweeping process. Suppose we have an $N \times N$ matrix $\mathbf{A}$:

$$\mathbf{A} = \begin{pmatrix} a_{11} & a_{12} & \cdots & a_{1N} \\ a_{21} & a_{22} & \cdots & a_{2N} \\ \vdots & \vdots & \ddots & \vdots \\ a_{N1} & a_{N2} & \cdots & a_{NN} \end{pmatrix}. \tag{12.91}$$

We want to find an operator $\mathbf{Q}$ such that when $\mathbf{A}$ is multiplied by $\mathbf{Q}$ all elements in the first column are zero except the element on the diagonal, i.e. we want to find $\mathbf{Q}$ such that:

$$\mathbf{QA} = \begin{pmatrix} a'_{11} & a'_{12} & \cdots & a'_{1N} \\ 0 & a'_{22} & \cdots & a'_{2N} \\ \vdots & \vdots & \ddots & \vdots \\ 0 & a'_{N2} & \cdots & a'_{NN} \end{pmatrix}. \tag{12.92}$$

This problem can be formulated slightly differently, suppose we denote the first columns of $\mathbf{A}$ by the vector $\mathbf{u}$:

$$\mathbf{u} \equiv \begin{pmatrix} a_{11} \\ a_{21} \\ \vdots \\ a_{N1} \end{pmatrix}. \tag{12.93}$$

The operator $\mathbf{Q}$ that we want to find maps this vector to a new vector which only has a nonzero component in the first element:

$$\mathbf{Qu} = \begin{pmatrix} a'_{11} \\ 0 \\ \vdots \\ 0 \end{pmatrix} = a'_{11} \hat{\mathbf{e}}_1 , \tag{12.94}$$

where $\hat{\mathbf{e}}_1$ is the unit vector in the $x_1$-direction:

$$\hat{\mathbf{e}}_1 \equiv \begin{pmatrix} 1 \\ 0 \\ \vdots \\ 0 \end{pmatrix}. \tag{12.95}$$

The desired operator $\mathbf{Q}$ can be found with a so-called Householder transformation. For a given unit vector $\hat{\mathbf{n}}$ the Householder transformation is defined by:

$$\mathbf{Q} \equiv \mathbf{I} - 2\hat{\mathbf{n}}\hat{\mathbf{n}}^T . \tag{12.96}$$

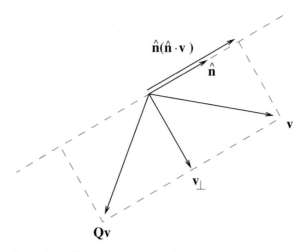

Fig. 12.7.   Geometrical interpretation of the Householder transformation.

**Problem b:** Show that the Householder transformation can be written as $\mathbf{Q} = \mathbf{I} - 2\mathbf{P}$, where $\mathbf{P}$ is the operator for projection along $\hat{\mathbf{n}}$.

**Problem c:** It follows from (12.3) that any vector $\mathbf{v}$ can be decomposed into a component along $\hat{\mathbf{n}}$ and a perpendicular component: $\mathbf{v} = \hat{\mathbf{n}}\,(\hat{\mathbf{n}} \cdot \mathbf{v}) + \mathbf{v}_\perp$. Show that after the Householder transformation the vector is given by:

$$\mathbf{Qv} = -\hat{\mathbf{n}}\,(\hat{\mathbf{n}} \cdot \mathbf{v}) + \mathbf{v}_\perp \ . \tag{12.97}$$

**Problem d:** Convince yourself that the Householder transformation of $\mathbf{v}$ is correctly shown in figure 12.7.

**Problem e:** Use (12.97) to show that $\mathbf{Q}$ does not change the length of a vector. Use this result to show that $a'_{11}$ in (12.94) is given by $a'_{11} = |\mathbf{u}|$.

With (12.94) this means that the Householder transformation should satisfy

$$\mathbf{Qu} = |\mathbf{u}|\,\hat{\mathbf{e}}_1 \ . \tag{12.98}$$

Our goal is now to find a unit vector $\hat{\mathbf{n}}$ such that this expression is satisfied.

**Problem f:** Use (12.96) to show that if $\mathbf{Q}$ satisfies the requirement (12.98) $\hat{\mathbf{n}}$ must satisfy the following equation:

$$2\hat{\mathbf{n}}\,(\hat{\mathbf{n}} \cdot \hat{\mathbf{u}}) = \hat{\mathbf{u}} - \hat{\mathbf{e}}_1 \ ; \tag{12.99}$$

in this expression $\hat{\mathbf{u}}$ is the unit vector in the direction $\mathbf{u}$.

**Problem g:** Equation (12.99) implies that $\hat{\mathbf{n}}$ is directed in the direction of the vector $\hat{\mathbf{u}} - \hat{\mathbf{e}}_1$, therefore $\hat{\mathbf{n}}$ can be written as $\hat{\mathbf{n}} = C(\hat{\mathbf{u}} - \hat{\mathbf{e}}_1)$, with $C$ an undetermined constant. Show that (12.98) implies that $C = 1/\sqrt{2[1 - (\hat{\mathbf{u}} \cdot \hat{\mathbf{e}}_1)]}$. Also show that this value of $C$ indeed leads to a vector $\hat{\mathbf{n}}$ that is of unit length.

This value of $C$ implies that the unit vector $\hat{\mathbf{n}}$ to be used in the Householder transformation (12.96) is given by

$$\hat{\mathbf{n}} = \frac{\hat{\mathbf{u}} - \hat{\mathbf{e}}_1}{\sqrt{2[1 - (\hat{\mathbf{u}} \cdot \hat{\mathbf{e}}_1)]}} \ . \tag{12.100}$$

To see how the Householder transformation can be used to render the matrix elements below the diagonal equal to zero apply the transformation $\mathbf{Q}$ to the linear equation $\mathbf{Ax} = \mathbf{y}$.

**Problem h:** Show that this leads to a new system of equations given by

$$\begin{pmatrix} |\mathbf{u}| & a'_{12} & \cdots & a'_{1N} \\ 0 & a'_{22} & \cdots & a'_{2N} \\ \vdots & \vdots & \ddots & \vdots \\ 0 & a'_{N2} & \cdots & a'_{NN} \end{pmatrix} \mathbf{x} = \mathbf{Qy} \ . \tag{12.101}$$

A second Householder transformation can now be applied to render all elements in the second column below the diagonal element $a'_{22}$ equal to zero. In this way, all the columns of $\mathbf{A}$ can be successively swiped. Note that in order to apply the Householder transformation one only needs to compute (12.100) and (12.96) to carry out a matrix multiplication. These operations can be carried out efficiently on computers.

# 13

---

# The Dirac delta function

## 13.1 Introduction of the delta function

In linear algebra, the identity matrix $\mathbf{I}$ plays a central role. This operator maps any vector $\mathbf{v}$ onto itself

$$\mathbf{Iv} = \mathbf{v} \,. \tag{13.1}$$

This expression can also be written in component form as

$$\sum_j I_{ij} v_j = v_i \,. \tag{13.2}$$

The identity matrix has diagonal elements that are equal to unity and its off-diagonal elements are equal to zero. This means that the elements of the identity matrix are equal to the Kronecker delta $I_{ij} = \delta_{ij}$, which is defined as:

$$\delta_{ij} \equiv \begin{cases} 1 & \text{for} \quad i = j \\ 0 & \text{for} \quad i \neq j \end{cases} \,. \tag{13.3}$$

Expression (13.2) shows that when the identity matrix $I_{ij}$ acts on all the components $v_j$ of a vector, it selects the component $v_i$. The question we address in this chapter is how can this idea be generalized to continuous functions instead of vectors. In other words, can we find a function $I(x_0, x)$ such that when it is integrated with a function $f(x)$ it selects that function as the location $x_0$:

$$\int I(x_0, x) f(x) dx = f(x_0) \,? \tag{13.4}$$

Note the resemblance between this expression and (13.2) for a vector. The vector $\mathbf{v}$ is replaced by the function $f$, the summation over $j$ is changed into the integration over $x$ and the index $i$ in (13.2) corresponds to the value $x_0$ in (13.4).

As a first guess for the operator $I(x_0, x)$ let us consider the following generalization of the definition (13.3) of the Kronecker delta to continuous functions:

$$I(x_0, x) \overset{?}{\equiv} \begin{cases} 1 & \text{for} \quad x_0 = x \\ 0 & \text{for} \quad x_0 \neq x \end{cases} . \tag{13.5}$$

This is not a very good guess. This is because when $I(x_0, x)$ is viewed as a function of $x$, $I(x_0, x)$ is almost everywhere equal to zero except when $x = x_0$. When we integrate $I(x_0, x)$ over $x$ the integrand is zero except at the point $x = x_0$, but this point gives a vanishing contribution to the integral because the 'width' of this point is equal to zero. (Mathematically one would say that a point has zero measure.) This means that the integral of $I(x_0, x)$ can only give a finite result when $I(x_0, x)$ is infinite at $x = x_0$.

As an improved guess let us therefore try the definition

$$I(x_0, x) \overset{?}{\equiv} \begin{cases} \infty & \text{for} \quad x_0 = x \\ 0 & \text{for} \quad x_0 \neq x \end{cases} . \tag{13.6}$$

This is not a very precise definition because it is not clear what we mean by '$\infty$'. However, we can learn something from this naive guess because it shows that $I(x_0, x)$ is not a well-behaved function. It is discontinuous at $x = x_0$ and its value is infinite at that point. It is clear from this that whatever definition of $I(x_0, x)$ we use, it will not lead to a well-behaved function.

A useful definition of $I(x_0, x)$ can be obtained from the boxcar function $B_a(x)$ which is defined as

$$B_a(x) \equiv \begin{cases} \dfrac{1}{2a} & \text{for} \quad |x| \leq a \\ \\ 0 & \text{for} \quad |x| > a \end{cases} . \tag{13.7}$$

This function is shown for three values of the parameter $a$ in figure 13.1. (A boxcar is the rectangular railroad car that is used for carrying freight, and the function $B_a(x)$ is called the boxcar function because it has the same rectangular shape as the boxcar used by the railways.)

**Problem a:** Show that

$$\int_{-\infty}^{\infty} B_a(x) dx = 1 . \tag{13.8}$$

Let us now center the boxcar at a location $x_0$, multiply it by $f(x)$ and integrate over $x$.

**Problem b:** Use the definition of the boxcar function to derive that

$$\int_{-\infty}^{\infty} B_a(x - x_0) f(x) dx = \frac{1}{2a} \int_{x_0-a}^{x_0+a} f(x) dx . \tag{13.9}$$

Pay attention to the limits of integration on the right hand side.

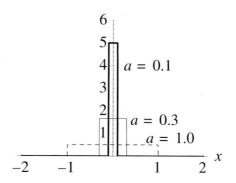

Fig. 13.1.   The boxcar function $B_a(x)$ for $a = 1.0$ (dashed line), $a = 0.3$ (thin solid line) and $a = 0.1$ (thick solid line).

The integral on the right hand side is nothing but the average of the function $f(x)$ over the interval $(x_0 - a, x_0 + a)$ because the pre-factor $1/2a$ corrects for the width of that interval. This is a very useful expression; as the parameter $a$ goes to zero the integral (13.9) gives the function at location $x_0$ because the limit $a \downarrow 0$ gives the mean value of $f(x)$ over the interval $(x_0 - 0, x_0 + 0)$. This means that

$$\lim_{a \downarrow 0} \int_{-\infty}^{\infty} B_a(x - x_0) f(x) dx = f(x_0) . \tag{13.10}$$

This means that this limit has the desired properties of the identity operator $I(x_0, x)$ for continuous functions. It is customary to denote this operator as $\delta(x - x_0)$ and to call it the *Dirac delta function*. (Usually one refers to this function simply as the 'delta function.') This means that the delta function satisfies the following property:

$$\boxed{\int_{-\infty}^{\infty} \delta(x - x_0) f(x) dx = f(x_0) .} \tag{13.11}$$

A comparison of this expression with (13.10) suggests that

$$\delta(x - x_0) ' = {}' \lim_{a \downarrow 0} B_a(x - x_0) . \tag{13.12}$$

Consider figure 13.1 again. It can be seen from that figure and definition (13.7) that as $a$ goes to zero, the value of the boxcar function becomes infinite and that the width of the boxcar goes to zero. In this sense the limit on the right hand side does not exist, and for this reason the = sign is placed between quotes. The word 'delta function' is really a misnomer because this 'function' is not a function at all. Its value is not defined and it is only nonzero in a region with measure zero. However, within the integral (13.11) the action of the delta function is well defined, it is an operator that selects the function value $f(x_0)$ at position $x_0$. The delta function is an example of a *distribution*. This is a mathematical object that is only defined within an integral. Note that the limit $a \downarrow 0$ of the integral (13.10) *is* well defined. This means that the properties of the delta function can only be meaningfully stated when the delta function is used in an integral.

Let us first give a formal proof that definition (13.12) indeed leads to a delta function with the desired property (13.11). It is clear from (13.9) that we only need to consider the function $f(x)$ in the interval $(x_0 - a, x_0 + a)$ for small values of $a$. Therefore it is useful to represent the function $f(x)$ by a Taylor series around the point $x_0$.

**Problem c:** Use (2.17) to show that $f(x)$ can be represented by the following Taylor series around the point $x_0$:

$$f(x) = \sum_{n=0}^{\infty} \frac{1}{n!} \frac{d^n f}{dx^n}(x_0)(x - x_0)^n$$

$$= f(x_0) + \frac{df}{dx}(x_0)(x - x_0) + \frac{1}{2}\frac{d^2 f}{dx^2}(x_0)(x - x_0)^2 + \cdots .$$
$$(13.13)$$

**Problem d:** When this is inserted in (13.9) each term gives a contribution proportional to $\int_{x_0 - a}^{x_0 + a}(x - x_0)^n dx$. Show that for odd values of $n$ this integral is equal to zero and that for even powers of $n$ it is given by

$$\int_{x_0 - a}^{x_0 + a} (x - x_0)^n dx = \frac{2}{n + 1} a^{n+1} .$$
$$(13.14)$$

**Problem e:** Use these results to show that

$$\int_{-\infty}^{\infty} B_a(x - x_0) f(x) dx = \sum_{n \text{ even}} \frac{1}{(n + 1)!} \frac{d^n f}{dx^n}(x_0) a^n$$

$$= f(x_0) + \frac{1}{6}\frac{d^2 f}{dx^2}(x_0) a^2$$

$$+ \frac{1}{120}\frac{d^4 f}{dx^4}(x_0) a^4 + \cdots .$$
$$(13.15)$$

In the limit $a \downarrow 0$ all the terms in this series vanish except the first term. This means that we have proven that (13.10) is indeed satisfied.

It should be noted that it is not necessary to define the delta function as the limit $a \downarrow 0$ of the boxcar function. One can also define the delta function using a Gaussian function with width $a$:

$$g_a(x) = \frac{1}{a\sqrt{\pi}} e^{-x^2/a^2} .$$
$$(13.16)$$

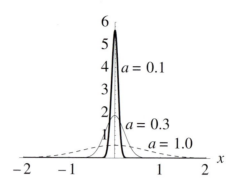

Fig. 13.2.    The Gaussian function $g_a(x)$ for $a = 1.0$ (dashed line), $a = 0.3$ (thin solid line) and $a = 0.1$ (thick solid line).

This function is shown for various values of $a$ in figure 13.2. A comparison of this figure with figure 13.1 shows that for small values of $a$ these functions have similar properties. One can indeed formally define the delta function using the Gaussian function (13.16) instead of the boxcar function. When the analysis of problem d is applied to this function, the first term is again given by $f(x_0)$. The higher-order terms have different coefficients because they follow from the integral $\int_{-\infty}^{\infty} (x - x_0)^n e^{-x^2/a^2} dx$ rather than the integral $\int_{x_0-a}^{x_0+a} (x - x_0)^n dx$. However, in the limit $a \downarrow 0$ these higher-order terms do not contribute. This example shows that the delta function can be defined as the limit of either boxcar functions or Gaussian functions. In fact the delta function can be defined as the limit of other types of functions as well.

## 13.2 Properties of the delta function

In the previous section the delta function was formally introduced. In this section we derive some properties of the delta function that are useful in a variety of applications.

**Problem a:** Apply the identity (13.11) to the function $f(x) = 1$ to derive that

$$\int_{-\infty}^{\infty} \delta(x - x_0)dx = 1 . \tag{13.17}$$

This expression states that the 'surface area' under the delta function is equal to zero. As the width of this function goes to zero, the value of the function must become infinite to ensure that (13.17) is satisfied. Note that according to (13.8) the boxcar function $B_a(x)$ that we used to define the delta function indeed satisfies this property.

**Problem b:** Show that the integral of the Gaussian function defined in (13.16) is also equal to unity:

$$\int_{-\infty}^{\infty} g_a(x)dx = 1 . \qquad (13.18)$$

In this derivation you can use that $\int_{-\infty}^{\infty} e^{-y^2} dy = \sqrt{\pi}$. This means that when the delta function is defined as the limit $a \downarrow 0$ of the Gaussian function $g_a(x)$ property (13.17) is indeed satisfied.

For the next property we consider the delta function $\delta\left(c(x - x_0)\right)$, where $c$ is a constant. Let us first consider the case in which $c$ is positive.

**Problem c:** Make the substitution $y = cx$ to derive the following identity:

$$\int_{-\infty}^{\infty} \delta\left(c(x - x_0)\right) f(x)dx = \int_{-\infty}^{\infty} \delta\left(y - cx_0\right) f(y/c)\frac{1}{c}dy . \quad (13.19)$$

**Problem d:** Carry out the $y$-integration using property (13.11) of the delta function to show that

$$\int_{-\infty}^{\infty} \delta\left(c(x - x_0)\right) f(x)dx = \frac{1}{c}f(x_0) \qquad \text{(positive } c) . \qquad (13.20)$$

**Problem e:** Carry out the same analysis for negative values of the constant $c$, and show that (13.19) in that case is given by

$$\int_{-\infty}^{\infty} \delta\left(c(x - x_0)\right) f(x)dx = \int_{+\infty}^{-\infty} \delta\left(y - cx_0\right) f(y/c)\frac{1}{c}dy . \quad (13.21)$$

Explain why the integration runs from $+\infty$ to $-\infty$.

**Problem f:** When the integration limits in the last integral are reversed, the integral obtains an additional minus sign. Carry out the $y$-integration and derive that

$$\int_{-\infty}^{\infty} \delta\left(c(x - x_0)\right) f(x)dx = -\frac{1}{c}f(x_0) \qquad \text{(negative } c) . \qquad (13.22)$$

For negative values of $c$, one can use that $-c = |c|$. For positive values of $c$, obviously $c = |c|$. This means that (13.20) and (13.21) can be combined in the single property

$$\int_{-\infty}^{\infty} \delta\left(c(x - x_0)\right) f(x)dx = \frac{1}{|c|}f(x_0) . \qquad (13.23)$$

Following (13.11) we also know that the right hand side of this expression can also be written as $(1/\left|c\right|) \int_{-\infty}^{\infty} \delta\left(x - x_0\right) f(x)dx$. A comparison with the left hand side of (13.23) implies that

$$\delta\left(c\left(x - x_0\right)\right) = \frac{1}{\left|c\right|}\delta\left(x - x_0\right) . \tag{13.24}$$

## 13.3 The delta function of a function

In some applications one arrives at an integral of the delta function in which the argument of the delta function is a function $g(x)$ rather than the integration variable $x$. This means that one needs to evaluate the integral $\int \delta\left(g(x)\right) f(x)dx$. We will encounter such an integral in (18.51). The delta function $\delta(x - x_0)$ only gives a nonzero contribution when $x$ is close to $x_0$. Therefore, the delta function $\delta\left(g(x)\right)$ only needs to be evaluated near the value of $x$ where $g(x) = 0$. Let us denote this value by $x_0$, so that

$$g(x_0) = 0 . \tag{13.25}$$

This point $x_0$ is shown in figure 13.3. Near the point $x_0$, the function $g(x)$ can be represented by a Taylor series:

$$g(x) = g(x_0) + \frac{dg}{dx}(x_0)(x - x_0) + \frac{1}{2}\frac{d^2g}{dx^2}(x_0)(x - x_0)^2 + \cdots . \tag{13.26}$$

Because $x_0$ gives the zero-crossing of $g(x)$, see (13.25), the first term is equal to zero. Ignoring the second order term and all higher order terms then gives the following first order approximation for the function $g(x)$ near its zero-crossing:

$$g(x) = \frac{dg}{dx}(x_0)(x - x_0) . \tag{13.27}$$

**Problem a:** Show that the right hand side of (13.27) describes the straight line that is tangent to the curve $g(x)$ at the zero crossing $x_0$. This tangent line is shown as the dashed line in figure 13.3.

**Problem b:** Using this relation one finds that

$$\int \delta\left(g(x)\right) f(x)dx = \int \delta\left(\frac{dg}{dx}(x_0)(x - x_0)\right) f(x)dx . \tag{13.28}$$

The derivative $dg/dx$ at the point $x_0$ can be considered to be a constant. Use the results of the previous section to derive that

$$\int \delta\left(g(x)\right) f(x)dx = \int \frac{1}{\left|\dfrac{dg}{dx}(x_0)\right|}\delta(x - x_0)f(x)dx , \tag{13.29}$$

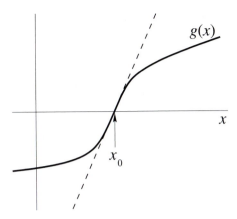

**Fig. 13.3.** The function $g(x)$ (thick solid line) with a zero-crossing at $x_0$ and the tangent line (dashed line) at that point.

and show that this gives

$$\int \delta\left(g(x)\right) f(x) dx = \frac{1}{\left|\dfrac{dg}{dx}(x_0)\right|} f(x_0) \,. \tag{13.30}$$

**Problem c:** Insert (13.11) into the right hand side of (13.30) to derive the following property of the delta function

$$\delta\left(g(x)\right) = \frac{1}{\left|\dfrac{dg}{dx}(x_0)\right|} \delta(x - x_0) \,, \tag{13.31}$$

where it must be kept in mind that $x_0$ denotes the zero-crossing of $g(x)$.

When the function $g(x)$ has more than one zero-crossing, the analysis of this section can be applied to each of these zero-crossings. The contribution of all the different zero-crossings must be added together because all the points where $g(x) = 0$ give a contribution to the integral. This means that when the zero-crossings are denoted by $x_i$ (so that $g(x_i) = 0$)

$$\delta\left(g(x)\right) = \sum_i \frac{1}{\left|\dfrac{dg}{dx}(x_i)\right|} \delta(x - x_i) \,. \tag{13.32}$$

## 13.4 The delta function in more dimensions

The delta function has been defined for functions of a single variable. Its definition can be extended to functions of more variables. As an example we consider here the delta function in three dimensions. The delta function is then defined as the product of the delta functions for each of the coordinates:

$$\delta(\mathbf{r} - \mathbf{r}_0) \equiv \delta(x - x_0)\,\delta(y - y_0)\,\delta(z - z_0) \ . \tag{13.33}$$

This definition of the delta function can be used in the integral $\int \delta(\mathbf{r} - \mathbf{r}_0)f(\mathbf{r})dV$.

**Problem a:** Write the volume integral as $dV = dxdydz$, insert (13.33) and carry out the integration over $x$ to show that

$$\int \delta(\mathbf{r} - \mathbf{r}_0)f(\mathbf{r})dV = \int \delta(y - y_0)\,\delta(z - z_0)\,f(x_0, y, z)dydz \ , \tag{13.34}$$

paying attention to the arguments of the function $f$.

**Problem b:** Carry out the $y$-integration and then the $z$-integration to derive that

$$\int \delta(\mathbf{r} - \mathbf{r}_0)f(\mathbf{r})dV = f(x_0, y_0, z_0) \ . \tag{13.35}$$

The right hand side of (13.35) is the function $f$ at location $\mathbf{r}_0$. This means that the multi-dimensional delta function defined in (13.33) satisfies the following property

$$\boxed{\int \delta(\mathbf{r} - \mathbf{r}_0)f(\mathbf{r})dV = f(\mathbf{r}_0) \ .} \tag{13.36}$$

This expression generalizes (13.11) to more dimensions.

## 13.5 The delta function on the sphere

Up to this point we have analyzed the delta function in Cartesian coordinates. When the delta function is used in a coordinate system that is not Cartesian, additional terms appear in the definition of the delta function. This is illustrated in this section with the delta function on a sphere. Suppose we define a function $f(\theta, \varphi)$ on the unit sphere. Using definition (13.11) the action of the delta function on the sphere as expressed in the angles $\theta$ and $\varphi$ can be written as

$$\int_0^{2\pi} \int_0^{\pi} \delta(\theta - \theta_0)\delta(\varphi - \varphi_0)f(\theta, \varphi)d\theta d\varphi = f(\theta_0, \varphi_0) \ . \tag{13.37}$$

Every point on the unit sphere can be characterized by the unit vector $\hat{\mathbf{r}}$ that points from the origin to that point. It follows from the first identity of (3.7) that for given angles $\theta$ and $\varphi$ this unit vector is given by

$$\hat{\mathbf{r}} = \begin{pmatrix} \sin\theta\cos\varphi \\ \sin\theta\sin\varphi \\ \cos\theta \end{pmatrix} ; \tag{13.38}$$

a similar definition holds for the unit vector $\hat{\mathbf{r}}_0$ that corresponds to the angles $\theta_0$ and $\varphi_0$.

In this section we rewrite (13.37) as an integration over the unit sphere. According to (3.35), the surface element on the sphere is given by $dS = r^2\sin\theta d\theta d\varphi$. On the unit sphere the radius is, by definition, given by $r = 1$, so that on the unit sphere $dS = \sin\theta d\theta d\varphi$.

**Problem a:** Show that (13.37) can be written as

$$\oint \frac{1}{\sin\theta}\delta(\theta - \theta_0)\delta(\varphi - \varphi_0)f(\theta,\varphi)dS = f(\theta_0,\varphi_0) , \tag{13.39}$$

where the symbol $\oint \cdots dS$ denotes the integration over the unit sphere.

We can consider the function $f$ to be a function of the angles $\theta$ and $\varphi$, but we can also see $f$ as a function of the unit vector $\hat{\mathbf{r}}$, which means that $f = f(\hat{\mathbf{r}})$. Therefore we can write the previous expression as

$$\oint \frac{1}{\sin\theta}\delta(\theta - \theta_0)\delta(\varphi - \varphi_0)f(\hat{\mathbf{r}})dS = f(\hat{\mathbf{r}}_0) . \tag{13.40}$$

Formally this can also be written as

$$\oint \delta(\hat{\mathbf{r}} - \hat{\mathbf{r}}_0)f(\hat{\mathbf{r}})dS = f(\hat{\mathbf{r}}_0) , \tag{13.41}$$

where $\delta(\hat{\mathbf{r}} - \hat{\mathbf{r}}_0)$ defines the delta function on the unit sphere.

**Problem b:** Show that

$$\delta(\hat{\mathbf{r}} - \hat{\mathbf{r}}_0) = \frac{1}{\sin\theta}\delta(\theta - \theta_0)\delta(\varphi - \varphi_0) . \tag{13.42}$$

This expression shows that when the delta function is computed in a non-Cartesian coordinate system (such as spherical coordinates defined on the unit sphere) additional terms appear in the delta function that account for the curvilinear character of the coordinate system. The terms that appear in the delta function compensate the terms in the Jacobian that account for the curvilinear character of the employed coordinate system.

## 13.6 The self energy of the electron

Up to this point the delta function has been treated as a mathematical tool. However, it is often used as a description of the concept of a point charge or point mass. Physically, the idea of a point mass is that a finite mass $M$ is concentrated at a certain point $\mathbf{r}_0$. The associated mass-density $\rho(\mathbf{r})$ is then equal to zero everywhere except at $\mathbf{r} = \mathbf{r}_0$. The integral of the mass-density must be equal to the total mass: $\int \rho(\mathbf{r})d^3r = M$, where $\int \cdots d^3\mathbf{r}$ denotes the three-dimensional volume integral.

**Problem a:** Verify that these properties are satisfied by the mass-density

$$\rho(\mathbf{r}) = M\delta(\mathbf{r} - \mathbf{r}_0) . \qquad (13.43)$$

For the moment we restrict our attention to the concept of a point mass and the associated gravitational field, but because of the equivalence of the laws for the gravitational field and the electrostatic field the results can be applied to a stationary electric field as well.

It was shown in section 7.2 that the gravitational field generated by a spherically symmetric body depends outside the body only on the total mass and not on the mass distribution within the body. For a mass centered at the origin ($\mathbf{r}_0 = 0$), the gravitational field is given by

$$\mathbf{g}(\mathbf{r}) = -\frac{GM}{r^2}\hat{\mathbf{r}} . \qquad (7.5)$$

This gravitational field is associated with a gravitational potential $V(\mathbf{r})$ that is related to the gravity field by the expression

$$\mathbf{g}(\mathbf{r}) = -\nabla V(\mathbf{r}) . \qquad (13.44)$$

**Problem b:** Use (7.5) to show that the gravitational potential for a point mass $M$ located in the origin is given by

$$V(\mathbf{r}) = -\frac{GM}{r} . \qquad (13.45)$$

(In the integration you encounter one integration constant. This integration constant follows from the requirement that the potential energy vanishes at infinity.)

In the example in this section the point mass serves as the source of the gravitational field. The response of any linear system to a source function plays a very important role in mathematical physics because a general source of the field can always be written as a superposition of delta functions. The response to a delta function excitation is called the

*Green's function*: this concept is treated in the chapters 17 and 18. We have seen in section 13.1 that the delta function is singular, and it is in fact so singular that is cannot be considered to be a function. Very often, the response to such a singular source function is also singular at the location of that source function. This means that the Green's function is usually singular at the point of excitation.

At this point we have computed the gravitational field and its potential energy for a point mass, and it appears that these can be computed without any problems. However, there is a complication. As shown in expression (1.53) of Jackson [42] the energy $E$ of a charge-density $\rho(\mathbf{r})$ that is placed in a potential $V(\mathbf{r})$ is given by

$$E = \frac{1}{2} \int \rho(\mathbf{r})V(\mathbf{r})d^3r \ . \tag{13.46}$$

(In ref. [42] this is derived for electrostatic energy, but the same expression holds for gravitational energy.)

**Problem c:** Use (13.43) and (13.45) to show that the gravitational energy of a point charge placed in the origin is infinite.

This means that a point mass has infinite energy, which indicates that the concept of a point mass is not without complications when the energy is concerned.

Let us consider what the energy is of a spherically symmetric mass distribution when the mass $M$ is homogeneously distributed in a sphere of radius $R$. The concept of the boxcar function as defined in (13.7) can easily be generalized to three dimensions by the following definition:

$$B_R(\mathbf{r}) \equiv \begin{cases} \dfrac{3}{4\pi R^3} & \text{for} \quad |\mathbf{r}| \leq R \\ 0 & \text{for} \quad |\mathbf{r}| > R \end{cases} . \tag{13.47}$$

**Problem d:** Show that $\int B_R(\mathbf{r})d^3r = 1$ and use this to show that the mass-density of such a mass distribution is given by

$$\rho(\mathbf{r}) = MB_R(\mathbf{r}) \ . \tag{13.48}$$

Outside the mass, the gravitational field is given by (7.5) and the associated potential energy is derived in (13.45). Inside the mass the gravitational field is given by (7.7).

**Problem e:** Use (7.7), (13.44) and the requirement that the potential is continuous everywhere to derive that the potential energy is given by

$$V(\mathbf{r}) \equiv \begin{cases} -\dfrac{GM}{2R^3}(3R^2 - r^2) & \text{for} \quad |\mathbf{r}| \leq R \\[4mm] -\dfrac{GM}{r} & \text{for} \quad |\mathbf{r}| > R \end{cases} \qquad (13.49)$$

**Problem f:** Use this expression and (13.46) to show that the gravitational energy is given by

$$E = -\frac{3GM^2}{5R} . \qquad (13.50)$$

Note that when the radius $R$ of the mass goes to zero, the gravitational energy becomes infinite; this is what we derived in problem c.

Let us consider this homogeneous mass distribution for the moment as a simplified model of the mass distribution of the Earth. It follows from (13.50) that the gravitational energy decreases without bound when the radius of the Earth is decreased. Since physical systems tend to minimize their energy, we can raise the question of why the radius of the Earth has not become smaller than its present-day value of about 6370 km. The fact that the potential energy is negative corresponds to the fact that mass always attracts itself. The decrease of the gravitational energy with decreasing radius thus corresponds to a gravitational collapse of the body. So we can phrase our question in a different way: why does the Earth not collapse to a black hole?

There are two effects that need to be considered. First, the gravitational force in the Earth leads to a compression of the material within the Earth. This compression induces elastic forces within the Earth. The radius of the Earth is dictated by the balance between the gravitational force and the elastic reactive force. If energy were used to describe this balance, one would state that the positive potential energy of the elastic deformation balances the unbounded negative growth of the gravitational energy as the radius is decreased. The second factor to take into account is that the gravitational fields that are treated in this section follow from Newton's law of gravitation. However, this law only holds for weak fields, and for very strong fields it should be replaced by the laws of general relativity [62]. It follows from (7.5) that the gravitational field of a point mass grows without bounds as one moves closer to the point mass. This means that ultimately Newton's law of gravitation ceases to be a good description of the field close to the point mass. This implies that although the concept of a point mass as described by a delta function is appealing, it is physically not without complications.

You may think that this is a purely academic issue. However, the same reasoning applies to electric point charges. By analogy with (13.50), it follows that the energy of a homogeneous charge distribution within a sphere of radius $R$ and a total charge $q$ is given by

$$E = + \frac{3q^2}{20\pi\varepsilon_0 R} \, . \tag{13.51}$$

The energy becomes infinite as the radius $R$ tends to zero. Note that the minus sign in (13.50) is replaced by a plus sign. This corresponds to the fact that equal masses attract each other while equal charges repel each other. Let us now consider the electron as a homogeneous charge $q$ with radius $R$. Expression (13.51) then states that an electron has an infinite energy.

**Problem g:** Show that the energy of the electron is minimized when the radius $R$ goes to infinity.

This means that an electron would grow beyond bounds in order to minimize its energy. Just as with the previous discussion of the gravitational energy of the Earth there are a number of reasons why this is not a physically accurate description. By analogy with the elastic forces in the Earth, one could argue there may be other forces acting within the electron that keep the charge together. However, the concept of an electron as a homogeneous charge distribution is physically not accurate. The charge of an electron is quantized, and the laws of electrostatics are not applicable to the description of the internal structure of the electron.

Nevertheless, the self energy of the electron is a long standing problem [97] because the quantum theory of the electron also predicts an infinite energy of the electron. The more advanced quantum field theory of the electron also predicts an infinite self energy of the electron, but in this case the singularity is logarithmic rather than algebraic [97]. This issue has both mathematical and philosophical aspects. Since the electron has a fixed quantized charge, one cannot consider an electron separately from its field. In fact, through quantum fluctuations the electron can interact with virtual particles that are generated in its field. This means that the energy of an electron that we observe is always a mixture of the energy the electron would have in the absence of its interaction with its field and the interaction with its field (and all virtual particles that may be generated in that field). This has led to the concept of mass-renormalization [76] in which one accounts for the fact that the energy of the electron in the absence of electromagnetic fields is different from the mass that we observe.

# 14

---

# Fourier analysis

Fourier analysis is concerned with the decomposition of signals into sine and cosine waves. This technique is of obvious relevance for spectral analysis where one decomposes a time signal into its different frequency components. As an example, the spectrum of a low-C on a soprano saxophone is shown in figure 14.1. However, the use of Fourier analysis goes far beyond this application because Fourier analysis can also be used to find solutions of differential equations and a large number of other applications. In this chapter the real Fourier transform on a finite interval is used as a starting point. From this the complex Fourier transform and the Fourier transform on an infinite interval are derived. At several stages of the analysis, the similarity between Fourier analysis and linear algebra will be made apparent.

## 14.1 The real Fourier series on a finite interval

Consider a function $f(x)$ that is defined on the interval $-L < x \leq L$. This interval is of length $2L$, and let us assume that $f(x)$ is periodic with period $2L$. This means that if one translates this function over a distance $2L$ the value does not change:

$$f(x + 2L) = f(x) \,. \tag{14.1}$$

We want to expand this function into a set of basis functions. Since $f(x)$ is periodic with period $2L$, these basis functions must be periodic with the same period.

**Problem a:** Show that the functions $\cos(n\pi x/L)$ and $\sin(n\pi x/L)$ with integer $n$ are periodic with period $2L$, i.e. show that these functions satisfy (14.1).

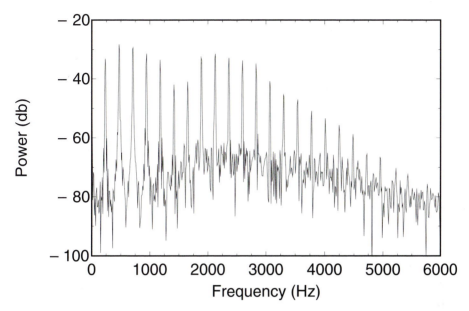

Fig. 14.1. The energy of the sound made by the author playing a low-C on his soprano saxophone as a function of frequency. The unit used for the horizontal axis is the hertz (the number of oscillations per second), the unit on the vertical axis is decibels (a logarithmic measure of energy).

The main statement of Fourier analysis is that one can write $f(x)$ as a superposition of these periodic sine and cosine waves:

$$f(x) = \frac{1}{2}a_0 + \sum_{n=1}^{\infty} a_n \cos\left(n\pi x/L\right) + \sum_{n=1}^{\infty} b_n \sin\left(n\pi x/L\right) . \qquad (14.2)$$

The factor $1/2$ with the coefficient $a_0$ has no special significance, and is used in order to simplify subsequent expressions. To show that (14.2) is actually true is not trivial. Providing this proof essentially amounts to showing that the functions $\cos\left(n\pi x/L\right)$ and $\sin\left(n\pi x/L\right)$ actually contain enough 'degrees of freedom' to describe $f(x)$. However, since $f(x)$ is a function of a continuous variable $x$ this function has infinitely many degrees of freedom and since there are infinitely many coefficients $a_n$ and $b_n$ counting the number of degrees of freedom does not work. Mathematically one would say that one needs to show that the set of functions $\cos\left(n\pi x/L\right)$ and $\sin\left(n\pi x/L\right)$ is a 'complete set.' We will not concern ourselves here with this proof, and simply start working with the Fourier series (14.2).

At this point it is not yet clear what the coefficients $a_n$ and $b_n$ are. In order to derive these coefficients one needs to use the following integrals:

$$\int_{-L}^{L} \cos^2\left(n\pi x/L\right) dx = \int_{-L}^{L} \sin^2\left(n\pi x/L\right) dx = L \qquad (n \geq 1) ; \qquad (14.3)$$

$$\int_{-L}^{L} \cos\left(n\pi x/L\right) \cos\left(m\pi x/L\right) dx = 0 \qquad \text{if} \qquad n \neq m ; \qquad (14.4)$$

$$\int_{-L}^{L} \sin\left(n\pi x/L\right) \sin\left(m\pi x/L\right) dx = 0 \qquad \text{if} \qquad n \neq m ; \qquad (14.5)$$

$$\int_{-L}^{L} \cos\left(n\pi x/L\right) \sin\left(m\pi x/L\right) dx = 0 \qquad \text{all} \qquad n, m . \qquad (14.6)$$

**Problem b:** Derive these identities. In doing so you need to use trigonometric identities such as $\cos\alpha\cos\beta = \left[\cos(\alpha + \beta) + \cos(\alpha - \beta)\right]/2$. If you have difficulty deriving these identities you should consult a textbook such as Boas [15].

**Problem c:** In order to find the coefficient $b_m$, multiply the Fourier expansion (14.2) by $\sin\left(m\pi x/L\right)$, integrate the result from $-L$ to $L$ and use (14.3)–(14.6) to evaluate the integrals. Show that this gives:

$$b_n = \frac{1}{L} \int_{-L}^{L} f(x) \sin\left(n\pi x/L\right) dx . \qquad (14.7)$$

**Problem d:** Use a similar analysis to show that:

$$a_n = \frac{1}{L} \int_{-L}^{L} f(x) \cos\left(n\pi x/L\right) dx . \qquad (14.8)$$

In deriving this result treat the cases $n \neq 0$ and $n = 0$ separately. It is now clear why the factor $1/2$ was introduced in the $a_0$ term of (14.2); without this factor expression (14.8) would have an additional factor 2 for $n = 0$.

There is a close relation between the Fourier series (14.2) and the coefficients given in the expressions above and the projection of a vector on a number of basis vectors in linear algebra as shown in section 12.1. To see this relation we restrict ourselves for simplicity to functions $f(x)$ that are odd functions of $x$: $f(-x) = -f(x)$, but this restriction is by no means essential. For these functions all coefficients $a_n$ are equal to zero. As an analog of a basis vector in linear algebra let us define the following basis function $u_n(x)$:

$$u_n(x) \equiv \frac{1}{\sqrt{L}} \sin\left(n\pi x/L\right) . \qquad (14.9)$$

An essential ingredient in the projection operators of section 12.1 is the inner product between vectors. It is also possible to define an inner product for functions, and for the present example the inner product of two functions $f(x)$ and $g(x)$ is defined as:

$$(f \cdot g) \equiv \int_{-L}^{L} f(x)g(x)dx \ . \tag{14.10}$$

**Problem e:** The basis functions $u_n(x)$ defined in (14.9) are the analog of a set of orthonormal basis vectors. To see this, use (14.3) and (14.5) to show that

$$(u_n \cdot u_m) = \delta_{nm} \ , \tag{14.11}$$

where $\delta_{nm}$ is the Kronecker delta.

This expression implies that the basis functions $u_n(x)$ are mutually orthogonal. If the norm of such a basis function is defined as $\|u_n\| \equiv \sqrt{(u_n \cdot u_n)}$, (14.11) implies that the basis functions are normalized (i.e. have norm 1). These functions are the generalization of orthogonal unit vectors to a function space. The (odd) function $f(x)$ can be written as a sum of the basis functions $u_n(x)$:

$$f(x) = \sum_{n=1}^{\infty} c_n u_n(x) \ . \tag{14.12}$$

**Problem f:** Take the inner product of (14.12) with $u_m(x)$ and show that $c_m = (u_m \cdot f)$. Use this to show that the Fourier expansion of $f(x)$ can be written as $f(x) = \sum_{n=1}^{\infty} u_n(x) (u_n \cdot f)$, and that on leaving out the explicit dependence on the variable $x$ the result is

$$f = \sum_{n=1}^{\infty} u_n (u_n \cdot f) \ . \tag{14.13}$$

This equation bears a close resemblance to the expression derived in section 12.1 for the projection of vectors. The projection of a vector $\mathbf{v}$ along a unit vector $\hat{\mathbf{n}}$ was shown to be

$$\mathbf{Pv} = \hat{\mathbf{n}} (\hat{\mathbf{n}} \cdot \mathbf{v}) \ . \tag{12.2}$$

A comparison with (14.13) shows that $u_n(x) (u_n \cdot f)$ can be interpreted as the projection of the function $f(x)$ on the function $u_n(x)$. To reconstruct the function, one must sum over the projections along all basis functions, hence the summation in (14.13). It is shown in (12.12) that in order to find the projection of the vector $\mathbf{v}$ onto the subspace spanned by a finite number of orthonormal basis vectors one simply has to sum the projections of the vector $\mathbf{v}$ on all the basis vectors that span the subspace: $\mathbf{Pv} = \sum_i \hat{\mathbf{n}}_i (\hat{\mathbf{n}}_i \cdot \mathbf{v})$. In a similar way, one can sum the Fourier series (14.13) over only a limited number of basis functions to obtain the projection of $f(x)$ on a limited number of basis functions:

$$f_{filtered} = \sum_{n=n_1}^{n_2} u_n (u_n \cdot f) \ . \tag{14.14}$$

In this expression it was assumed that only values $n_1 \leq n \leq n_2$ have been used. The projected function is called $f_{filtered}$ because this projection is really a filtering operation.

**Problem g:** Show that the functions $u_n(x)$ are sinusoidal waves with wavelength $\lambda = 2L/n$.

This means that restricting the $n$-values in the sum (14.14) amounts to using only wavelengths between $2L/n_2$ and $2L/n_1$ for the projected function. Since only certain wavelengths are used, this projection really acts as a filter that allows only certain wavelengths in the filtered function.

It is the filtering property that makes the Fourier transform so useful for filtering data sets by excluding wavelengths that are unwanted. In fact, the Fourier transform forms the basis of digital filtering techniques that have many applications in science and engineering, see for example the books of Claerbout [19] and Robinson and Treitel [75].

## 14.2 The complex Fourier series on a finite interval

In the theory of the preceding section there is no reason why the function $f(x)$ should be real. Although the basis functions $\cos(n\pi x/L)$ and $\sin(n\pi x/L)$ are real, the Fourier sum (14.2) can be complex because the coefficients $a_n$ and $b_n$ can be complex. The equation of de Moivre gives the relation between these basis functions and complex exponential functions:

$$e^{in\pi x/L} = \cos(n\pi x/L) + i \sin(n\pi x/L) \ . \tag{14.15}$$

This expression can be used to rewrite the Fourier series (14.2) using the basis functions $e^{in\pi x/L}$ rather than sine and cosines.

**Problem a:** Replace $n$ by $-n$ in (14.15) to show that:

$$\left. \begin{aligned} \cos(n\pi x/L) &= \frac{1}{2} \left( e^{in\pi x/L} + e^{-in\pi x/L} \right) , \\[2mm] \sin(n\pi x/L) &= \frac{1}{2i} \left( e^{in\pi x/L} - e^{-in\pi x/L} \right) . \end{aligned} \right\} \tag{14.16}$$

**Problem b:** Insert this relation in the Fourier series (14.2) to show that this Fourier series can also be written as:

$$f(x) = \sum_{n=-\infty}^{\infty} c_n e^{in\pi x/L} , \qquad (14.17)$$

with the coefficients $c_n$ given by:

$$\left. \begin{array}{ll} c_n = (a_n - ib_n)/2 & \text{for } n > 0 , \\ c_n = (a_{|n|} + ib_{|n|})/2 & \text{for } n < 0 , \\ c_0 = a_0/2 . \end{array} \right\} \qquad (14.18)$$

Note that the absolute value $|n|$ is used for $n < 0$.

**Problem c:** Explain why the $n$-summation in (14.17) extends from $-\infty$ to $\infty$ rather than from 0 to $\infty$.

**Problem d:** Relations (14.7) and (14.8) can be used to express the coefficients $c_n$ in the function $f(x)$. Treat the cases $n > 0$, $n < 0$ and $n = 0$ separately to show that for all values of $n$ the coefficient $c_n$ is given by:

$$c_n = \frac{1}{2L} \int_{-L}^{L} f(x) e^{-in\pi x/L} dx . \qquad (14.19)$$

The sum (14.17) with (14.19) constitutes the complex Fourier transform over a finite interval. Again, there is a close analogy with the projections of vectors shown in section 12.1. Before we can explore this analogy, the inner product between two complex functions $f(x)$ and $g(x)$ needs to be defined. This inner product is not given by $(f \cdot g) = \int f(x)g(x)dx$. The reason for this is that the length of a vector is defined by $\|\mathbf{v}\|^2 = (\mathbf{v} \cdot \mathbf{v})$, and a straightforward generalization of this expression to functions using the inner product given above would give for the norm of the function: $\|f\|^2 = (f \cdot f) = \int f^2(x)dx$. However, when $f(x)$ is purely imaginary this would lead to a negative norm. This can be avoided by defining the inner product of two complex functions by:

$$(f \cdot g) \equiv \int_{-L}^{L} f^*(x)g(x)dx , \qquad (14.20)$$

where the asterisk denotes the complex conjugate.

**Problem e:** Show that with this definition the norm of $f(x)$ is given by $\|f\|^2 = (f \cdot f) = \int |f(x)|^2 dx$.

With this inner product the norm of the function is guaranteed to be positive. Now that we have an inner product, the analogy with the projections in linear algebra can be explored. In order to do this, define the following basis functions:

$$u_n(x) \equiv \frac{1}{\sqrt{2L}} e^{in\pi x/L} .$$  (14.21)

**Problem f:** Show that these functions are orthonormal with respect to the inner product (14.20), i.e. show that:

$$(u_n \cdot u_m) = \delta_{nm} .$$  (14.22)

Pay special attention to the normalization of these functions; i.e. to the case $n = m$.

**Problem g:** Expand $f(x)$ in these basis functions, $f(x) = \sum_{n=-\infty}^{\infty} \gamma_n u_n(x)$ and show that $f(x)$ can be written as:

$$f = \sum_{n=-\infty}^{\infty} u_n (u_n \cdot f) .$$  (14.23)

**Problem h:** Make the comparison between this expression and the expressions for the projections of vectors in section 12.1.

## 14.3 The Fourier transform on an infinite interval

In several applications, one wants to compute the Fourier transform of a function that is defined on an infinite interval. This amounts to taking the limit $L \to \infty$. However, a simple inspection of (14.19) shows that one cannot simply take the limit $L \to \infty$ of the expressions of the previous section because in that limit $c_n = 0$. In order to define the Fourier transform for an infinite interval define the variable $k$ by:

$$k \equiv \frac{n\pi}{L} .$$  (14.24)

An increment $\Delta n$ corresponds to an increment $\Delta k$ given by $\Delta k = \pi \Delta n/L$. In the summation over $n$ in the Fourier expansion (14.17), $n$ is incremented by unity: $\Delta n = 1$. This corresponds to an increment $\Delta k = \pi/L$ of the variable $k$. In the limit $L \to \infty$ this increment goes to zero, which implies that the summation over $n$ should be replaced by an integration over $k$:

$$\sum_{n=-\infty}^{\infty} (\cdots) \to \frac{\Delta n}{\Delta k} \int_{-\infty}^{\infty} (\cdots) \, dk = \frac{L}{\pi} \int_{-\infty}^{\infty} (\cdots) \, dk \qquad \text{as } L \to \infty .$$  (14.25)

**Problem a:** Explain the presence of the factor $\Delta n/\Delta k$ and prove the last identity.

This is not enough to generalize the Fourier transform of the previous section to an infinite interval. As noted earlier, the coefficients $c_n$ vanish in the limit $L \to \infty$. Also note that the integral on the right hand side of (14.25) is multiplied by $L/\pi$, and this coefficient is infinite in the limit $L \to \infty$. Both complications can be solved by defining the following function:

$$F(k) \equiv \frac{L}{\pi} c_n \ , \tag{14.26}$$

where the relation between $k$ and $n$ is given by (14.24).

**Problem b:** Show that with the replacements (14.25) and (14.26) the limit $L \to \infty$ of the complex Fourier transform (14.17) and (14.19) can be taken and that the result can be written as:

$$f(x) = \int_{-\infty}^{\infty} F(k) e^{ikx} dk \ , \tag{14.27}$$

$$F(k) = \frac{1}{2\pi} \int_{-\infty}^{\infty} f(x) e^{-ikx} dx \ . \tag{14.28}$$

## 14.4 The Fourier transform and the delta function

In this section the Fourier transform of the delta function is treated. This is not only useful in a variety of applications, but it will also establish the relation between the Fourier transform and the closure relation introduced in section 12.1. Consider the delta function centered at $x = x_0$:

$$f(x) = \delta(x - x_0) \ . \tag{14.29}$$

**Problem a:** Show that the Fourier transform $F(k)$ of this function is given by:

$$F(k) = \frac{1}{2\pi} e^{-ikx_0} \ . \tag{14.30}$$

**Problem b:** Show that this implies that the Fourier transform of the delta function $\delta(x)$ centered at $x = 0$ is a constant. Determine this constant.

**Problem c:** Use (14.27) to show that

$$\delta(x - x_0) = \frac{1}{2\pi} \int_{-\infty}^{\infty} e^{ik(x-x_0)} dk \ . \tag{14.31}$$

**Problem d:** Use a similar analysis to derive that

$$\delta(k - k_0) = \frac{1}{2\pi} \int_{-\infty}^{\infty} e^{-i(k-k_0)x} dx \ . \tag{14.32}$$

These expressions are very useful in a number of applications. Again, there is close analogy between these expressions and the projection of vectors introduced in section 12.1. To establish this connection let us define the following basis functions:

$$u_k(x) \equiv \frac{1}{\sqrt{2\pi}} e^{ikx} , \qquad (14.33)$$

and use the inner product defined in (14.20) with the integration limits extending from $-\infty$ to $\infty$.

**Problem e:** Show that (14.32) implies that

$$(u_k \cdot u_{k_0}) = \delta(k - k_0) . \qquad (14.34)$$

This implies that the functions $u_k(x)$ form an orthonormal set, because this relation generalizes (14.22) to a continuous basis of functions.

**Problem f:** Use (14.31) to derive that:

$$\int_{-\infty}^{\infty} u_k(x) u_k^*(x_0) \, dk = \delta(x - x_0) . \qquad (14.35)$$

This expression is the counterpart of the closure relation (12.13) introduced in section 12.1 for finite-dimensional vector spaces. Note that the delta function $\delta(x - x_0)$ plays the role of the identity operator $\mathbf{I}$ with components $I_{ij} = \delta_{ij}$ in (12.13) and that the summation $\sum_{i=1}^{N}$ over the basis vectors is replaced by an integration $\int_{-\infty}^{\infty} dk$ over the basis functions. Both differences are due to the fact that we are dealing in this section with an infinite-dimensional function space rather than a finite-dimensional vector space. Also note that in (14.35) the complex conjugate of $u_k(x_0)$ is taken. The reason for this is that for complex unit vectors $\hat{\mathbf{n}}$ the transpose in the completeness relation (12.13) should be replaced by the Hermitian conjugate. This involves taking the complex conjugate as well as taking the transpose.

## 14.5 Changing the sign and scale factor

In the Fourier transformation (14.27) from the wavenumber domain $(k)$ to the position domain $(x)$, the exponent has a plus sign, $e^{+ikx}$, and the coefficient multiplying the integral is given by 1. In other texts on Fourier transforms you may encounter a different sign in the exponent and different scale factors are sometimes used. For example, the exponent in the Fourier transformation from the wavenumber domain to the position domain may have a minus sign, $e^{-ikx}$, and there may be a scale factor such as $1/\sqrt{2\pi}$. It turns out that there is a freedom in choosing the sign of the exponential as well as in the scaling of the Fourier transform. We will first study the effect of a scaling parameter on the Fourier transform.

**Problem a:** Let the function $F(k)$ defined in (14.28) be related to a new function $\tilde{F}(k)$ by a scaling with a scale factor $C$: $F(k) = C\tilde{F}(k)$. Use (14.27) and (14.28) to show that:

$$f(x) = C \int_{-\infty}^{\infty} \tilde{F}(k)e^{ikx}dk , \qquad (14.36)$$

$$\tilde{F}(k) = \frac{1}{2\pi C} \int_{-\infty}^{\infty} f(x)e^{-ikx}dx . \qquad (14.37)$$

These expressions are completely equivalent to the original Fourier transform pair (14.27) and (14.28). The constant $C$ is completely arbitrary. This implies that one may take any multiplication constant for the Fourier transform; the only restriction is that the product of the coefficients for Fourier transform and the backward transform is equal to $1/2\pi$.

**Problem b:** Prove this last statement.

In the literature, notably in quantum mechanics, one often encounters the Fourier transform pair using the value $C = 1/\sqrt{2\pi}$. This leads to the Fourier transform pair:

$$f(x) = \frac{1}{\sqrt{2\pi}} \int_{-\infty}^{\infty} \tilde{F}(k)e^{ikx}dk , \qquad (14.38)$$

$$\tilde{F}(k) = \frac{1}{\sqrt{2\pi}} \int_{-\infty}^{\infty} f(x)e^{-ikx}dx . \qquad (14.39)$$

This normalization not only has the advantage that the multiplication factors for the forward and backward transformations are identical $(1/\sqrt{2\pi})$, but the constants are also identical to the constant used in (14.33) to create a set of orthonormal functions.

Next we investigate a change in the sign of the exponent in the Fourier transform. To do this, we will use the function $\tilde{F}(k)$ defined by: $\tilde{F}(k) = F(-k)$.

**Problem c:** Change the integration variable $k$ in (14.27) to $-k$ and show that the Fourier transform pair (14.27) and (14.28) is equivalent to:

$$f(x) = \int_{-\infty}^{\infty} \tilde{F}(k)e^{-ikx}dk , \qquad (14.40)$$

$$\tilde{F}(k) = \frac{1}{2\pi} \int_{-\infty}^{\infty} f(x)e^{ikx}dx . \qquad (14.41)$$

Note that these expressions only differ from the earlier expressions by the sign of the exponents. This means that there is a freedom in the choice of this sign. It does not matter which sign convention you use. Any choice of the sign and the multiplication constant for the Fourier transform can be used as long as:

> (i) The product of the constants for the forward and backward transform is equal to $1/2\pi$ and (ii) the sign of the exponent for the forward and the backward transformation is opposite.

In this book, the Fourier transform pair (14.27) and (14.28) will generally be used for the Fourier transform from the space $(x)$ domain to the wavenumber $(k)$ domain.

Of course, the Fourier transform can also be used to transform a function in the time $(t)$ domain to the frequency $(\omega)$ domain. Perhaps illogically the following convention will used in this book for this Fourier transform pair:

$$f(t) = \int_{-\infty}^{\infty} F(\omega)e^{-i\omega t}\,d\omega \;, \tag{14.42}$$

$$F(\omega) = \frac{1}{2\pi} \int_{-\infty}^{\infty} f(t)e^{i\omega t}\,dt \;. \tag{14.43}$$

The reason for this choice is that the combined Fourier transform from the $(x,t)$-domain to the $(k,\omega)$-domain that is obtained by combining (14.27) and (14.42) is given by:

$$f(x,t) = \int\!\!\int_{-\infty}^{\infty} F(k,\omega)e^{i(kx-\omega t)}\,dk\,d\omega \;. \tag{14.44}$$

The function $e^{i(kx-\omega t)}$ in this integral describes a wave that moves for positive values of $k$ and $\omega$ in the direction of increasing values of $x$. To see this, let us assume that we are at a crest of this wave and that we follow the motion of the crest over a time $\Delta t$ and we want to find the distance $\Delta x$ that the crest has moved in that time interval. If we follow a wave crest, the phase of the wave is constant, and hence $kx - \omega t$ is constant.

**Problem d:** Show that this implies that $\Delta x = c\Delta t$, with $c$ given by $c = \omega/k$. Why does this imply that the wave moves with velocity $c$?

The exponential in the double Fourier transform (14.44) therefore describes for positive values of $\omega$ and $k$ a wave travelling in the positive direction with velocity $c = \omega/k$. However, note that this is no proof that we should use the Fourier transform (14.44) and not a transform with a different choice of the sign in the exponent. In fact, one should realize that in the Fourier transform (14.44) one needs to integrate over *all* values of $w$ and $k$ so that negative values of $w$ and $k$ contribute to the integral as well.

**Problem e:** Use (14.28) and (14.43) to derive the inverse of the double Fourier transform (14.44).

## 14.6 The convolution and correlation of two signals

There are different ways in which one can combine signals to create a new signal. In this section the convolution and correlation of two signals is treated. For the sake of argument the signals are taken to be functions of time, and the Fourier transform pair (14.42) and (14.43) is used for the forward and inverse Fourier transforms. Suppose a function $f(t)$ has a Fourier transform $F(w)$ defined by (14.42) and another function $h(t)$ has a similar Fourier transform $H(w)$:

$$h(t) = \int_{-\infty}^{\infty} H(w)e^{-iwt}dw . \tag{14.45}$$

The two Fourier transforms $F(w)$ and $H(w)$ can be multiplied in the frequency domain, and we want to find out what the Fourier transform of the product $F(w)H(w)$ is in the time domain.

**Problem a:** Show that:

$$F(w)H(w) = \frac{1}{(2\pi)^2} \int\!\!\!\int_{-\infty}^{\infty} f(t_1)h(t_2)e^{iw(t_1+t_2)}dt_1dt_2 . \tag{14.46}$$

**Problem b:** Show that after a Fourier transformation this function corresponds in the time domain to:

$$\int_{-\infty}^{\infty} F(w)H(w)e^{-iwt}dw = \frac{1}{(2\pi)^2} \int\!\!\!\int\!\!\!\int_{-\infty}^{\infty} f(t_1)h(t_2)e^{iw(t_1+t_2-t)}dt_1dt_2dw .$$
$$\tag{14.47}$$

**Problem c:** Use the representation (14.31) of the delta function to carry out the integration over $w$ and show that this gives:

$$\int_{-\infty}^{\infty} F(w)H(w)e^{-iwt}dw = \frac{1}{2\pi} \int\!\!\!\int_{-\infty}^{\infty} f(t_1)h(t_2)\delta(t_1 + t_2 - t)dt_1dt_2 .$$
$$\tag{14.48}$$

**Problem d:** The integration over $t_1$ can now be carried out. Do this, and show that after renaming the variable $t_2$ as $\tau$ the result can be written as:

$$\int_{-\infty}^{\infty} F(\omega)H(\omega)e^{-i\omega t}d\omega = \frac{1}{2\pi}\int_{-\infty}^{\infty} f(t-\tau)h(\tau)d\tau = \frac{1}{2\pi}(f*h)(t) .$$
(14.49)

The $\tau$-integral in the middle term is called the *convolution* of the functions $f$ and $h$; this operation is denoted by the symbol $(f*h)$. Equation (14.49) states that a multiplication of the spectra of two functions in the frequency domain corresponds to the convolution of these functions in the time domain. For this reason, (14.49) is called the *convolution theorem*. This theorem is schematically indicated in the following diagram:

$$f(t) \longleftrightarrow F(\omega) ,$$
$$h(t) \longleftrightarrow H(\omega) ,$$
$$\frac{1}{2\pi}(f*h) \longleftrightarrow F(\omega)H(\omega) .$$

Note that in the convolution theorem, a scale factor $1/2\pi$ is present on the left hand side. This scale factor depends on the choice of the scale factors that one uses in the Fourier transform, see section 14.5.

**Problem e:** Use a change of the integration variable to show that the convolution of $f$ and $h$ can be written in the following two ways:

$$(f*h)(t) = \int_{-\infty}^{\infty} f(t-\tau)h(\tau)d\tau = \int_{-\infty}^{\infty} f(\tau)h(t-\tau)d\tau . \quad (14.50)$$

**Problem f:** In order to see what the convolution theorem looks like when a different scale factor is used in the Fourier transform, define $F(\omega) = C\tilde{F}(\omega)$, and a similar scaling for $H(\omega)$. Show that with this choice of the scale factors, the Fourier transform of $\tilde{F}(\omega)\tilde{H}(\omega)$ is in the time domain given by $(1/2\pi C)(f*h)(t)$. Hint: first determine the scale factor that one needs to use in the transformation from the frequency domain to the time domain.

The convolution of two time series plays a very important role in exploration geophysics. Suppose one carries out a seismic experiment in which one uses a source such as dynamite or a vibrator to generate waves that propagate through the Earth. Let the source signal in the frequency domain be given by $S(\omega)$. The waves reflect at layers in the Earth and are recorded by geophones. In the ideal case, the source signal would have the shape of a simple spike, and the waves reflected by all the reflectors would show up as a sequence of individual spikes. In that case the recorded data

would indicate the true reflectors in the Earth. Let the signal $r(t)$ recorded
in this ideal case have a Fourier transform $R(\omega)$ in the frequency domain.
The problem that one faces is that a real seismic source is often not very
impulsive. If the recorded data $d(t)$ have a Fourier transform $D(\omega)$ in the
frequency domain, then this Fourier transform is given by

$$D(\omega) = R(\omega)S(\omega) \,. \qquad (14.51)$$

One is only interested in $R(\omega)$ which is the Earth response in the frequency domain, but in practice one records the product $R(\omega)S(\omega)$. In the
time domain this is equivalent to saying that one has recorded the con-
volution $\int_{-\infty}^{\infty} r(\tau)s(t-\tau)d\tau$ of the Earth response with the source signal,
but that one is only interested in the Earth response $r(t)$. One would like
to 'undo' this convolution; this process is called *deconvolution*. Carrying
out the deconvolution seems trivial in the frequency domain. According to
(14.51) one only needs to divide the data in the frequency domain by the
source spectrum $S(\omega)$ to obtain $R(\omega)$. The problem is that in practice one
often does not know the source spectrum $S(\omega)$. This makes seismic deconvolution a difficult process; see the collection of articles compiled by Webster [96]. It has been strongly argued by Ziolkowski [105] that the seismic
industry should make a larger effort to record the source signal accurately.

   The convolution of two signals was obtained in this section by taking
the product $F(\omega)H(\omega)$ and carrying out a Fourier transformation back
to the time domain. The same steps can be taken by multiplying $F(\omega)$
by the complex conjugate $H^*(\omega)$ and applying a Fourier transformation
to go the time domain.

**Problem g:** Take steps similar to those in the derivation of the convolution to show that

$$\int_{-\infty}^{\infty} F(\omega)H^*(\omega)e^{-i\omega t}d\omega = \frac{1}{2\pi}\int_{-\infty}^{\infty} f(t+\tau)h^*(\tau)d\tau \,. \qquad (14.52)$$

The right hand side of this expression is called the *correlation* of the functions $f(t)$ and $h^*(t)$. Note that this expression is very similar to the convolution theorem (14.50). This result implies that the Fourier transform
of the product of a function and the complex conjugate in the frequency
domain corresponds to the correlation in the time domain. Note again the
constant $1/2\pi$ on the right hand side. This constant again depends on the
scale factors used in the Fourier transform.

**Problem h:** Set $t = 0$ in (14.52) and let the function $h(t)$ be equal to
   $f(t)$. Show that this gives:

$$\int_{-\infty}^{\infty} |F(\omega)|^2\, d\omega = \frac{1}{2\pi}\int_{-\infty}^{\infty} |f(t)|^2\, dt \,. \qquad (14.53)$$

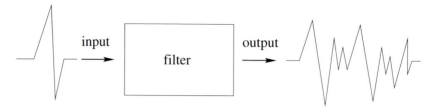

Fig. 14.2.   Schematic representation of a linear filter.

This is known as *Parseval's theorem*. To see its significance, note that $\int_{-\infty}^{\infty} |f(t)|^2 \, dt = (f \cdot f)$, with the inner product of (14.20) with $t$ as the integration variable and with the integration extending from $-\infty$ to $\infty$. Since $\sqrt{(f \cdot f)}$ is the norm of $f$ measured in the time domain, and since $\int_{-\infty}^{\infty} |F(\omega)|^2 \, d\omega$ is the square of the norm of $F$ measured in the frequency domain, Parseval's theorem states that with this definition of the norm, the norm of a function is equal in the time domain and in the frequency domain (up to the scale factor $1/2\pi$).

## 14.7 Linear filters and the convolution theorem

Let us consider a linear system that has an output signal $o(t)$ when it is given an input signal $i(t)$, see figure 14.2. There are numerous examples of this kind of system. As an example, consider a damped harmonic oscillator that is driven by a force; this system is described by the differential equation $\ddot{x} + 2\beta\dot{x} + \omega_0^2 x = F/m$, where the dot denotes a time derivative. The force $F(t)$ can be seen as the input signal, and the response $x(t)$ of the oscillator can be seen as the output signal. The relation between the input signal and the output signal is governed by the characteristics of the system under consideration; in this example it is the physics of the damped harmonic oscillator that determines the relation between the input signal $F(t)$ and the output signal $x(t)$.

Note that we have not defined yet what a *linear* filter is. A filter is linear when an input $c_1 i_1(t) + c_2 i_2(t)$ leads to an output $c_1 o_1(t) + c_2 o_2(t)$, in which $o_1(t)$ is the output corresponding to the input $i_1(t)$ and $o_2(t)$ is the output the input $i_2(t)$.

**Problem a:** Can you think of another example of a linear filter?

**Problem b:** Can you think of a system that has one input signal and one output signal, in which these signals are related through a nonlinear relation? This would be an example of a nonlinear filter; the theory of this section would not apply to such a filter.

It is possible to determine the output $o(t)$ for any input $i(t)$ if the output to a delta function input is known. Consider the special input signal $\delta(t-\tau)$ that consists of a delta function centered at $t = \tau$. Since a delta function has 'zero-width' (if it has a width at all) such an input function is very impulsive. Let the output for this particular input be denoted by $g(t, \tau)$. Since this function is the response at time $t$ to an impulsive input at time $\tau$ this function is called the *impulse response*:

> *The impulse response function $g(t, \tau)$ is the output of the system at time $t$ due to an impulsive input at time $\tau$.*

How can the impulse response be used to find the response to an arbitrary input function? Any input function can be written as:

$$i(t) = \int_{-\infty}^{\infty} \delta(t - \tau) i(\tau) d\tau . \qquad (14.54)$$

This identity follows from the definition of the delta function. However, we can also look at this expression from a different point of view. The integral on the right hand side of (14.54) can be seen as a superposition of infinitely many delta functions $\delta(t - \tau)$. Each delta function when considered as a function of $t$ is centered at time $\tau$. Since we integrate over $\tau$ these different delta functions are superposed to construct the input signal $i(t)$. Each of the delta functions in the integral (14.54) is multiplied by $i(\tau)$. This term plays the role of a coefficient that gives a weight to the delta function $\delta(t - \tau)$.

At this point it is crucial to use that the filter is linear. Since the response to the input $\delta(t-\tau)$ is the impulse response $g(t, \tau)$, and since the input can be written as the superposition (14.54) of delta function input signals $\delta(t - \tau)$, the output can be written as the same superposition of impulse response signals $g(t, \tau)$:

$$o(t) = \int_{-\infty}^{\infty} g(t, \tau) i(\tau) d\tau . \qquad (14.55)$$

**Problem c:** Carefully compare (14.54) and (14.55). Note the similarity and make sure you understand the reasoning that has led to (14.55).

You may find this 'derivation' of (14.55) rather vague. The notion of the impulse response will be treated in much greater detail in chapter 17 because it plays a crucial role in mathematical physics.

At this point we make another assumption about the system. Apart from its linearity we will also assume it is *invariant for translations in time*. This is a complex way of saying that we assume that the properties of the filter do not change with time. This is the case for the damped harmonic oscillator described at the beginning of this section. However, this oscillator would not be invariant for translations in time if the damping parameter were a function of time as well: $\beta = \beta(t)$. In that case, the system would give different responses when the same input is used at different times.

When the properties of the filter do not depend on time, the impulse response $g(t, \tau)$ depends only on the *difference* $t - \tau$. To see this, consider the damped harmonic oscillator again. The response at a certain time depends only on the time that has lapsed between the excitation at time $\tau$ and the time of observation $t$. Therefore, for a time-invariant filter:

$$g(t, \tau) = g(t - \tau) \, . \tag{14.56}$$

Inserting this in (14.55) shows that for a linear time-invariant filter the output is given by the convolution of the input with the impulse response:

$$o(t) = \int_{-\infty}^{\infty} g(t - \tau)i(\tau)d\tau = (g * i)\,(t) \, . \tag{14.57}$$

**Problem d:** Let the Fourier transform of $i(t)$ be given by $I(\omega)$, the Fourier transform of $o(t)$ by $O(\omega)$ and the Fourier transform of $g(t)$ by $G(\omega)$. Use (14.57) to show that these Fourier transforms are related by:

$$O(\omega) = 2\pi G(\omega)I(\omega) \, . \tag{14.58}$$

Expressions (14.57) and (14.58) are key results in the theory in linear time-invariant filters. The first of these states that one only needs to know the response $g(t)$ to a single impulse to compute the output of the filter to *any* input signal $i(t)$. Equation (14.58) has two important consequences. First, if one knows the Fourier transform $G(\omega)$ of the impulse response, one can compute the Fourier transform $O(\omega)$ of the output. An inverse Fourier transform then gives the output $o(t)$ in the time domain.

**Problem e:** Show that $G(\omega)e^{-i\omega t}$ is the response of the system to the input signal $e^{-i\omega t}$.

This means that if one knows the response of the filter to the harmonic signal $e^{-i\omega t}$ at any frequency, one knows $G(\omega)$ and the response to any input signal can be determined.

The second important consequence of (14.58) is that the output at frequency $\omega$ depends only on the input and impulse response at the same frequency $\omega$, and not on other frequencies. This last property does not hold for nonlinear systems, because for these different frequency components of the input signal are mixed by the nonlinearity of the system. An example of this phenomenon is the Earth's climate which has variations which contain frequency components that cannot be explained by periodic variations in the orbital parameters in the Earth, but which are due to the nonlinear character of the climate response to the amount of energy received by the Sun [80].

The fact that a filter can be used either by specifying its Fourier transform $G(\omega)$ (or equivalently the response to an harmonic input $e^{-i\omega t}$) or by prescribing the impulse response $g(t)$ implies that a filter can be designed either in the frequency domain or in the time domain. In section 14.8 the action of a filter in the time domain is described. A Fourier transform then leads to a compact description of the filter response in the frequency domain. In section 14.9 the converse route is taken; the filter is designed in the frequency domain, and a Fourier transform is used to derive an expression for the filter in the time domain.

As a last reminder it should be mentioned that although the theory of linear filters is introduced here for filters that act in the time domain, the theory is equally valid for filters in the spatial domain. In the latter case, the wavenumber $k$ plays the role that the angular frequency played in this section. Since there may be more than one spatial dimension, the theory for the spatial domain must be generalized to include higher-dimensional spatial Fourier transforms. However, this does not change the principles involved.

## 14.8  The dereverberation filter

As an example of a filter that is derived in the time domain we consider here the description of reverberations in marine seismics. Suppose a seismic survey is carried out at sea. In such an experiment a ship tows a string of hydrophones that record the pressure variations just below the surface of the water, see figure 14.3. Since the pressure vanishes at the surface of the water, the surface of the water totally reflects pressure waves and the reflection coefficient for reflection at the water surface is equal to $-1$. Let the reflection coefficient for waves reflecting upwards from the sea bed be denoted by $r$. Since the contrast between the water and the solid earth below is not small, this reflection coefficient can be considerable.

**Problem a:** Give a physical argument why this reflection coefficient must be smaller or equal than unity: $r \leq 1$.

Since the reflection coefficient of the sea bed is not small, waves can bounce back and forth repeatedly between the water surface and the sea bed. These reverberations are an unwanted artifact in seismic experiments. This is because a wave that has bounced back and forth in the water layer can be misinterpreted on a seismic section as a reflector in the Earth. For this reason one wants to eliminate these reverberations from seismic data.

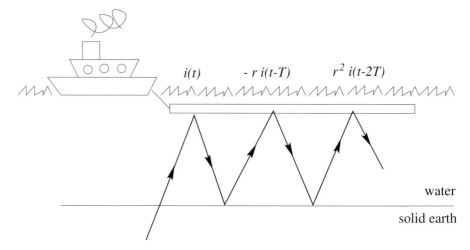

Fig. 14.3.   The generation of reverberations in a marine seismic experiment.

Suppose the wavefield recorded by the hydrophones in the absence of reverberations is denoted by $i(t)$. Let the time it takes for a wave to travel from the water surface to the sea bed and back be denoted by $T$; this time is called the two-way travel time.

**Problem b:** Show that a wave that has bounced back and forth once is given by $-ri(t-T)$. Hint: determine the amplitude of this wave from the reflection coefficients it encounters on its path and account for the time delay due to the bouncing up and down once in the water layer.

**Problem c:** Generalize this result to a wave that bounces back and forth $n$-times in the water layer and show that the signal $o(t)$ recorded by the hydrophones is given by:

$$o(t) = i(t) - r\, i(t - T) + r^2\, i(t - 2T) + \cdots$$

or

$$o(t) = \sum_{n=0}^{\infty} (-r)^n\, i(t - nT) , \qquad (14.59)$$

see figure 14.3.

The notation $i(t)$ and $o(t)$ that was introduced in the previous section is deliberately used here. The action of the reverberation in the water layer is seen as a linear filter. The input of the filter $i(t)$ is the wavefield that would have been recorded if the waves did not bounce back and forth in the water layer. The output is the wavefield that results from the reverberations in the water layer. In a marine seismic experiment one records the wavefield $o(t)$ while one would like to know the signal $i(t)$ that contains just the reflections from below the water bottom. The process of removing the reverberations from the signal is called 'dereverberation.' The aim of this section is to derive a dereverberation filter that allows us to extract the input $i(t)$ from the recorded output $o(t)$.

**Problem d:** Can you see a way to determine $i(t)$ from (14.59) when $o(t)$ is given?

**Problem e:** It may not be obvious that (14.59) describes a linear filter of the form (14.57) that maps the input $i(t)$ onto the output $o(t)$. Show that (14.59) can be written in the form (14.57) with the impulse response $g(t)$ given by:

$$g(t) = \sum_{n=0}^{\infty} (-r)^n \, \delta(t - nT) , \qquad (14.60)$$

with $\delta(t)$ the Dirac delta function.

**Problem f:** Show that $g(t)$ is indeed the impulse response. In other words, show that if a delta function is incident as a primary arrival at the water surface the reverberations within the water layer lead to the signal (14.60).

You have probably discovered that it is not simple to solve problem d. However, the problem becomes much simpler if the analysis is carried out in the frequency domain. Let the Fourier transforms of $i(t)$ and $o(t)$, as defined by the transform (14.43), be denoted by $I(\omega)$ and $O(\omega)$ respectively. It follows from (14.59) that one needs to find the Fourier transform of $i(t - nT)$.

**Problem g:** According to the definition (14.43) the Fourier transform of $i(t - \tau)$ is given by $(1/2\pi) \int_{-\infty}^{\infty} i(t - \tau) e^{i\omega t} dt$. Use a change of the integration variable to show that the Fourier transform of $i(t - \tau)$ is given by $I(\omega) e^{i\omega \tau}$.

What you have derived here is the *shift property* of the Fourier transform: a translation of a function over a time $\tau$ corresponds in the frequency domain to a multiplication by $e^{i\omega \tau}$

$$\left. \begin{array}{ccc} i(t) & \longleftrightarrow & I(\omega) , \\ i(t - \tau) & \longleftrightarrow & I(\omega) e^{i\omega \tau} . \end{array} \right\} \qquad (14.61)$$

**Problem h:** Apply a Fourier transform to (14.59) for the output, use the shift property (14.61) for each term and show that the output in the frequency domain is related to the Fourier transform of the input by the following expression:

$$O(\omega) = \sum_{n=0}^{\infty} (-r)^n\, e^{i\omega nT} I(\omega) \ . \tag{14.62}$$

**Problem i:** Use the theory of section 14.7 to show that the filter that describes the generation of reverberations is given in the frequency domain by:

$$G(\omega) = \frac{1}{2\pi} \sum_{n=0}^{\infty} (-r)^n\, e^{i\omega nT} \ . \tag{14.63}$$

**Problem j:** Since we know that the reflection coefficient $r$ is less than or equal to 1 (see problem a), this series is guaranteed to converge. Sum this series to show that

$$G(\omega) = \frac{1}{2\pi} \frac{1}{1 + re^{i\omega T}} \ . \tag{14.64}$$

This is a very useful result because it implies that the output and the input in the frequency domain are related by

$$O(\omega) = \frac{1}{1 + re^{i\omega T}}\, I(\omega) \ . \tag{14.65}$$

Note that the action of the reverberation leads in the frequency domain to a simple division by $(1 + re^{i\omega T})$. Note also that (14.65) has a form similar to (2.40) which accounts for the reverberation of waves between two stacks of reflectors. This resemblance is no coincidence because the physics of waves bouncing back and forth between two reflectors is similar.

**Problem k:** The goal of this section was to derive a dereverberation filter that produces $i(t)$ when $o(t)$ is given. Use (14.65) to derive a dereverberation filter in the frequency domain.

The dereverberation filter you have just derived is very simple in the frequency domain; it only involves a multiplication of every frequency component $O(\omega)$ by a scalar. Since multiplication is a simple and efficient procedure it is attractive to carry out dereverberation in the frequency domain. The dereverberation filter you have just derived was developed originally by Backus [6].

The simplicity of the dereverberation filter hides a nasty complication. If the reflection coefficient $r$ and the two-way travel time $T$ are known exactly and if the sea bed is exactly horizontal there is no problem with the dereverberation filter. However, in practice one only has *estimates* of these quantities. Let these estimates be denoted by $r'$ and $T'$ respectively. The reverberations lead in the frequency domain to a division by $1+re^{i\omega T}$, while the dereverberation filter based on the estimated parameters leads to a multiplication with $1 + r'e^{i\omega T'}$. The net effect of the generation of the reverberations and the subsequent dereverberation is thus given in the frequency domain by a multiplication by

$$\frac{1+r'e^{i\omega T'}}{1+re^{i\omega T}} \ .$$

**Problem l:** Show that when the reflection coefficients are close to unity and when the estimate of the travel time is not accurate $(T' \neq T)$ the term given above differs appreciably from unity. Explain why this implies that the dereverberation does not work well.

In practice one faces not only the problem that the estimates of the reflection coefficients and the two-way travel time may be inaccurate, but in addition the sea bed may not be exactly flat and there may be variations in the reflection coefficient along the sea bed. In which case the performance of the dereverberation filter can be significantly degraded.

## 14.9 Design of frequency filters

In this section we consider the problem in which a time series $i(t)$ is recorded and is contaminated with high-frequency noise. The aim of this section is to derive a filter in the time domain that removes the frequency components with a frequency greater than a cut-off frequency $\omega_0$ from the time series. Such a filter is called a low-pass filter because only frequency components lower than the threshold $\omega_0$ pass the filter.

**Problem a:** Show that this filter is given in the frequency domain by:

$$G(\omega) = \begin{cases} 1 & \text{if} & |\omega| \leq \omega_0 \\ 0 & \text{if} & |\omega| > \omega_0 \end{cases} . \tag{14.66}$$

**Problem b:** Explain why the absolute value of the frequency should be used in this expression.

**Problem c:** Show that this filter is given in the time domain by

$$g(t) = \int_{-\omega_0}^{\omega_0} e^{-i\omega t} \, d\omega \ . \tag{14.67}$$

**Problem d:** Carry out the integration over frequency to derive that the filter is explicitly given by

$$g(t) = 2\omega_0 \operatorname{sinc}(\omega_0 t) , \qquad (14.68)$$

where the sinc function is defined by

$$\operatorname{sinc} x \equiv \frac{\sin x}{x} . \qquad (14.69)$$

**Problem e:** Sketch the impulse response (14.68) of the low-pass filter as a function of time. Determine the behavior of the filter for $t = 0$ and show that the first zero-crossing of the filter is at time $t = \pm\pi/\omega_0$.

The zero-crossing of the filter is of fundamental importance. It implies that the width of the impulse response in the time domain is given by $2\pi/\omega_0$.

**Problem f:** Show that the width of the filter in the frequency domain is given by $2\omega_0$.

This means that when the the cut-off frequency $\omega_0$ is increased, the width of the filter in the frequency domain increases but the width of the filter in the time domain decreases. A large width of the filter in the frequency domain corresponds to a small width of the filter in the time domain and vice versa.

**Problem g:** Show that the product of the width of the filter in the time domain and the width of the same filter in the frequency domain is given by $4\pi$.

The significance of this result is that this product is independent of frequency. This implies that the filter cannot be arbitrarily peaked in both the time domain and the frequency domain. This effect has pronounced consequences since it is the essence of the *uncertainty relation of Heisenberg* which states that the position and momentum of a particle can never be known exactly; more details can be found in the book of Merzbacher [55].

The filter (14.68) does not actually have very desirable properties; it has two basic problems. The first problem is that the filter decays only slowly with time. This means that the length of the filter in the time domain is very great, and hence the convolution of a time series with the filter is numerically a rather inefficient process. This can be solved by making the cut-off of the filter in the frequency domain more gradual than the frequency cut-off defined in (14.66), for example by using the filter $G(\omega) = (1 + |\omega|/\omega_0)^{-n}$ with $n$ a positive integer.

**Problem h:** Does this filter have the steepest cut-off for low values of $n$ or for high values of $n$? Hint: make a plot of $G(\omega)$ as a function of $\omega$.

The second problem is that the filter is not *causal*. This means that when a function is convolved with the filter (14.68), the output of the filter depends on the value of the input at later times, i.e. the filter output depends on the input in the future.

**Problem i:** Show that this is the case, and that the output depends on the input of earlier times only when $g(t) = 0$ for $t < 0$.

A causal filter can be designed by using the theory of analytic functions described in chapter 15. The design of filters is quite an art; details can be found for example in the books of Robinson and Treitel [75] or Claerbout [19].

## 14.10 Linear filters and linear algebra

There is a close analogy between the theory of linear filters of section 14.7 and the eigenvector decomposition of a matrix in linear algebra as treated in section 12.4. To see this we will use the same notation as in section 14.7 and use the Fourier transform (14.45) to write the output of the filter in the time domain as:

$$o(t) = \int_{-\infty}^{\infty} O(\omega)e^{-i\omega t}d\omega . \tag{14.70}$$

**Problem a:** Use expression (14.58) to show that this can be written as

$$o(t) = 2\pi \int_{-\infty}^{\infty} G(\omega)I(\omega)e^{-i\omega t}d\omega , \tag{14.71}$$

and carry out an inverse Fourier transformation of $I(\omega)$ to find the following expression

$$o(t) = \int\!\!\int_{-\infty}^{\infty} G(\omega)e^{-i\omega t}e^{i\omega \tau}i(\tau)d\omega d\tau . \tag{14.72}$$

In order to establish the connection with linear algebra we introduce by analogy with (14.33) the following basis functions:

$$u_\omega(t) \equiv \frac{1}{\sqrt{2\pi}}e^{-i\omega t} , \tag{14.73}$$

and the inner product

$$(f \cdot g) \equiv \int_{-\infty}^{\infty} f^*(t)g(t)dt . \tag{14.74}$$

**Problem b:** Show that these basis functions are orthonormal for this inner product in the sense that

$$(u_\omega \cdot u_{\omega'}) = \delta(\omega - \omega') . \tag{14.75}$$

**Problem c:** These functions play the same role as the eigenvectors in section 12.4. To which expression in section 12.4 does the above expression correspond?

**Problem d:** Show that (14.72) can be written as

$$o(t) = \int_{-\infty}^{\infty} G(\omega) u_\omega(t) \left( u_\omega \cdot i \right) d\omega . \tag{14.76}$$

This expression should be compared with

$$\mathbf{Ap} = \sum_{n=1}^{N} \lambda_n \hat{\mathbf{v}}^{(n)} \left( \hat{\mathbf{v}}^{(n)} \cdot \mathbf{p} \right) . \tag{12.44}$$

The integration over frequency plays the same role as the summation over eigenvectors in (12.44). Expression (14.76) can be seen as a description for the operator $g(t)$ in the time domain that maps the input function $i(t)$ onto the output $o(t)$.

**Problem e:** Use (14.57), (14.74) and (14.76) to show that:

$$g(t - \tau) = 2\pi \int_{-\infty}^{\infty} G(\omega) u_\omega(t) u_\omega^*(\tau) d\omega . \tag{14.77}$$

There is a close analogy between this expression and the dyadic decomposition of a matrix into its eigenvectors and eigenvalues derived in section 12.4.

**Problem f:** To see this connection show that (12.45) can be written in component form as:

$$A_{ij} = \sum_{n=1}^{N} \lambda_n \hat{v}_i^{(n)} \hat{v}_j^{(n)T} . \tag{14.78}$$

The sum over eigenvalues in (14.78) corresponds to the integration over frequency in (14.77). In section 12.4 linear algebra in a finite-dimensional vector space was treated. In such a space there is a finite number of eigenvalues. In this section, a function space with infinitely many degrees of freedom is analyzed; it will come as no surprise that for this reason the sum over a finite number of eigenvalues should be replaced by an integration over the continuous variable $\omega$. The index $i$ in (14.78) corresponds to the variable $t$ in (14.77) while the index $j$ corresponds to the variable $\tau$.

**Problem g:** Establish the connection between *all* variables in (14.77) and (14.78). Show specifically that $G(\omega)$ plays the role of eigenvalue $\lambda_n$ and $u_\omega$ plays the role of eigenvector. Which operation in (14.77) corresponds to the transpose that is taken of the second eigenvector in (14.78)?

You may wonder why the function $u_\omega(t) = e^{-i\omega t}/\sqrt{2\pi}$, defined in (14.73), and not some other function plays the role of the eigenvector of the impulse response operator $g(t - \tau)$. To see why this is so we have to understand what a linear filter actually does. Let us first consider the example of the reverberation filter of section 14.8. According to (14.59) the reverberation filter is given by:

$$o(t) = i(t) - r\, i(t-T) + r^2\, i(t-2T) + \cdots . \qquad (14.59)$$

It follows from this expression that what the filter really does is to take the input $i(t)$, translate it over a time $nT$ to a new function $i(t - nT)$, multiply each term with $(-r)^n$ and sum over all values of $n$. This means that the filter is a combination of three operations: (i) translation in time, (ii) multiplication and (iii) summation over $n$. The same conclusion holds for any general time-invariant linear filter.

**Problem h:** Use a change of the integration variable to show that the action of a time-invariant linear filter as given in (14.57) can be written as

$$o(t) = \int_{-\infty}^{\infty} g(\tau) i(t - \tau) d\tau . \qquad (14.79)$$

The function $i(t - \tau)$ is the function $i(t)$ translated over a time $\tau$. This translated function is multiplied by $g(\tau)$ and an integration over all values of $\tau$ is carried out. This means that in general the action of a linear filter can be seen as a combination of translation in time, multiplication and integration over all translations $\tau$. How can this be used to explain that the correct eigenfunctions to be used are $u_\omega(t) = e^{-i\omega t}/\sqrt{2\pi}$? The answer does not lie in the multiplication because *any* function is an eigenfunction of the operator that carries out multiplication by a constant, i.e. $af(t) = \lambda f(t)$ for every function $f(t)$. This means that the translation operator is the reason that the eigenfunctions are $u_\omega(t) = e^{-i\omega t}/\sqrt{2\pi}$. Let the operator that carries out a translation over a time $\tau$ be denoted by $T_\tau$:

$$T_\tau f(t) \equiv f(t - \tau) . \qquad (14.80)$$

**Problem i:** Show that the functions $u_\omega(t)$ defined in (14.73) are the eigenfunctions of the translation operator $T_\tau$, i.e. show that $T_\tau u_\omega(t) = \lambda u_\omega(t)$. Express the eigenvalue $\lambda$ of the translation operator in terms of the translation time $\tau$.

**Problem j:** Compare this result with the shift property of the Fourier transform that was derived in (14.61).

This means that the functions $u_\omega(t)$ are the eigenfunctions to be used for the eigenfunction decomposition of a linear time-invariant filter, because these functions are eigenfunctions of the translation operator.

**Problem k:** You identified in problem e the eigenvalues of the filter with $G(\omega)$. Show that this interpretation is correct: in other words show that when the filter $g$ acts on the function $u_\omega(t)$ the result can be written as $G(\omega)u_\omega(t)$. Hint: go back to problem e of section 14.7.

This analysis shows that the Fourier transform, which uses the functions $e^{-i\omega t}$, is so useful because these functions are the eigenfunctions of the translation operator. However, this also points to a limitation of the Fourier transform. Consider a linear filter that is not time-invariant, i.e. a filter in which the output does not depend only on the *difference* between the input time $\tau$ and the output time $t$. Such a filter satisfies the general equation (14.55) rather than the convolution integral (14.57). The action of a filter that is not time-invariant *cannot* in general be written as a combination of the following operations: multiplication, translation and integration. This means that for such a filter the functions $e^{-i\omega t}$ that form the basis of the Fourier transform are not the appropriate eigenfunctions. The upshot of this is that in practice the Fourier transform is only useful for systems that are time-dependent, or in general that are translationally invariant in the coordinate that is used.

# 15

---

# Analytic functions

In this chapter we consider complex functions in the complex plane. The reason for doing this is that the requirement that the function 'behaves well' (this is defined later) imposes remarkable constraints on such complex functions. Since these constraints coincide with some of the laws of physics, the theory of complex functions has a number of important applications in mathematical physics. In this chapter complex functions $h(z)$ are treated that are decomposed into a real and imaginary parts:

$$h(z) = f(z) + ig(z) \; ; \tag{15.1}$$

hence the functions $f(z)$ and $g(z)$ are assumed to be real. The complex number $z$ will frequently be written as $z = x + iy$, so that $x = \Re e(z)$ and $y = \Im m(z)$, where $\Re e$ and $\Im m$ denote the real and imaginary part respectively.

## 15.1 The theorem of Cauchy–Riemann

Let us first consider a real function $F(x)$ of a real variable $x$. The derivative of such a function is defined by the rule

$$\frac{dF}{dx} = \lim_{\Delta x \to 0} \frac{F(x + \Delta x) - F(x)}{\Delta x} \; . \tag{15.2}$$

In general there are two ways in which $\Delta x$ can approach zero: from above and from below. For a function that is differentiable it does not matter whether $\Delta x$ approaches zero from above or from below. If the limits $\Delta x \downarrow 0$ and $\Delta x \uparrow 0$ do give a different result it is a sign that the function does not behave well, it has a kink and the derivative is not unambiguously defined, see figure 15.1.

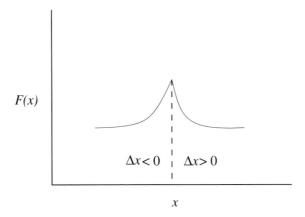

Fig. 15.1.   A function $F(x)$ that is not differentiable.

For complex functions the derivative is defined in the same way as in (15.1) for real functions:

$$\frac{dh}{dz} = \lim_{\Delta z \to 0} \frac{h(z + \Delta z) - h(z)}{\Delta z} . \tag{15.3}$$

For real functions, $\Delta x$ could approach zero in two ways: from below and from above. However, the limit $\Delta z \to 0$ in (15.3) can be taken in infinitely many ways. As an example see figure 15.2 where some different paths that one can use to let $\Delta z$ approach zero are sketched. These do not always give the same result.

**Problem a:** Consider the function $h(z) = e^{1/z}$. Using the definition (15.3) compute $dh/dz$ at the point $z = 0$ when $\Delta z$ approaches zero: (i) from the positive real axis, (ii) from the negative real axis, (iii) from the positive imaginary axis and (iv) from the negative imaginary axis.

You have discovered that for some functions the result of the limit $\Delta z$ depends critically on the path that one uses in the limit process. The derivative of such a function is not defined unambiguously. However, for many functions the value of the derivative does *not* depend on the way that $\Delta z$ approaches zero. When these functions and their derivatives are also finite, they are called *analytic functions.* The requirement that the derivative does not depend on the way in which $\Delta z$ approaches zero imposes a strong constraint on the real and imaginary parts of the complex function. To see this we will let $\Delta z$ approach zero along the real axis and along the imaginary axis.

**Problem b:** Consider a complex function of the form (15.1) and compute the derivative $dh/dz$ by setting $\Delta z = \Delta x$ with $\Delta x$ a real

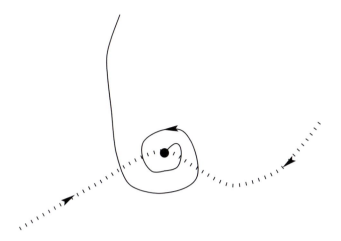

Fig. 15.2.   Examples of paths along which the limit $\Delta z \to 0$ can be taken.

number. (Hence $\Delta z$ approaches zero along the real axis.) Show that
the derivative is given by $dh/dz = \partial f/\partial x + i\partial g/\partial x$.

**Problem c:** Compute the derivative $dh/dz$ also by setting $\Delta z = i\Delta y$
with $\Delta y$ a real number. (Hence $\Delta z$ approaches zero along the imag-
inary axis.) Show that the derivative is given by $dh/dz = \partial g/\partial y - i\partial f/\partial y$.

**Problem d:** When $h(z)$ is analytic these two expressions for the derivative are by definition equal. Show that this implies that:

$$\frac{\partial f}{\partial x} = \frac{\partial g}{\partial y} , \tag{15.4}$$

$$\frac{\partial g}{\partial x} = -\frac{\partial f}{\partial y} . \tag{15.5}$$

These are puzzling expressions since (15.4) and (15.5) imply that the real
and imaginary parts of an analytic complex functions are not independent
of each other: they are coupled by the constraints imposed by the equations above. Expressions (15.4) and (15.5) are called the Cauchy–Riemann
relations.

**Problem e:** Use these relations to show that both $f(x,y)$ and $g(x,y)$
are harmonic functions. These are functions for which the Laplacian
vanishes:

$$\nabla^2 f = \nabla^2 g = 0 . \tag{15.6}$$

Hence we have found not only that $f$ and $g$ are coupled to each other,
in addition the functions $f$ and $g$ must be harmonic functions. This is

exactly the reason why this theory is so useful in mathematical physics because harmonic functions arise in several applications, see the examples in the coming sections. However, we have not found all the properties of harmonic functions yet.

**Problem f:** Show that:
$$(\nabla f \cdot \nabla g) = 0 \ . \qquad (15.7)$$

Since the gradient of a function is perpendicular to the lines where the function is constant this implies that the curves where $f$ is constant and where $g$ is constant intersect each other at a fixed angle.

**Problem g:** Determine this angle.

**Problem h:** Verify properties (15.4)–(15.7) explicitly for the function $h(z) = z^2$. Also sketch the lines in the complex plane where $f = \Re e(h)$ and $g = \Im m(h)$ are constant.

We have still not fully explored all the properties of analytic functions. Let us consider a line integral $\oint_C h(z)dz$ along a closed contour $C$ in the complex plane.

**Problem i:** Use the property $dz = dx + idy$ to deduce that:
$$\oint_C h(z)dz = \oint_C \mathbf{v} \cdot d\mathbf{r} + i \oint_C \mathbf{w} \cdot d\mathbf{r} \ , \qquad (15.8)$$

where $d\mathbf{r} = \begin{pmatrix} dx \\ dy \end{pmatrix}$ and the vectors $\mathbf{v}$ and $\mathbf{w}$ are defined by:

$$\mathbf{v} = \begin{pmatrix} f \\ -g \end{pmatrix}, \qquad \mathbf{w} = \begin{pmatrix} g \\ f \end{pmatrix} \ . \qquad (15.9)$$

Note that we are now using $x$ and $y$ in a dual role as the real and imaginary parts of a complex number, as well as the Cartesian coordinates in a plane. In the following problem we will (perhaps confusingly) use the notation $z$ both for a complex number in the complex plane, as well as for the familiar $z$-coordinate in a three-dimensional Cartesian coordinate system.

**Problem j:** Show that the Cauchy–Riemann relations (15.4)–(15.5) imply that the $z$-component of the *curl* of $\mathbf{v}$ and $\mathbf{w}$ vanishes: $(\nabla \times \mathbf{v})_z = (\nabla \times \mathbf{w})_z = 0$, and use (15.8) and the theorem of Stokes (8.2) to show that when $h(z)$ is analytic everywhere within the contour $C$:

$$\oint_C h(z)dz = 0, \text{where } h(z) \text{ is analytic within } C \ . \qquad (15.10)$$

This means that the line integral of a complex functions along *any* contour which encloses a region of the complex plane where that function is analytic is equal to zero. We will make extensive use of this property in chapter 16 where we treat integration in the complex plane.

## 15.2 The electric potential

Analytic functions are often useful in the determination of the electric field and the potential for two-dimensional problems. The electric field satisfies the field equation (7.13): $(\nabla \cdot \mathbf{E}) = \rho(\mathbf{r})/\varepsilon_0$. In free space the charge-density vanishes, hence $(\nabla \cdot \mathbf{E}) = 0$. The electric field is related to the potential $V$ through the relation

$$\mathbf{E} = -\nabla V \ . \tag{15.11}$$

**Problem a:** Show that in free space the potential is a harmonic function:

$$\nabla^2 V(x, y) = 0 \ . \tag{15.12}$$

We can exploit the theory of analytic functions by noting that the real and imaginary parts of analytic functions both satisfy (15.12). This implies that if we take $V(x, y)$ to be the real part of a complex analytic function $h(x + iy)$, (15.12) is automatically satisfied.

**Problem b:** If follows from (15.11) that the electric field is perpendicular to the lines where $V$ is constant. Show that this implies that the electric field lines are also perpendicular to the lines $V = const$. Use the theory of the previous section to argue that the field lines are the lines where the imaginary part of $h(x + iy)$ is constant.

This means that we receive a bonus by expressing the potential $V$ as the real part of a complex analytic function, because the field lines simply follow from the requirement that $\Im m(h) = const$.

Suppose we want to know the potential in the half-space $y \geq 0$ when we have specified the potential on the $x$-axis. (Mathematically this means that we want to solve the equation $\nabla^2 V = 0$ for $y \geq 0$ when $V(x, y = 0)$ is given.) If we can find an analytic function $h(x + iy)$ such that on the $x$-axis (where $y = 0$) the real part of $h$ is equal to the potential, we have solved our problem because the real part of $h$ by definition satisfies the required boundary condition and it satisfies the field equation (15.12).

**Problem c:** Consider a potential that is given on the $x$-axis by

$$V(x, y = 0) = V_0 e^{-x^2/a^2} \ . \tag{15.13}$$

Show that on the $x$-axis this function can be written as $V = \Re e(h)$ with

$$h(z) = V_0 e^{-z^2/a^2} . \tag{15.14}$$

**Problem d:** This means that we can determine the potential and the field lines throughout the half-plane $y \geq 0$. Use the theory of this section to show that the potential is given by

$$V(x, y) = V_0 e^{(y^2 - x^2)/a^2} \cos \left( \frac{2xy}{a^2} \right) . \tag{15.15}$$

**Problem e:** Verify explicitly that this solution satisfies the boundary condition at the $x$-axis and that is satisfies the field equation (15.12).

**Problem f:** Show that the field lines are given by the relation

$$e^{(y^2 - x^2)/a^2} \sin \left( \frac{2xy}{a^2} \right) = const. \tag{15.16}$$

**Problem g:** Sketch the field lines and the lines where the potential is constant in the half-space $y \geq 0$.

In this derivation we have extended the solution $V(x, y)$ into the upper half-plane by identifying it with an analytic function in the half-plane which has on the $x$-axis a real part which equals the potential on the $x$-axis. Note that we found the solution to this problem without explicitly solving the partial differential equation (15.12) which governs the potential. The approach we have taken is called *analytic continuation* since we continue an analytic function from one region (the $x$-axis) into the upper half-plane. Analytic continuation turns out to be a very unstable process. This can be verified explicitly for this example.

**Problem h:** Sketch the potential $V(x, y)$ as a function of $x$ for the values $y = 0$, $y = a$ and $y = 10a$. What is the wavelength of the oscillations in the $x$-direction of the potential $V(x, y)$ for these values of $y$? Show that when we slightly perturb the constant $a$ the perturbation of the potential increases for increasing values of $y$. This implies that when we slightly perturb the boundary condition the solution is more perturbed as we move further away from that boundary.

## 15.3 Fluid flow and analytic functions

As a second application of the theory of analytic functions we consider fluid flow. At the end of section 5.2 we saw that the streamlines of the flow can be determined by solving the differential equations $d\mathbf{r}/dt = \mathbf{v}(\mathbf{r})$

for the velocity field (5.10)–(5.11). This requires the solution of a system of nonlinear differential equations which is very difficult. Here the theory of analytic functions is used to solve this problem in a simple way. We consider once again a fluid that is incompressible $((\nabla \cdot \mathbf{v}) = 0)$ and specialize to the special case in which the vorticity of the flow vanishes:

$$\nabla \times \mathbf{v} = 0 . \tag{15.17}$$

Such a flow is called *irrotational* because it does not rotate (see sections 6.3 and 6.4). The requirement (15.17) is automatically satisfied when the velocity is the gradient of a scalar function $f$:

$$\mathbf{v} = \nabla f . \tag{15.18}$$

**Problem a:** Show this.

The function $f$ plays the same role for the velocity field $\mathbf{v}$ as the electric potential $V$ does for the electric field $\mathbf{E}$. For this reason, flow with a vorticity equal to zero is called *potential flow*.

**Problem b:** Show that the requirement that the flow is incompressible implies that
$$\nabla^2 f = 0 . \tag{15.19}$$

We now specialize to the special case of incompressible and irrotational flow in two dimensions. In that case we can again use the theory of analytic functions to describe the flow field. Once we can identify the potential $f(x, y)$ with the real part of an analytic function $h(x + iy)$ we know that (15.19) must be satisfied.

**Problem c:** Consider the velocity field (5.9) due to a point source at $\mathbf{r} = 0$. Show that this flow field follows from the potential

$$f(x, y) = \frac{V}{2\pi} \ln r , \tag{15.20}$$

where $r = \sqrt{x^2 + y^2}$.

**Problem d:** Verify explicitly that this potential satisfies (15.19) except for $r = 0$. What is the physical reason why (15.19) is not satisfied at $r = 0$?

We now want to identify the potential $f(x, y)$ with the real part of an analytic function $h(x + iy)$. We know already that the flow follows from the requirement that the velocity is the gradient of the potential, hence it follows by taking the gradient of the real part of $h$. The curves $f =$

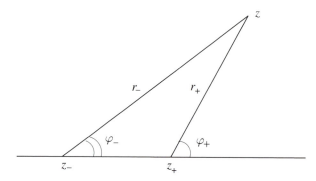

Fig. 15.3. Definition of the geometric variables for the fluid flow with a source and a sink.

*const.* are perpendicular to the flow because $\mathbf{v} = \nabla f$ is perpendicular to these curves. However, it was shown in section 15.1 that the curves $g = \Im m(h) = const.$ are perpendicular to the curves $f = \Re e(h) = const.$ This means that the flow lines are given by the curves $g = \Im m(h) = const.$ In order to use this, we first have to find an analytic function with a real part given by (15.20). The simplest guess is to replace $r$ in (15.20) by the complex variable $z$:

$$h(z) = \frac{V}{2\pi} \ln z \ . \tag{15.21}$$

**Problem e:** Verify that for complex $z$ the real part of this function indeed satisfies (15.20) and that the imaginary part $g$ of this function is given by:

$$g = \frac{V}{2\pi} \varphi \ , \tag{15.22}$$

where $\varphi = \arctan(y/x)$ is the argument of the complex number. Hint: use the representation $z = re^{i\varphi}$ for the complex numbers.

**Problem f:** Sketch the lines $g(x, y) = const.$ and verify that these lines indeed represent the flow lines in this flow.

Now we consider the more complicated problem in section 5.2 in which the flow has a source at $\mathbf{r}_+ = (L, 0)$ and a sink at $\mathbf{r}_- = (-L, 0)$. The velocity field is given by (5.10) and (5.11) and our goal is to determine the flow lines without solving the differential equation $d\mathbf{r}/dt = \mathbf{v}(\mathbf{r})$. The points $\mathbf{r}_+$ and $\mathbf{r}_-$ can be represented in the complex plane by the complex numbers $z_+ = L + i0$ and $z_- = -L + i0$ respectively. For the source at $\mathbf{r}_+$ flow is represented by the potential $(V/2\pi) \ln |z - z_+|$; this follows from the solution (15.21) for a single source by moving this source to $z = z_+$.

**Problem g:** Using a similar reasoning determine the contribution to the potential $f$ by the sink at $\mathbf{r}_-$. Show that the potential of the total flow is given by:

$$f(z) = \frac{V}{2\pi} \ln \left( \frac{|z - z_+|}{|z - z_-|} \right) . \tag{15.23}$$

**Problem h:** Express this potential in $x$ and $y$, compute the gradient and verify that this potential indeed gives the flow of (5.10)–(5.11). You may find figure 15.3 helpful.

We have found that the potential $f$ is the real part of the complex function

$$h(z) = \frac{V}{2\pi} \ln \left( \frac{z - z_+}{z - z_-} \right) . \tag{15.24}$$

**Problem i:** Write $z - z_\pm = r_\pm e^{i\varphi_\pm}$ (with $r_\pm$ and $\varphi_\pm$ defined in figure 15.3), take the imaginary part of (15.24), and show that $g = \Im m(h)$ is given by:

$$g = \frac{V}{2\pi} (\varphi_+ - \varphi_-) . \tag{15.25}$$

**Problem j:** Show that the streamlines of the flow are given by the relation

$$\arctan \left( \frac{y}{x - L} \right) - \arctan \left( \frac{y}{x + L} \right) = const. \tag{15.26}$$

A plot of the streamlines can thus be drawn by making a contour plot of the function on the left hand side of (15.26). This treatment is definitely much simpler than solving the differential equation $d\mathbf{r}/dt = \mathbf{v}(\mathbf{r})$.

# 16

_____

# Complex integration

In chapter 15 the properties of analytic functions in the complex plane were treated. One of the key results is that the contour integral of a complex function is equal to zero when the function is analytic everywhere in the area of the complex plane enclosed by that contour, see (15.10). From this it follows that the integral of a complex function along a closed contour is only nonzero when the function is not analytic in the area enclosed by the contour. Functions that are not analytic come in different types. In this section complex functions are considered that are not analytic only at isolated points. These points where the function is not analytic are called the *poles* of the function.

## 16.1 Nonanalytic functions

When a complex function is analytic at a point $z_0$, it can be expanded in a Taylor series around that point. This implies that within a certain region around $z_0$ the function can be written as:

$$h(z) = \sum_{n=0}^{\infty} a_n (z - z_0)^n . \tag{16.1}$$

Note that in this sum only positive powers of $(z - z_0)$ appear.

**Problem a:** Show that the function $h(z) = \sin(z)/z$ can be written as a Taylor series around the point $z_0 = 0$ of the form (16.1) and determine the coefficients $a_n$.

Not all functions can be represented in a series of the form (16.1). As an example consider the function $h(z) = e^{1/z}$. The function is not analytic at the point $z = 0$ (why?). Expanding the exponential leads to the expansion

$$h(z) = e^{1/z} = 1 + \frac{1}{z} + \frac{1}{2!}\frac{1}{z^2} + \cdots = \sum_{n=0}^{\infty} \frac{1}{n!}\frac{1}{z^n} . \tag{16.2}$$

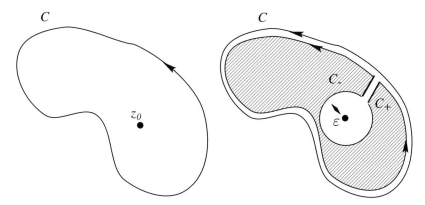

Fig. 16.1. Definition of the contours for the contour integration.

In this case an expansion in *negative* powers of $z$ is needed to represent this function. Of course, each of the terms $1/z^n$ is for $n \geq 1$ singular at $z = 0$; this reflects the fact that the function $e^{1/z}$ has a pole at $z = 0$. In this section we consider complex functions which can be expanded around a point $z_0$ as an infinite sum of *integer* powers of $(z - z_0)$:

$$h(z) = \sum_{n=-\infty}^{\infty} a_n(z - z_0)^n \ . \tag{16.3}$$

However, it should be noted that not every function can be represented as such a sum.

**Problem b:** Can you think of one?

## 16.2 The residue theorem

It was argued at the beginning of this chapter that the integral of a complex function around a closed contour in the complex plane is only nonzero when the function is not analytic at some point in the area enclosed by the contour. In this section we derive the value of the contour integral. Let us integrate a complex function $h(z)$ along a contour $C$ in the complex plane that encloses a pole of the function at the point $z_0$, see the left hand panel of figure 16.1. Note that the integration is carried out in the counterclockwise direction. It is assumed that around the point $z_0$ the function $h(z)$ can be expanded in a power series of the form (16.3). It is our goal to determine the value of the contour integral $\oint_C h(z)dz$.

The first step in the determination of the contour integral is to recognize that within the shaded area in the right hand panel of figure 16.1 the function $h(z)$ is analytic because we have assumed that $h(z)$ is only

nonanalytic at the point $z_0$. By virtue of the identity (15.10) this implies
that

$$\oint_{C^*} h(z)dz = 0 , \tag{16.4}$$

where the path $C^*$ consists of the contour $C$, a small circle with radius
$\varepsilon$ around $z_0$ and the paths $C^+$ and $C^-$ in the right hand panel of figure
16.1.

**Problem a:** Show that the integrals along $C^+$ and $C^-$ do not give a net
contribution to the total integral:

$$\int_{C^+} h(z)dz + \int_{C^-} h(z)dz = 0 . \tag{16.5}$$

Hint: note the opposite directions of integration along the paths $C^+$
and $C^-$.

**Problem b:** Use this result and (16.4) to show that the integral along
the original contour $C$ is identical to the integral along the small
circle $C_\varepsilon$ around the point where $h(z)$ is not analytic:

$$\oint_C h(z)dz = \oint_{C_\varepsilon} h(z)dz . \tag{16.6}$$

**Problem c:** The integration along $C$ is in the counterclockwise direction. Is the integration along $C_\varepsilon$ in the clockwise or in the counterclockwise direction?

Expression (16.6) is very useful because the integral along the small circle
can be evaluated by using that close to $z_0$ the function $h(z)$ can be written
as the series (16.3). When one does this the integration path $C_\varepsilon$ needs to
be parameterized. This can be achieved by writing the points on the path
$C_\varepsilon$ as

$$z = z_0 + \varepsilon e^{i\varphi} , \tag{16.7}$$

with $\varphi$ running from 0 to $2\pi$ since $C_\varepsilon$ is a complete circle.

**Problem d:** Use (16.3), (16.6) and (16.7) to derive that

$$\oint_C h(z)dz = \sum_{n=-\infty}^{\infty} ia_n\varepsilon^{(n+1)} \int_0^{2\pi} e^{i(n+1)\varphi}d\varphi . \tag{16.8}$$

This is very useful because it expresses the contour integral in the coefficients $a_n$ of the expansion (16.3). It turns out that only the coefficient
$a_{-1}$ gives a nonzero contribution.

**Problem e:** Show by direct integration that:

$$\int_0^{2\pi} e^{im\varphi} d\varphi = \begin{cases} 0 & \text{for } m \neq 0 \\ 2\pi & \text{for } m = 0 \end{cases} . \qquad (16.9)$$

**Problem f:** Use this result to derive that only the term $n = -1$ contributes to the sum on the right hand side of (16.8) and that

$$\oint_C h(z)dz = 2\pi i a_{-1} \qquad (16.10)$$

It may seem surprising that only the term $n = -1$ contributes to the sum on the right hand side of (16.8). However, we could have anticipated this result because we had already discovered that the contour integral does not depend on the precise choice of the integration path. It can be seen that in the sum (16.8) each term is proportional to $\varepsilon^{(n+1)}$. Since we know that the integral does not depend on the choice of the integration path, and hence on the size of the circle $C_\varepsilon$, one would expect that only terms which do not depend on the radius $\varepsilon$ contribute. This is only the case when $n + 1 = 0$, hence only for the term $n = -1$ is the contribution independent of the size of the circle. It is indeed only this term that gives a nonzero contribution to the contour integral.

The coefficient $a_{-1}$ is usually called the *residue* and is denoted by the symbol Res $h(z_0)$ rather than $a_{-1}$. However, remember that there is nothing mysterious about the residue, it is simply defined by the definition

$$\text{Res } h(z_0) \equiv a_{-1} . \qquad (16.11)$$

With this definition the result (16.10) can trivially be written as

$$\oint_C h(z)dz = 2\pi i \text{Res } h(z_0) \qquad \text{(counterclockwise direction)} . \qquad (16.12)$$

This may appear to be a rather uninformative rewrite of (16.10) but it is the form (16.12) that you will find in the literature. The identity (16.12) is called the *residue theorem*.

Of course the residue theorem is only useful when one can determine the coefficient $a_{-1}$ in the expansion (16.3). You can find in section 2.12 of Butkov [18] an overview of methods for computing the residue. Here we will present the two most widely used ones. The first method is to determine the power series expansion (16.3) of the function explicitly.

**Problem g:** Determine the power series expansion of the function $h(z) = \sin(z)/z^4$ around $z = 0$ and use this expansion to find the residue.

Unfortunately, this method does not always work. For some special functions other tricks can be used. Here we consider functions with a simple pole; these are functions where the terms for $n < -1$ do not contribute in the expansion (16.3):

$$h(z) = \sum_{n=-1}^{\infty} a_n (z - z_0)^n \ . \tag{16.13}$$

An example of such a function is $h(z) = \cos(z)/z$. The residue at the point $z_0$ follows by 'extracting' the coefficient $a_{-1}$ from (16.13).

**Problem h:** Multiply (16.13) by $(z - z_0)$, and take the limit $z \to z_0$ to show that:

$$\text{Res } h(z_0) = \lim_{z \to z_0} (z - z_0) h(z) \qquad \text{(simple pole)} \ . \tag{16.14}$$

However, remember that this recipe only works for functions with a simple pole, it gives the wrong answer (infinity) when applied to a function which has nonzero coefficients $a_n$ for $n < -1$ in (16.3).

In the treatment in this section we have considered an integration in the counterclockwise direction around the pole $z_0$.

**Problem i:** Repeat the derivation of this section for a contour integration in the *clockwise* direction around the pole $z_0$ and show that in that case

$$\oint_C h(z) dz = -2\pi i \text{Res } h(z_0) \qquad \text{(clockwise direction)} \ . \tag{16.15}$$

Find out in which step of the derivation the minus sign is picked up!

**Problem j:** It may happen that a contour encloses not a single pole but a number of poles at points $z_j$. Find for this case a contour similar to the contour $C^*$ in the right hand panel of figure 16.1 to show that the contour integral is equal to the *sum* of the contour integrals around the individual poles $z_j$. Use this to show that for this situation:

$$\oint_C h(z) dz = 2\pi i \sum_j \text{Res } h(z_j) \qquad \text{(counterclockwise direction)} \ .$$

$$\tag{16.16}$$

## 16.3 Solving integrals without knowing the primitive function

The residue theorem has some applications to integrals that do not contain a complex variable at all! As an example consider the integral

$$I = \int_{-\infty}^{\infty} \frac{1}{1+x^2} dx \ . \tag{16.17}$$

If you know that $1/(1+x^2)$ is the derivative of $\arctan x$ it is not difficult to solve this integral:

$$I = [\arctan x]_{-\infty}^{\infty} = \frac{\pi}{2} - \left(-\frac{\pi}{2}\right) = \pi \ . \tag{16.18}$$

Now suppose you did not know that $\arctan x$ is the primitive function of $1/(1+x^2)$. In that case you would be unable to see that the integral (16.17) is equal to $\pi$. Complex integration offers a way to obtain the value of this integral in a systematic fashion.

First note that the path of integration in the integral can be viewed as the real axis in the complex plane. Nothing prevents us from viewing the real function $1/(1+x^2)$ as a complex function $1/(1+z^2)$ because on the real axis $z$ is equal to $x$. This means that

$$I = \int_{C_{real}} \frac{1}{1+z^2} dz \ , \tag{16.19}$$

where $C_{real}$ denotes the real axis as integration path. At this point we cannot yet apply the residue theorem because the integration is not over a closed contour in the complex plane, see figure 16.2. Let us close the integration path using the circular path $C_R$ in the upper half-plane with radius $R$, see figure 16.2. At the end of the calculation we will let $R$ go to infinity so that the integration over the semicircle moves to infinity.

**Problem a:** Show that

$$I = \int_{C_{real}} \frac{1}{1+z^2} dz = \oint_C \frac{1}{1+z^2} dz - \int_{C_R} \frac{1}{1+z^2} dz \ . \tag{16.20}$$

The circular integral is over the closed contour in figure 16.2. What we have done is that we have closed the contour and subtracted the integral over the semicircle that we added to obtain a closed integration path. This is the general approach in the application of the residue theorem; one adds segments to the integration path to obtain an integral over a closed contour in the complex plane, and corrects for the segments that one has added to the path. Obviously, this is only useful when the integral over the segment that one has added can be computed easily or when

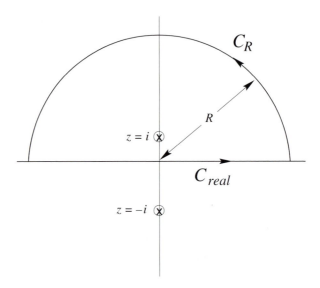

Fig. 16.2.  Definition of the integration paths in the complex plane.

it vanishes. In this example the integral over the semicircle vanishes as $R \to \infty$. This can be seen from the following estimations:

$$\left| \int_{C_R} \frac{1}{1+z^2} dz \right| \leq \int_{C_R} \left| \frac{1}{1+z^2} \right| |dz| \leq \int_{C_R} \frac{1}{|z|^2 - 1} |dz| = \frac{\pi R}{R^2 - 1} \to 0$$

$$\text{as } R \to \infty . \quad (16.21)$$

The estimate (16.21) implies that the last integral in (16.20) vanishes in the limit $R \to \infty$. This means that

$$I = \oint_C \frac{1}{1+z^2} dz . \qquad (16.22)$$

Now we are in the position to apply the residue theorem because we have reduced the problem to the evaluation of an integral along a closed contour in the complex plane. We know from section 16.2 that this integral is determined by the poles of the function that is integrated within the contour.

**Problem b:** Show that the function $1/(1+z^2)$ has poles for $z = +i$ and $z = -i$.

Only the pole at $z = +i$ is within contour $C$, see figure 16.2. Since $1/(1+z^2) = 1/[(z-i)(z+i)]$ this pole is simple (why?).

**Problem c:** Use (16.14) to show that for the pole at $z = i$ the residue is given by: Res $= 1/2i$.

**Problem d:** Use the residue theorem (16.12) to deduce that

$$I = \int_{-\infty}^{\infty} \frac{1}{1+x^2} dx = \pi .$$
(16.23)

This value is identical to the value obtained at the beginning of this section by using that the primitive of $1/(1+x^2)$ is equal to $\arctan x$. Note that the analysis is very systematic and that we did not need to 'know' the primitive function.

In the treatment of this problem there is no reason why the contour should be closed in the upper half-plane. The estimate (16.21) holds equally well for a semicircle in the lower half-plane.

**Problem e:** Repeat the analysis of this section with the contour closed in the lower half-plane. Use that now the pole at $z = -i$ contributes and take into account that the sense of integration is in the clockwise rather than the anticlockwise direction. Show that this leads to the same result (16.23) that was obtained by closing the contour in the upper half-plane.

In the evaluation of the integral (16.17) there was freedom whether to close the contour in the upper half-plane or in the lower half-plane. This is not always the case. To see this consider the integral

$$J = \int_{-\infty}^{\infty} \frac{\cos (x - \pi/4)}{1 + x^2} dx .$$
(16.24)

Since $e^{ix} = \cos x + i \sin x$ this integral can be written as

$$J = \Re \left( \int_{-\infty}^{\infty} \frac{e^{i(x-\pi/4)}}{1 + x^2} dx \right) ,$$
(16.25)

where $\Re(\cdots)$ again denotes the real part. We want to evaluate this integral by closing this integration path with a semicircle either in the upper half-plane or in the lower half-plane. Due to the term $e^{ix}$ in the integral we now have no real choice in this issue. The decision whether to close the integral in the upper half-plane or in the lower half-plane is dictated by the requirement that the integral over the semicircle vanishes as $R \to \infty$. This can only happen when the integrand vanishes (faster than $1/R$) as $R \to \infty$. Let $z$ be a point in the complex plane on the semicircle $C_R$ that we use for closing the contour. On the semicircle $z$ can be written as $z = Re^{i\varphi}$, where in the upper half-plane $0 \leq \varphi < \pi$ and in the lower half-plane $\pi \leq \varphi < 2\pi$.

**Problem f:** Use this representation of $z$ to show that

$$\left| e^{iz} \right| = e^{-R \sin \varphi} .$$
(16.26)

**Problem g:** Show that in the limit $R \to \infty$ this term only goes to zero when $z$ is in the upper half-plane.

This means that the integral over the semicircle only vanishes when we close the contour in the upper half-plane. Using steps similar to those in (16.21) one can show that the integral over a semicircle in the upper half-plane vanishes as $R \to \infty$.

**Problem h:** Take exactly the same steps as in the derivation of (16.23) and show that

$$\int_{-\infty}^{\infty} \frac{\cos(x - \pi/4)}{1 + x^2} dx = \frac{\pi}{\sqrt{2}\, e} . \tag{16.27}$$

Note that this integral is equal to a combination of the three irrational numbers $\pi$, $\sqrt{2}$ and $e$!

**Problem i:** Determine the integral $\int_{-\infty}^{\infty}[\sin(x-\pi/4)/(1+x^2)]dx$ without doing any additional calculations. Hint: look carefully at (16.25) and spot $\sin(x - \pi/4)$.

It is important to note that in finding the intergrals in this section we did not need to know the primitive function of the function that we integrated. This is the reason why complex integration is such a powerful technique; it allows us to solve many integrals without knowing their primitive functions. The structure of the singularities of the integrand in the complex plane and the behavior of the function on a bounding contour carry sufficient information to integrate the function. This rather miraculous property is due to the fact that according to (15.6) in regions where the integrand is analytic it satisfies Laplace's equation: $\nabla^2 f = \nabla^2 g = 0$ (where $f$ and $g$ are the real and imaginary parts of the integrand respectively). The singularities of the integrand act as sources or sinks for $f$ and $g$. We have learned already that the specification of the sources and sinks ($\rho$) plus boundary conditions is sufficient to solve Poisson's equation: $\nabla^2 V = \rho$. In the same way, the specification of the singularities (the sources and sinks) of a complex function plus its values on a bounding contour (the boundary conditions) are sufficient to compute the integral of this complex function.

## 16.4 Response of a particle in syrup

Up to this point, contour integration has been applied to mathematical problems. However, this technique does have important applications in physical problems. In this section we consider a particle with mass $m$ on which a force $f(t)$ is acting. The particle is suspended in syrup, which damps the velocity of the particle, and it is assumed that this damping force is proportional to the velocity $v(t)$ of the particle. The equation of motion of the particle is given by

$$m\frac{dv}{dt} + \beta v = f(t) ,\qquad(16.28)$$

where $\beta$ is a parameter that determines the strength of the damping of the motion by the fluid. The question we want to solve is: what is the velocity $v(t)$ for a given force $f(t)$?

We solve this problem using a Fourier transform technique. The Fourier transform of $v(t)$ is denoted by $V(\omega)$. The velocity in the frequency domain is related to the velocity in the time domain by the relation:

$$v(t) = \int_{-\infty}^{\infty} V(\omega)e^{-i\omega t}d\omega ,\qquad(16.29)$$

and its inverse
$$V(\omega) = \frac{1}{2\pi}\int_{-\infty}^{\infty} v(t)e^{+i\omega t}dt .\qquad(16.30)$$

The force $f(t)$ is Fourier transformed using the same expressions; in the frequency domain it is denoted by $F(\omega)$.

**Problem a:** Use the definitions of the Fourier transform to show that the equation of motion (16.28) is given in the frequency domain by

$$-i\omega m V(\omega) + \beta V(\omega) = F(\omega) .\qquad(16.31)$$

Comparing this with the original equation (16.28) we can see why the Fourier transform is so useful. The original expression (16.28) is a differential equation, while (16.31) is an algebraic equation. Since algebraic equations are much easier to solve than differential equations we have made considerable progress.

**Problem b:** Solve the algebraic equation (16.31) for $V(\omega)$ and use the Fourier transform (16.29) to derive that

$$v(t) = \frac{i}{m}\int_{-\infty}^{\infty}\frac{F(\omega)e^{-i\omega t}}{\left(\omega + \dfrac{i\beta}{m}\right)}d\omega .\qquad(16.32)$$

Now we have an explicit relation between the velocity $v(t)$ in the time domain and the force $F(\omega)$ in the frequency domain. This is not quite what we want since we want to find the relation between the velocity and the force $f(t)$ in the time domain.

**Problem c:** Use the inverse Fourier transform (16.30) (but for the force) to show that

$$v(t) = \frac{i}{2\pi m} \int_{-\infty}^{\infty} f(t') \int_{-\infty}^{\infty} \frac{e^{-i\omega(t-t')}}{\left(\omega + \dfrac{i\beta}{m}\right)} \, d\omega \, dt' \, . \qquad (16.33)$$

This equation looks messy, but we can simplify it by writing it as

$$v(t) = \frac{i}{2\pi m} \int_{-\infty}^{\infty} f(t') I(t - t') \, dt' \, , \qquad (16.34)$$

with

$$I(t - t') = \int_{-\infty}^{\infty} \frac{e^{-i\omega(t-t')}}{\left(\omega + \dfrac{i\beta}{m}\right)} \, d\omega \, . \qquad (16.35)$$

The problem we now face is to evaluate this last integral. For this we use complex integration. The integration variable is now called $\omega$ rather than $z$ but this does not change the principles. We close the contour by adding a semicircle either in the upper half-plane or in the lower half-plane to the integral (16.35) along the real axis. On a semicircle with radius $R$ the complex number $\omega$ can be written as $\omega = Re^{i\varphi}$.

**Problem d:** Show that $\left| e^{-i\omega(t-t')} \right| = e^{-R(t'-t)\sin\varphi}$.

**Problem e:** The integral along the semicircle should vanish in the limit $R \to \infty$. Use the result of problem d to show that the contour should be closed in the upper half-plane for $t < t'$ and in the lower half-plane for $t > t'$, see figure 16.3.

**Problem f:** Show that the integrand in (16.35) has one pole at the negative imaginary axis at $\omega = -i\beta/m$ and that the residue at this pole is given by

$$Res = e^{-(\beta/m)(t-t')} \, . \qquad (16.36)$$

**Problem g:** Use these results and the theorems derived in section 16.2 to show that:

$$I(t - t') = \begin{cases} 0 & \text{for } t < t' \\ -2\pi i e^{-(\beta/m)(t-t')} & \text{for } t > t' \end{cases} \, . \qquad (16.37)$$

Hint: treat the cases $t < t'$ and $t > t'$ separately.

Let us first consider (16.37). One can see from (16.33) that $I(t - t')$ is a function which describes the effect of a force acting at time $t'$ on the

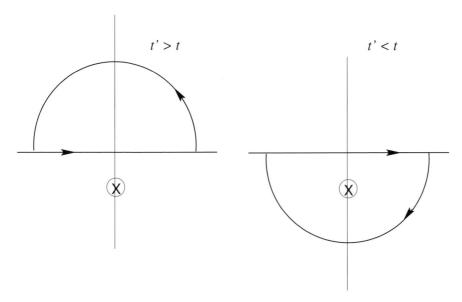

Fig. 16.3. The poles in the complex plane and the closure of contours for $t' > t$ (left) and $t' < t$ (right).

velocity at time $t$. Expression (16.37) tells us that this effect is zero when $t < t'$. In other words, this expression tells us that the force $f(t')$ has no effect on the velocity $v(t)$ when $t < t'$. This is equivalent to saying that the force only affects the velocity at *later* times. In this sense, (16.37) can be seen as an expression of *causality*; the cause $f(t')$ only influences the effect $v(t)$ for later times.

**Problem h:** Insert (16.37) and (16.35) in (16.33) to show that

$$v(t) = \frac{1}{m} \int_{-\infty}^{t} e^{-(\beta/m)(t-t')} f(t')dt' \ . \tag{16.38}$$

Pay particular attention to the limits of integration.

**Problem i:** Convince yourself that this expression states that the force (the 'cause') only has an influence on the velocity (the 'effect') for *later* times.

You may be very happy that for this problem we have managed to give a proof of the causality principle. However, there is a problem hidden in the analysis. Suppose we switch off the damping parameter $\beta$, i.e. we remove the syrup from the problem. One can easily see that setting $\beta = 0$ in the final result (16.38) poses no problem. However, suppose that we had set $\beta = 0$ at the start of the problem.

**Problem j:** Show that in that case the pole in figure 16.3 is located on
the *real* axis.

This implies that it is not clear how this pole affects the response. In
particular, it is not clear whether this pole gives a nonzero contribution
for $t < t'$ (as it would if we consider it to lie in the upper half-plane)
or for $t > t'$ (as it would if we consider it to lie in the lower half-plane).
This is a disconcerting result since it implies that causality only follows
from the analysis when the problem contains some dissipation. This is
not an artifact of the analysis using complex integration that we have
used. What we are encountering here is a manifestation of the problem
that in the absence of dissipation the laws of physics are symmetric for
time-reversal, whereas the world around us seems to move in only one
time direction This is the poorly resolved issue of the 'arrow of time'
[22][68][70].

# 17

_____

# Green's functions: principles

Green's functions play a very important role in mathematical physics. The Green's function plays a similar role to that of the impulse response for linear filters that was treated in section 14.7. The general idea is that if one knows the response of a system to a delta-function input, the response of the system to any input can be reconstructed by superposing the response to the delta-function input in an appropriate manner. However, the use of Green's functions suffers from the same limitation as the use of the impulse response for linear filters: since the superposition principle underlies the use of Green's functions they are only useful for systems that are linear. Excellent treatments of Green's functions can be found in the book of Barton [8] which is completely devoted to Green's functions and also in Butkov [18]. Since the principles of Green's functions are so fundamental, the first section of this chapter is not presented as a set of problems.

## 17.1 The girl on a swing

In order to become familiar with Green's functions let us consider the example of a girl on a swing which is pushed by her mother, see figure 17.1. When the amplitude of the swing is not too large, the motion of the swing is described by the equation of a harmonic oscillator that is driven by an external force $F(t)$ which is a general function of time:

$$\ddot{x} + \omega_0^2 x = F(t)/m . \qquad (17.1)$$

The eigenfrequency of the oscillator is denoted by $\omega_0$. It is not easy to solve this equation for a general driving force. For simplicity, we will solve (17.1) for the special case that the mother gives a single push to her daughter. The push is given at time $t = 0$, has duration $\Delta$ and the

Fig. 17.1.   The girl on a swing.

magnitude of the force is denoted by $F_0$. This means that:

$$F(t) = \begin{cases} 0 & \text{for} \quad t < 0 \\ F_0 & \text{for} \quad 0 \leq t < \Delta \\ 0 & \text{for} \quad \Delta \leq t \end{cases} . \qquad (17.2)$$

We will look here for a *causal* solution. This is another way of saying that we are looking for a solution in which the cause (the driving force) precedes the effect (the motion of the oscillator). This means that we require that the oscillator does not move before for $t < 0$. For $t \geq \Delta$ the driving force vanishes and the solution is given by a linear combination of $\cos(\omega_0 t)$ and $\sin(\omega_0 t)$. For $0 \leq t < \Delta$ the solution can be found by making the substitution $x = y + F_0/m\omega_o^2$.

When $t < 0$ and $t \geq \Delta$ the function $x(t)$ satisfies the differential equation $\ddot{x} + \omega_0^2 x = 0$. This differential equation has general solutions of the form $x(t) = A\cos(\omega_0 t) + B\sin(\omega_0 t)$. For $t < 0$ the displacement vanishes, hence the constants $A$ and $B$ vanish. This determines the solution for $t < 0$ and $t \geq \Delta$. For $0 \leq t < \Delta$ the displacement satisfies the differential equation $\ddot{x}+\omega_0^2 x = F_0/m$. The solution can be found by writing $x(t) = F_0/m\omega_0^2 + y(t)$. The function $y(t)$ then satisfies the equation $\ddot{y}+\omega_0^2 y = 0$, which has the general solution $C\cos(\omega_0 t)+D\sin(\omega_0 t)$. This means that the general solution $x(t)$ of the oscillator is given by

$$x(t) = \begin{cases} 0 & \text{for} \quad t < 0 \\ F_0/m\omega_0^2 + C\cos(\omega_0 t) + D\sin(\omega_0 t) & \text{for} \quad 0 \le t < \Delta \\ A\cos(\omega_0 t) + B\sin(\omega_0 t) & \text{for} \quad \Delta \le t \end{cases},$$
(17.3)

where $A$, $B$, $C$ and $D$ are integration constants which are not yet known.

These integration constants follow from the requirement that the motion $x(t)$ of the oscillator is at all times continuous and that the velocity $\dot{x}(t)$ of the oscillator is at all times continuous. The last condition follows from the consideration that when the force is finite, the acceleration is finite and the velocity is therefore continuous. The requirement that both $x(t)$ and $\dot{x}(t)$ are continuous at $t = 0$ and at $t = \Delta$ leads to the following equations:

$$\left. \begin{aligned} F_0/m\omega_0^2 + C &= 0, \\ \omega_0 D &= 0, \\ F_0/m\omega_0^2 + C\cos(\omega_0\Delta) + D\sin(\omega_0\Delta) &= A\cos(\omega_0\Delta) + B\sin(\omega_0\Delta), \\ -C\sin(\omega_0\Delta) + D\cos(\omega_0\Delta) &= -A\sin(\omega_0\Delta) + B\cos(\omega_0\Delta). \end{aligned} \right\}$$
(17.4)

These equations are four linear equations for the four unknown integration constants $A$, $B$, $C$ and $D$. The upper two equations can be solved directly for the constants $C$ and $D$ to give the values $C = -(F_0/m\omega_0^2)$ and $D = 0$. These values for $C$ and $D$ can then be inserted into the lower two equations. Solving these equations for the constants $A$ and $B$ gives the values $A = -(F_0/m\omega_0^2)[1 - \cos(\omega_0\Delta)]$ and $B = (F_0/m\omega_0^2)\sin(\omega_0\Delta)$. Inserting these values of the constants into (17.3) shows that the motion of the oscillator is given by:

$$x(t) = \begin{cases} 0 & \text{for} \quad t < 0 \\ (F_0/m\omega_0^2)[1 - \cos(\omega_0 t)] & \text{for} \quad 0 \le t < \Delta. \\ (F_0/m\omega_0^2)\{\cos[\omega_0(t - \Delta)] - \cos(\omega_0 t)\} & \text{for} \quad \Delta \le t \end{cases}$$
(17.5)

This is the solution for a push of duration $\Delta$ delivered at time $t = 0$. Suppose now that the push is very short compared to the period of the oscillator, then $\omega_0\Delta \ll 1$. In that case one can use a Taylor expansion in $\omega_0\Delta$ for the term $\cos[\omega_0(t - \Delta)]$ in (17.5). This can be achieved by using that $\cos[\omega_0(t - \Delta)] = \cos(\omega_0 t)\cos(\omega_0\Delta) - \sin(\omega_0 t)\sin(\omega_0\Delta)$ and by using the Taylor expansions $\sin x = x - x^3/6 + O(x^5)$ and $\cos x = 1 - x^2/2 + O(x^4)$ for $\sin(\omega_0\Delta)$ and $\cos(\omega_0\Delta)$. Retaining only the term of order $(\omega_0\Delta)$ and ignoring terms of higher order in $(\omega_0\Delta)$ shows that for an impulsive push ($\omega_0 \Delta \ll 1$) the solution is given by:

$$x(t) = \begin{cases} 0 & \text{for} \quad t < 0 \\ (F_0/m\omega_0^2)\,(\omega_0\Delta)\sin(\omega_0 t) & \text{for} \quad t > \Delta \end{cases} . \tag{17.6}$$

We will not bother anymore with the solution between $0 \le t < \Delta$ because in the limit $\Delta \to 0$ this interval is of vanishing duration.

At this point we have all the ingredients needed to determine the response of the oscillator for a general driving force $F(t)$. Suppose we divide the time-axis in intervals of duration $\Delta$. In the $i$th interval, the force is given by $F_i = F(t_i)$, where $t_i$ is the time of the $i$th interval. We know from (17.6) the response to a force of duration $\Delta$ at time $t = 0$. The response to a force $F_i$ at time $t_i$ follows by replacing $F_0$ by $F_i$ and by replacing $t$ by $t - t_i$. Making these replacements it thus follows that the response to a force $F_i$ delivered over a time interval $\Delta$ at time $t_i$ is given by:

$$x(t) = \begin{cases} 0 & \text{for} \quad t < t_i \\ \dfrac{1}{m\omega_0}\sin\left[\omega_0\left(t - t_i\right)\right]\,F(t_i)\Delta & \text{for} \quad t > t_i \end{cases} . \tag{17.7}$$

This is the response due to the force acting at time $t_i$ only. To obtain the response to the full force $F(t)$ one should sum over the forces delivered at all times $t_i$. In the language of the girl on the swing one would say that (17.6) gives the motion of the swing for a single impulsive push, and that (17.7) gives the response of the swing to a sequence of pushes given by the mother. Since the differential equation (17.1) is linear we can use the superposition principle which states that the response to the superposition of two pushes is the sum of the response to the individual pushes. (In the language of section 14.7 we would say that the swing is a linear system.) This means that when the swing receives a number of pushes at different times $t_i$ the response can be written as the sum of the responses to every individual push. With (17.7) this gives:

$$x(t) = \sum_{t_i < t} \frac{1}{m\omega_0}\sin\left[\omega_0\left(t - t_i\right)\right]\,F(t_i)\Delta . \tag{17.8}$$

Note that in (17.7) the response to a push at time $t$ before the push at time $t_i$ vanishes. For this reason one only needs in (17.8) to sum over the pushes at *earlier* times because the pushes at later times give a vanishing contribution. For this reason the summation extended to times $t \ge t_i$.

Suppose now that the swing is not given a finite number of impulse pushes but instead that the driving force is a continuous function. This case can be handled by taking the limit $\Delta \to 0$. The summation in (17.8) then needs to be replaced by an integration. This can naturally

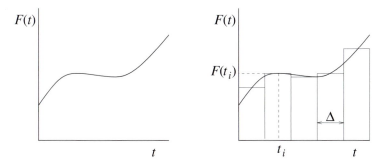

Fig. 17.2. A continuous function (left) and an approximation to this function that is constant within finite intervals (right).

be achieved because the duration $\Delta$ is then equal to the infinitesimal interval $dt$ used in the integration. What we are really doing here is replacing the continuous function $F(t)$ by a function that is constant within every interval $\Delta$ at times $t_i$, see figure 17.2, and then taking the limit where the width of the intervals goes to zero, i.e. $\Delta \to 0$. A similar treatment may be familiar to you from the theory of integration. When the limit $\Delta \to 0$ is taken the summation over $t_i$ can be replaced by an integration: $\sum_{t_i} (\cdots) \Delta \to \int (\cdots) d\tau$. The integration variable $\tau$ plays the role of the summation variable $t_i$ and the time-interval $\Delta$ is replaced $d\tau$. The response of the oscillator to a continuous force $F(t)$ is then given by

$$x(t) = \int_{-\infty}^{t} \frac{1}{m\omega_0} \sin\left[\omega_0 \left(t - \tau\right)\right] F(\tau) d\tau \ . \tag{17.9}$$

The integration is only carried out over times $\tau < t$ because the summation (17.8) extends only over the times $t_i < t$.

With a slight change in notation this result can be written as:

$$x(t) = \int_{-\infty}^{\infty} G(t, \tau) F(\tau) d\tau \ , \tag{17.10}$$

with

$$G(t, \tau) = \begin{cases} 0 & \text{for} \quad t < \tau \\ \dfrac{1}{m\omega_0} \sin\left[\omega_0 \left(t - \tau\right)\right] & \text{for} \quad t > \tau \end{cases} . \tag{17.11}$$

The function $G(t, \tau)$ in (17.11) is called the *Green's function* of the harmonic oscillator. Note that (17.10) is very similar to the response of a linear filter described by (14.55). This is not surprising; in both examples the response of a linear system to an impulsive input was determined, and it will be no surprise that the results are identical. In fact, the Green's function is defined as the response of a linear system to a delta-function input. Although Green's functions are often presented in a rather abstract way, one should remember that:

> *The Green's function of a system is nothing but the impulse response of the system, i.e. it is the response of the system to a delta-function excitation.*

## 17.2 You have seen Green's functions before!

Although the concept of a Green's function may appear to be new to you, you have already seen ample examples of Green's functions. One example is the electric field generated by a point charge $q$ at the origin which was treated in section 5.1:

$$\mathbf{E}(\mathbf{r}) = \frac{q\hat{\mathbf{r}}}{4\pi\varepsilon_0 r^2} \, . \tag{5.2}$$

Since this is the electric field generated by a delta-function charge at the origin, it is very closely related to the Green's function for this problem. The field equation (5.13) of the electric field is invariant for translations in space. This is a complex way of saying that the electric field depends only on the *relative* positions of the point charge and the point of observation.

**Problem a:** Show that this implies that the electric field at location $\mathbf{r}$ due to a point charge at location $\mathbf{r}'$ is given by:

$$\mathbf{E}(\mathbf{r}) = \frac{q}{4\pi\varepsilon_0} \frac{(\mathbf{r} - \mathbf{r}')}{|\mathbf{r} - \mathbf{r}'|^3} \, . \tag{17.12}$$

Now suppose that we don't have a single point charge, but instead a system of point charges $q_i$ at locations $\mathbf{r}_i$. Since the field equation is linear, the electric field generated by a sum of point charges is the sum of the fields generated by each point charge:

$$\mathbf{E}(\mathbf{r}) = \sum_i \frac{q_i}{4\pi\varepsilon_0} \frac{(\mathbf{r} - \mathbf{r}_i)}{|\mathbf{r} - \mathbf{r}_i|^3} \, . \tag{17.13}$$

**Problem b:** To which expression of the previous section does this equation correspond?

Just as in the previous section we now make the transition from a finite number of discrete inputs (either pushes of the swing or point charges) to an input function that is a continuous function (either the applied force to the oscillator as a function of time or a continuous electric charge). Let the electric charge per unit volume be denoted by $\rho(\mathbf{r})$. This means that the electric charge in a volume $dV$ is given by $\rho(\mathbf{r})dV$.

**Problem c:** Replace the sum in (17.13) by an integration over volume and use the appropriate charge for each volume element $dV$ to show that the electric field for a continuous charge distribution is given by:

$$\mathbf{E}(\mathbf{r}) = \iiint \frac{\rho(\mathbf{r}')}{4\pi\varepsilon_0} \frac{(\mathbf{r} - \mathbf{r}')}{|\mathbf{r} - \mathbf{r}'|^3} dV' \, , \tag{17.14}$$

where the volume integration is over $\mathbf{r}'$.

**Problem d:** Show that this implies that the electric field can be written as

$$\mathbf{E}(\mathbf{r}) = \iiint \mathbf{G}(\mathbf{r}, \mathbf{r}')\rho(\mathbf{r}')dV' \, , \tag{17.15}$$

with the Green's function given by

$$\mathbf{G}(\mathbf{r}, \mathbf{r}') = \frac{1}{4\pi\varepsilon_0} \frac{(\mathbf{r} - \mathbf{r}')}{|\mathbf{r} - \mathbf{r}'|^3} \, . \tag{17.16}$$

Note that this Green's function has the same form as the electric field for a point charge given in (17.12).

**Problem e:** Show that the Green's function is only a function of the relative distance $\mathbf{r} - \mathbf{r}'$. Explain why the integral (17.15) can be seen as a three-dimensional convolution: $\mathbf{E}(\mathbf{r}) = \iiint \mathbf{G}(\mathbf{r} - \mathbf{r}')\rho(\mathbf{r}')dV'$.

The main purpose of this section is not to show you that you had seen an example of a Green's function before. Rather, it provides an example showing that the Green's function is not necessarily a function of time and that it is not necessarily a scalar function; the Green's function (17.16) depends only on the position and not on time and it describes a vector field rather than a scalar. The most important thing to remember is that the Green's function is the impulse response of a linear system.

**Problem f:** You have seen another Green's function before if you worked through section 16.4 where the response of a particle in syrup was treated. Find the Green's function in that section and spot the expressions equivalent to (17.10) and (17.11).

## 17.3 The Green's function as impulse response

You may have found the derivation of the Green's function in section 17.1 rather complex. The reason for this is that in (17.3) the motion of the swing was determined before the push ($t < 0$), during the push ($0 < t < \Delta$) and after the push ($t > \Delta$). The requirement that the displacement $x$ and the velocity $\dot{x}$ were continuous then led to the system of equations (17.4) with four unknowns. However, in the end we took the limit $\Delta \to 0$ and did not use the solution for time $0 < t < \Delta$. This suggests that this method of solution is unnecessarily complicated. This is indeed the case. In this section an alternative derivation of the Green's function (17.11) is given which is based directly on the idea that the Green's function $G(t, \tau)$ describes the motion of the oscillator due to a delta-function force at time $\tau$:

$$\ddot{G}(t, \tau) + \omega_0^2 G(t, \tau) = \frac{1}{m} \delta(t - \tau) . \tag{17.17}$$

**Problem a:** For $t \neq \tau$ the delta function vanishes and the right hand side of this expression is equal to zero. We are looking for the *causal* Green's function, i.e. the solution where the cause (the force) precedes the effect (the motion of the oscillator). Show that these conditions imply that for $t \neq \tau$ the Green's function is given by:

$$G(t, \tau) = \begin{cases} 0 & \text{for} \quad t < \tau \\ A \cos[\omega_0(t - \tau)] + B \sin[\omega_0(t - \tau)] & \text{for} \quad t > \tau \end{cases}, \tag{17.18}$$

where $A$ and $B$ are unknown integration constants.

The integration constants follow from the conditions at $t = \tau$. Since we have two unknown parameters we need to impose two conditions. The first condition is that the displacement of the oscillator is continuous at $t = \tau$. If this were not the case the velocity of the oscillator would be infinite at that moment.

**Problem b:** Show that the requirement of continuity of the Green's function at $t = \tau$ implies that $A = 0$.

The second condition requires more care. We derive the second condition first mathematically and then explore its physical meaning. The second condition follows by integrating (17.17) over $t$ from $\tau - \varepsilon$ to $\tau + \varepsilon$ and by taking the limit $\varepsilon \downarrow 0$. Integrating (17.17) in this way gives:

$$\int_{\tau-\varepsilon}^{\tau+\varepsilon} \ddot{G}(t, \tau) dt + \omega_0^2 \int_{\tau-\varepsilon}^{\tau+\varepsilon} G(t, \tau) dt = \frac{1}{m} \int_{\tau-\varepsilon}^{\tau+\varepsilon} \delta(t - \tau) dt . \tag{17.19}$$

**Problem c:** Show that the right hand side of (17.19) is equal to $1/m$, regardless of the value of $\varepsilon$.

**Problem d:** Show that the absolute value of the second term on the left hand side of (17.19) is smaller than $2\varepsilon\omega_0^2 \max(G)$, where $\max(G)$ is the maximum of $G$ over the integration interval. Since the Green's function is finite this means that the middle term vanishes in the limit $\varepsilon \downarrow 0$.

**Problem e:** Show that the first term on the left hand side of (17.19) is equal to $\dot{G}(t = \tau + \varepsilon, \tau) - \dot{G}(t = \tau - \varepsilon, \tau)$. This quantity will be denoted by $[\dot{G}(t, \tau)]_{t=\tau-\varepsilon}^{t=\tau+\varepsilon}$.

**Problem f:** Show that in the limit $\varepsilon \downarrow 0$ (17.19) gives:

$$\left[ \dot{G}(t, \tau) \right]_{t=\tau-\varepsilon}^{t=\tau+\varepsilon} = \frac{1}{m} \, . \tag{17.20}$$

**Problem g:** Show that this condition together with the continuity of $G$ implies that the integration constants in (17.18) have the values $A = 0$ and $B = 1/m\omega_0$, i.e. that the Green's function is given by:

$$G(t, \tau) = \begin{cases} 0 & \text{for} \quad t < \tau \\ \dfrac{1}{m\omega_0} \sin \left[ \omega_0 \left( t - \tau \right) \right] & \text{for} \quad t > \tau \end{cases} . \tag{17.21}$$

A comparison with (17.11) shows that the Green's function derived in this section is identical to the Green's function derived in section 17.1. Note that the solution was obtained here without invoking the motion of the oscillator during the moment of excitation. This would have been very difficult because the duration of the excitation (a delta function) is equal to zero, if it can be defined at all.

There is, however, something strange about the derivation in this section. In section 17.1 the solution was found by requiring that the displacement $x$ and its first derivative $\dot{x}$ were continuous at all times. As used in problem b the first condition is also met by the solution (17.21). However, the derivative $\dot{G}$ is *not* continuous at $t = \tau$.

**Problem h:** Which of the equations that you derived above states that the first derivative is not continuous?

**Problem i:** $G(t, \tau)$ denotes the displacement of the oscillator. Show that expression (17.20) states that the velocity of the oscillator changes discontinuously at $t = \tau$.

**Problem j:** Give a physical reason why the velocity of the oscillator was continuous in the first part of section 17.1 and why the velocity is discontinuous for the Green's function derived in this section. Hint: how large is the force needed to produce a finite jump in the velocity of a particle when the force is applied over a time-interval of length zero (the width of the delta-function excitation).

How can we reconcile this result with the solution obtained in section 17.1?

**Problem k:** Show that the change in the velocity in the solution $x(t)$ in (17.7) is proportional to $F(t_i)\Delta$, i.e. that

$$[\dot{x}]_{t_i-\varepsilon}^{t_i+\varepsilon} = \frac{1}{m} F(t_i)\Delta .$$  (17.22)

This means that the change in the velocity depends on the strength of the force times the duration of the force. The physical reason for this is that the change in the velocity depends on the integral of the force over time divided by the mass of the particle.

**Problem l:** Derive this last statement directly from Newton's law ($F = ma$).

When the force is finite and when $\Delta \to 0$, the jump in the velocity is zero and the velocity is continuous. However, when the force is infinite (as is the case for a delta function), the jump in the velocity is nonzero and the velocity is discontinuous.

In many applications the Green's function is the solution of a differential equation with a delta function as the excitation. This implies that some derivative, or combination of derivatives, of the Green's function are equal to a delta function at the point (or time) of excitation. This usually has the effect that the Green's function or its derivative is not a continuous function. The delta function in the differential equation usually leads to a singularity in the Green's function or its derivative.

## 17.4 The Green's function for a general problem

In this section, the theory of Green's functions is treated in a more abstract fashion. Every linear differential equation for a function $u$ with a source term $F$ can be written symbolically as:

$$Lu = F .$$  (17.23)

For example in (17.1) for the girl on the swing, $u$ is the displacement $x(t)$ while $L$ is a differential operator given by

$$L = m\frac{d^2}{dt^2} + m\omega_0^2 ,$$  (17.24)

where it is understood that a differential operator acts term by term on the function to the right of the operator.

**Problem a:** Find the differential operator $L$ and the source term $F$ for the electrical field treated in section 17.2 from the field equation (5.13).

In the notation used in this section, the Green's function depends on the position vector $\mathbf{r}$, but the results derived here are equally valid for a Green's function that depends only on time or on position and time. In general, the differential equation (17.23) must be supplemented with boundary conditions to give an unique solution. In this section the position of the boundary is denoted by $\mathbf{r}_B$ and it is assumed that the function $u$ has the value $u_B$ at the boundary:

$$u(\mathbf{r}_B) = u_B . \tag{17.25}$$

Let us first find a single solution to the differential equation (17.23) without bothering about boundary conditions. We follow the same treatment as in section 14.7 where in (14.54) the input of a linear function was written as a superposition of delta functions. In the same way, the source function can be written as:

$$F(\mathbf{r}) = \int \delta(\mathbf{r} - \mathbf{r}')F(\mathbf{r}')dV' . \tag{17.26}$$

This expression follows from the properties of the delta function. One can interpret this expression as an expansion of the function $F(\mathbf{r})$ in delta functions because the integral (17.26) describes a superposition of delta functions $\delta(\mathbf{r} - \mathbf{r}')$ centered at $\mathbf{r} = \mathbf{r}'$; each of these delta functions is given a weight $F(\mathbf{r}')$. We want to use a Green's function to construct a solution. The Green's function $G(\mathbf{r}, \mathbf{r}')$ is the response at location $\mathbf{r}$ due to a delta-function source at location $\mathbf{r}'$, i.e. the Green's function satisfies:

$$LG(\mathbf{r}, \mathbf{r}') = \delta(\mathbf{r} - \mathbf{r}') . \tag{17.27}$$

The response to the input $\delta(\mathbf{r} - \mathbf{r}')$ is given by $G(\mathbf{r}, \mathbf{r}')$, and the source functions can be written as a superposition of these delta functions with weight $F(\mathbf{r}')$. This suggests that a solution of (17.23) is given by a superposition of Green's functions $G(\mathbf{r}, \mathbf{r}')$, where each Green's function has the same weight factor as the delta function $\delta(\mathbf{r} - \mathbf{r}')$ in the expansion (17.26) of $F(\mathbf{r})$ in delta functions. This means that the solution of (17.23) is given by:

$$u_P(\mathbf{r}) = \int G(\mathbf{r}, \mathbf{r}')F(\mathbf{r}')dV' . \tag{17.28}$$

**Problem b:** If you worked through section 14.7 discuss the relation between this expression and (14.55) for the output of a linear function.

It is crucial to understand at this point that we have used three steps to arrive at (17.28): (i) the source function is written as a superposition of delta functions, (ii) the response of the system to each delta-function input is defined and (iii) the solution is written as the same superposition of Green's function as was used in the expansion of the source function in delta functions:

$$\delta(\mathbf{r} - \mathbf{r}') \quad \longleftrightarrow \quad F(\mathbf{r}) \stackrel{(i)}{=} \int \delta(\mathbf{r} - \mathbf{r}')F(\mathbf{r}')dV'$$
$$\Downarrow (ii) \qquad\qquad\qquad \Downarrow \qquad\qquad\qquad\qquad\qquad (17.29)$$
$$G(\mathbf{r}, \mathbf{r}') \quad \longleftrightarrow \quad u_P(\mathbf{r}) \stackrel{(iii)}{=} \int G(\mathbf{r}, \mathbf{r}')F(\mathbf{r}')dV'$$

**Problem c:** Although this reasoning may sound plausible, we have not proved that $u_P(\mathbf{r})$ in (17.28) actually is a solution of the differential equation (17.23). Give a proof that this is indeed the case by letting the operator $L$ act on (17.28) and by using (17.27) for the Green's function. Hint: the operator $L$ acts on $\mathbf{r}$ while the integration is over $\mathbf{r}'$, the operator can thus be taken inside the integral.

It should be noted that we have not solved our problem yet, because $u_P$ does not necessarily satisfy the boundary conditions. In fact, (17.28) is just one of the many possible solutions to (17.23). It is a particular solution of the inhomogeneous equation (17.23), and this is the reason why the subscript $P$ is used. Equation (17.23) is called an *inhomogeneous equation* because the right hand side is nonzero. When the right hand side is zero one speaks of the *homogeneous equation*. This implies that a solution $u_H$ of the homogeneous equation satisfies

$$Lu_H = 0 . \qquad\qquad (17.30)$$

**Problem d:** In general one can add a solution of the homogeneous equation (17.30) to a particular solution, and the result still satisfies the inhomogeneous equation (17.23). Give a proof of this statement by showing that the function $u = u_P + u_H$ is a solution of (17.23). In other words show that the general solution of (17.23) is given by:

$$u(\mathbf{r}) = u_H(\mathbf{r}) + \int G(\mathbf{r}, \mathbf{r}')F(\mathbf{r}')dV' . \qquad\qquad (17.31)$$

**Problem e:** The problem is that we still need to enforce the boundary conditions (17.25). This can be achieved by requiring that the solution $u_H$ satisfies specific boundary conditions at $\mathbf{r}_B$. Insert (17.31) into the boundary conditions (17.25) and show that the required solution $u_H$ of the homogeneous equation must satisfy the following boundary conditions:

$$u_H(\mathbf{r}_B) = u_B(\mathbf{r}_B) - \int G(\mathbf{r}_B, \mathbf{r}')F(\mathbf{r}')dV' . \qquad\qquad (17.32)$$

This is all we need to solve the problem. What we have shown is that:

*The total solution (17.31) is given by the sum of the particular solu-*
*tion (17.28) plus a solution of the homogeneous equation (17.30) that*
*satisfies the boundary condition (17.32).*

This construction may appear to be very complex to you. However, you
should realize that the main complexity is the treatment of the boundary
condition. In many problems, the boundary condition dictates that the
function vanishes at the boundary ($u_B = 0$) and the Green's function
also vanishes at the boundary. It follows from (17.31) that in that case
the boundary condition for the homogeneous solution is $u_H(\mathbf{r}_B) = 0$.
This boundary condition is satisfied by the solution $u_H(\mathbf{r}) = 0$, which
implies that in that case one can dispense with the addition of $u_H$ to the
particular solution $u_P(\mathbf{r})$.

**Problem f:** Suppose that the boundary conditions do not prescribe the
value of the solution at the boundary but that instead of (17.25)
the normal derivative of the solution is prescribed by the boundary
conditions:

$$\frac{\partial u}{\partial n}(\mathbf{r}_B) = \hat{\mathbf{n}} \cdot \nabla u(\mathbf{r}_B) = w_B , \qquad (17.33)$$

where $\hat{\mathbf{n}}$ is the unit vector perpendicular to the boundary. How
should the theory in this section be modified to accommodate this
boundary condition?

The theory in this section is rather abstract. In order to make the issues
at stake more explicit the theory is applied in the next section to the
calculation of the temperature in the Earth.

## 17.5 Radiogenic heating and the Earth's temperature

As an application of the use of the Green's function we consider in this
section the calculation of the temperature in the Earth and specifically the
effect of the decay of radioactive elements in the crust on the temperature
in the Earth. Several radioactive elements such as $U_{235}$ do not fit well in
the lattice of mantle rocks. For this reason, these elements are expelled
from the material in the Earth's mantle and they accumulate in the crust.
Radioactive decay of these elements then leads to the production of heat
at the place where these elements have accumulated.

As a simplified example of this problem we assume that the temperature $T$ and the radiogenic heating $Q$ depend only on depth and that we
can ignore the sphericity of the Earth. In addition, we assume that the
radiogenic heating does not depend on time and that we consider only
the equilibrium temperature.

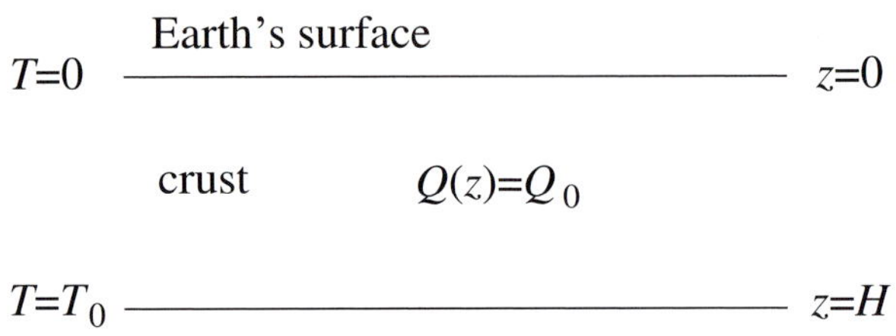

Earth's surface

$T=0$  ———————————————— $z=0$

crust            $Q(z)=Q_0$

$T=T_0$ ———————————————— $z=H$

Fig. 17.3.   Definition of the geometric variables and boundary conditions for the temperature in the crust.

**Problem a:** Show that these assumptions imply that the temperature is only a function of the $z$-coordinate: $T = T(z)$.

The temperature field satisfies the heat equation derived in section 10.4:

$$\frac{\partial T}{\partial t} = \kappa \nabla^2 T + Q \ . \tag{10.31}$$

**Problem b:** Use this expression to show that for the problem in this section the temperature field satisfies

$$\frac{d^2 T}{dz^2} = -\frac{Q(z)}{\kappa} \ . \tag{17.34}$$

This equation can be solved when the boundary conditions are specified. The thickness of the crust is denoted by $H$, see figure 17.3. The temperature is assumed to vanish at the Earth's surface. In addition, it is assumed that at the base of the crust the temperature has a fixed value $T_0$.[*] This implies that the boundary conditions are:

$$T(z = 0) = 0 \ , \qquad T(Z = H) = T_0 \ . \tag{17.35}$$

In this section we solve the differential equation (17.34) with the boundary conditions (17.35) using the Green's function technique described in the previous section. Analogously to (17.28) we first determine a particular solution $T_P$ of the differential equation (17.34) and worry about the boundary conditions later. The Green's function $G(z, z')$ to be used is the temperature at depth $z$ due to delta-function heating at depth $z'$:

$$\frac{d^2 G(z, z')}{dz^2} = \delta(z - z') \ . \tag{17.36}$$

---

[*] Geophysically this is an oversimplified boundary condition because in reality the temperature in the Earth is determined by the radiogenic heating everywhere in the Earth and by the heat that was formed during the Earth's formation.

**Problem c:** Use the theory of the previous section to show that the following function satisfies the heat equation (17.34):

$$T_P(z) = -\frac{1}{\kappa} \int_0^H G(z, z') Q(z') dz' . \tag{17.37}$$

Before further progress can be made it is necessary to find the Green's function, i.e. to solve the differential equation (17.36). In order to do this the boundary conditions for the Green's function need to be specified. In this example we will use a Green's function that vanishes at the endpoints of the depth interval:

$$G(z = 0, z') = G(Z = H, z') = 0 . \tag{17.38}$$

**Problem d:** Use (17.36) to show that for $z \neq z'$ the Green's function satisfies the differential equation $d^2 G(z, z')/dz^2 = 0$ and use this to show that the Green's function which satisfies the boundary conditions (17.38) must be of the form

$$G(z, z') = \begin{cases} \beta z & \text{for } z < z' \\ \gamma(z - H) & \text{for } z > z' \end{cases} , \tag{17.39}$$

with $\beta$ and $\gamma$ constants that need to be determined.

**Problem e:** Since there are two unknown constants, two conditions are needed. The first condition is that the Green's function is continuous for $z = z'$. Use the theory of section 17.3 and the differential equation (17.36) to show that the second requirement is:

$$\lim_{\varepsilon \downarrow 0} \left[ \frac{dG(z, z')}{dz} \right]_{z=z'-\varepsilon}^{z=z'+\varepsilon} = 1 , \tag{17.40}$$

i.e. that the first derivative makes a unit jump at the point of excitation.

**Problem f:** Apply these two conditions to the solution (17.39) to determine the constants $\beta$ and $\gamma$ and show that the Green's function is given by:

$$G(z, z') = \begin{cases} -\dfrac{1}{H}(H - z')z & \text{for } z < z' \\ \\ -\dfrac{1}{H}z'(H - z) & \text{for } z > z' \end{cases} . \tag{17.41}$$

In this notation the two regions $z < z'$ and $z > z'$ are separated. Note, however, that the solution in the two regions has a highly symmetric form. In the literature you will find that a solution such as (17.41) is often rewritten by defining $z_>$ to be the maximum of $z$ and $z'$ and $z_<$ to be the minimum of $z$ and $z'$:

$$\left. \begin{array}{l} z_> \equiv \max(z, z') , \\ z_< \equiv \min(z, z') . \end{array} \right\} \tag{17.42}$$

**Problem g:** Show that in this notation the Green's function (17.41) can be written as:

$$G(z, z') = -\frac{1}{H}(H - z_>)z_< . \tag{17.43}$$

For simplicity we assume that the radiogenic heating is constant in the crust:

$$Q(z) = Q_0 \qquad \text{for} \qquad 0 < z < H . \tag{17.44}$$

**Problem h:** Show that the particular solution (17.37) for this heating function is given by

$$T_P(z) = \frac{Q_0 H^2}{2\kappa}\frac{z}{H}\left(1 - \frac{z}{H}\right). \tag{17.45}$$

**Problem i:** Show that this particular solution satisfies the boundary conditions

$$T_P(z = 0) = T_P(z = H) = 0 . \tag{17.46}$$

**Problem j:** This means that this solution does not satisfy the boundary conditions (17.35) of our problem. Use the theory in section 17.4 to show that to obtain this solution we must add a solution $T_H(z)$ of the homogeneous equation $d^2 T_H/dz^2 = 0$ which satisfies the boundary conditions $T_H(z = 0) = 0$ and $T_H(z = H) = T_0$.

**Problem k:** Show that the solution of the homogeneous equation is given by $T_H(z) = T_0 z/H$ and that the total solution is given by

$$T(z) = T_0\frac{z}{H} + \frac{Q_0 H^2}{2\kappa}\frac{z}{H}\left(1 - \frac{z}{H}\right). \tag{17.47}$$

**Problem l:** Verify explicitly that this solution satisfies the differential equation (17.34) with the boundary conditions (17.35).

As shown in expression (10.29) the conductive heat flow is given by $\mathbf{J} = -\kappa\nabla T$. Since the problem is one-dimensional the heat flow is given by

$$J = -\kappa\frac{dT}{dz} . \tag{17.48}$$

**Problem m:** Compute the heat flow at the top ($z = 0$) and at the bottom ($z = H$) of the crust. Assuming that $T_0$ and $Q_0$ are both positive, does the heat flow at these locations increase or decrease because of the radiogenic heating $Q_0$? Give a physical interpretation of this result.

The derivation in this section used a Green's function which satisfies the boundary conditions (17.38) rather than the boundary conditions (17.35) of the temperature field. However, there is no particular reason why one should use these boundary conditions for the Green's function. To wit, one might think one could avoid the step of adding a solution $T_H(z)$ of the homogeneous equation by using a Green's function $\tilde{G}$ that satisfies the differential equation (17.39) and an inhomogeneous boundary condition at the base of the crust:

$$\tilde{G}(z = 0, z') = 0 , \qquad \tilde{G}(z = H, z') = H . \qquad (17.49)$$

(The boundary value at the base of the crust is set equal to $H$ because the Green's function has the dimension of length (see (17.41)) and the crustal thickness $H$ is the only length-scale in the problem.)

**Problem n:** Go through the same steps as you did earlier in this section by constructing the Green's function $\tilde{G}(z, z')$, computing the corresponding particular solution $\tilde{T}_P(z)$, verifying whether the boundary conditions (17.35) are satisfied by this particular solution and if necessary adding a solution of the homogeneous equation in order to satisfy the boundary conditions. Show that this again leads to (17.47).

The lesson to be learned from this section is that usually one needs to add a solution of the homogeneous equation to a particular solution in order to satisfy the boundary conditions. However, suppose that the boundary conditions of the temperature field were also homogeneous: $T = (z = 0) = T(z = H) = 0$. In that case the particular solution (17.45) that was constructed using a Green's function that satisfies the homogeneous boundary conditions (17.38) satisfies the boundary conditions of the full problem as well. This implies that it only pays to use a Green's function that satisfies the boundary conditions of the full problem when these boundary conditions are *homogeneous*, i.e. when the function itself vanishes ($T = 0$) or when the normal gradient of the function vanishes ($\partial T / \partial n = 0$) or when a linear combination of these quantities vanishes ($aT + b\partial T / \partial n = 0$). In all other cases one cannot avoid adding a solution of the homogeneous equation in order to satisfy the boundary conditions and the most efficient procedure is usually to use the Green's function that can most easily be computed.

## 17.6 Nonlinear systems and Green's functions

Up to this point, Green's functions have been applied to linear systems. The definition of a linear system was introduced in section 14.7. Suppose that a force $F_1$ leads to a response $x_1$ and that a force $F_2$ leads to a response $x_2$. A system is linear when the response to the linear combination $c_1 F_1 + c_2 F_2$ (with $c_1$ and $c_2$ constants) is the superposition response $c_1 x_1 + c_2 x_2$.

**Problem a:** Show that this definition implies that the response to the input times a constant is given by the response multiplied by the same constant. In other words show that for a linear system an input that is twice as large leads to a response that is twice as large.

**Problem b:** Show that the definition of linearity given above implies that the response to the sum of two force functions is the sum of the responses to the individual force functions.

This last property reflects that a linear system satisfies the *superposition principle* which states that for a linear system one can superpose the response to a sum of force functions.

Not every system is linear, and we exploit here the extent to which Green's functions are useful for nonlinear systems. As an example we consider the Verhulst equation:

$$\dot{x} = x - x^2 + F(t) . \tag{17.50}$$

This equation has been used in mathematical biology to describe the growth of a population. Suppose that only the term $x$ were present on the right hand side. In that case the solution would be given by $x(t) = Ce^t$. This means that the first term on the right hand side accounts for the exponential population growth which is due to the fact that the number of offspring is proportional to the size of the population. However, a population cannot grow indefinitely; when a population is too large limited resources restrict the growth, and this is accounted for by the $-x^2$ term on the right hand side. The term $F(t)$ accounts for external influences on the population. For example, a mass-extinction could be described by a strongly negative forcing function $F(t)$. We will consider first the solution for the case that $F(t) = 0$. Since the population size is positive we consider only positive solutions $x(t)$.

**Problem c:** Show that for the case $F(t) = 0$ the change of variable $y = 1/x$ leads to the linear equation $\dot{y} = 1 - y$. Solve this equation and show that the general solution of (17.50) (with $F(t) = 0$) is given by:

$$x(t) = \frac{1}{Ae^{-t} + 1} , \tag{17.51}$$

with $A$ an integration constant.

**Problem d:** Use this solution to show that any solution of the unforced equation goes to 1 for infinite times:

$$\lim_{t\to\infty} x(t) = 1 . \tag{17.52}$$

In other words, the population of the unforced Verhulst equation always converges to the same population size. Note that when the force vanishes after a finite time, the solution after that time must satisfy (17.51) which implies that the long-time limit is then also given by (17.52).

Now, consider the response to a delta-function excitation at time $t_0$ with strength $F_0$. The associated response $g(t, t_0)$ thus satisfies

$$\dot{g} - g + g^2 = F_0 \, \delta(t - t_0) . \tag{17.53}$$

Since this function is the impulse response of the system, the notation $g$ is used in order to bring out the resemblance with the Green's functions used earlier. We consider only causal solutions, i.e. we require that $g(t, t_0)$ vanishes for $t < t_0$: $g(t, t_0) = 0$ for $t < t_0$. For $t > t_0$ the solution satisfies the Verhulst equation without force, hence the general form is given by (17.51). The only task remaining is to find the integration constant $A$. This constant follows by a treatment similar to the analysis of section 17.3.

**Problem e:** Integrate (17.53) over $t$ from $t_0 - \varepsilon$ to $t_0 + \varepsilon$, take the limit $\varepsilon \downarrow 0$ and show that this leads to the following requirement for the discontinuity in $g$:

$$\lim_{\varepsilon \downarrow 0} [g(t, t_0)]_{t_0 - \varepsilon}^{t_0 + \varepsilon} = F_0 . \tag{17.54}$$

**Problem f:** Use this condition to show that the constant $A$ in the solution (17.51) is given by $A = (1/F_0 - 1)e^{t_0}$ and that the solution is given by:

$$g(t, t_0) = \begin{cases} 0 & \text{for} \quad t < t_0 \\ \dfrac{F_0}{(1 - F_0)e^{-(t-t_0)} + F_0} & \text{for} \quad t > t_0 \end{cases} . \tag{17.55}$$

At this point you should be suspicious of interpreting $g(t, t_0)$ as a Green's function. An important property of linear systems is that the response is proportional to the force. However, the solution $g(t, t_0)$ in (17.55) is not proportional to the strength $F_0$ of the force.

Let us now check whether we can use the superposition principle. Suppose the force function is the superposition of a delta-function force $F_1$ at $t = t_1$ and a delta-function force $F_2$ at $t = t_2$:

$$F(t) = F_1 \delta(t - t_1) + F_2 \delta(t - t_2) . \tag{17.56}$$

By analogy with (17.10) you might think that a Green's-function-type solution is given by:

$$x^{Green}(t) = \frac{F_1}{(1 - F_1)e^{-(t-t_1)} + F_1} + \frac{F_2}{(1 - F_2)e^{-(t-t_2)} + F_2} , \qquad (17.57)$$

for times larger than both $t_1$ and $t_2$. You can verify by direct substitution that this function is not a solution of the differential equation (17.50). However, this process is rather tedious and there is a simpler way to see that the function $x^{Green}(t)$ violates the differential equation (17.50).

**Problem g:** To see this, show that the solution $x^{Green}(t)$ has the following long-time behavior:

$$\lim_{t\to\infty} x^{Green}(t) = 2 . \qquad (17.58)$$

This limit is at odds with the limit (17.52) that every solution of the differential equation (17.50) should satisfy when the force vanishes after a certain finite time. This proves that $x^{Green}(t)$ is not a solution of the Verhulst equation.

This implies that the Green's function technique introduced in the previous sections cannot be used for a nonlinear equation such as the forced Verhulst equation. The reason for this is that Green's functions are based on the superposition principle; by knowing the response to a delta-function force and by writing a general force as a superposition of delta functions one can construct a solution by making the corresponding superposition of Green's functions, see (17.29). However, solutions of a nonlinear equation such as the Verhulst equation do not satisfy the principle of superposition. This implies that Green's function cannot be used effectively to construct the behavior of nonlinear systems. It is for this reason that Green's function are in practice only used for constructing the response of *linear* systems.

# 18

---

# Green's functions: examples

In the previous chapter the basic theory of Green's function was introduced. In this chapter a number of examples of Green's functions are shown that are often used in mathematical physics.

## 18.1 The heat equation in $N$ dimensions

In this section we consider once again the heat equation as introduced in section 10.4:

$$\frac{\partial T}{\partial t} = \kappa \nabla^2 T + Q . \tag{10.31}$$

First we construct a Green's function for this equation in $N$ space dimensions. The reason for this is that the analysis for $N$ dimensions is just as easy (or difficult) as the analysis for only one spatial dimension.

The heat equation is invariant for translations in both space and time. For this reason the Green's function $G(\mathbf{r},t;\mathbf{r}_0,t_0)$ that gives the temperature at location $\mathbf{r}$ and time $t$ due to a delta-function heat source at location $\mathbf{r}_0$ and time $t_0$ depends only on the relative distance $\mathbf{r} - \mathbf{r}_0$ and the relative time $t - t_0$.

**Problem a:** Show that this implies that $G(\mathbf{r},t;\mathbf{r}_0,t_0) = G(\mathbf{r} - \mathbf{r}_0, t - t_0)$.

Since the Green's function depends only on $\mathbf{r} - \mathbf{r}_0$ and $t - t_0$ it suffices to construct the simplest solution by considering the special case of a source at $\mathbf{r}_0 = 0$ at time $t_0 = 0$. This means that we construct the Green's function $G(\mathbf{r},t)$ that satisfies:

$$\frac{\partial G(\mathbf{r},t)}{\partial t} - \kappa \nabla^2 G(\mathbf{r},t) = \delta(\mathbf{r})\delta(t) . \tag{18.1}$$

This Green's function can most easily be constructed by carrying out a spatial Fourier transform. Using the Fourier transform (14.27) for each

269

of the $N$ spatial dimensions one finds that the Green's function has the following Fourier expansion:

$$G(\mathbf{r}, t) = \frac{1}{(2\pi)^N} \int g(\mathbf{k}, t) e^{i\mathbf{k}\cdot\mathbf{r}} d^N k \ . \tag{18.2}$$

Note that the Fourier transform is only carried out over the spatial dimensions and not over time. This implies that $g(\mathbf{k}, t)$ is a function of time as well. The differential equation that $g$ satisfies can be obtained by inserting the Fourier representation (18.2) in the differential equation (18.1). In doing this we also need the Fourier representation of $\nabla^2 G(\mathbf{r},t)$.

**Problem b:** Show by applying the Laplacian to the Fourier integral (18.1) that:

$$\nabla^2 G(\mathbf{r}, t) = \frac{-1}{(2\pi)^N} \int k^2 g(\mathbf{k}, t) e^{i\mathbf{k}\cdot\mathbf{r}} d^N k \ . \tag{18.3}$$

**Problem c:** As a last ingredient we need the Fourier representation of the delta function on the right hand side of (18.1). This multi-dimensional delta function is a shorthand notation for $\delta(\mathbf{r}) = \delta(x_1)\delta(x_2)\cdots\delta(x_N)$. Use the Fourier representation (14.31) of the delta function to show that:

$$\delta(\mathbf{r}) = \frac{1}{(2\pi)^N} \int e^{i\mathbf{k}\cdot\mathbf{r}} d^N k \ . \tag{18.4}$$

**Problem d:** Insert these results into the differential equation (18.1) of the Green's function to show that $g(\mathbf{k}, t)$ satisfies the differential equation

$$\frac{\partial g(\mathbf{k}, t)}{\partial t} + \kappa k^2 g(\mathbf{k}, t) = \delta(t) \ . \tag{18.5}$$

We have made considerable progress. The original equation (18.1) was a partial differential equation, whereas (18.5) is an ordinary differential equation for $g$ because only a time derivative is taken. In fact, you saw this equation before when you read section 16.4 which dealt with the response of a particle in syrup. Equation (18.5) is equivalent to the equation of motion (16.28) for a particle in syrup when the forcing force is a delta function.

**Problem e:** Use the theory of section 17.3 to show that the causal solution of (18.5) is given by:

$$g(\mathbf{k}, t) = e^{-\kappa k^2 t} \ . \tag{18.6}$$

This solution can be inserted into the Fourier representation (18.2) of the Green's function, and this gives:

$$G(\mathbf{r}, t) = \frac{1}{(2\pi)^N} \int e^{-\kappa k^2 t + i\mathbf{k} \cdot \mathbf{r}} d^N k \ . \tag{18.7}$$

The Green's function can be found by solving this Fourier integral. Before we do this, let us pause to consider the solution (18.6) for the Green's function in the wavenumber–time domain. The function $g(\mathbf{k}, t)$ gives the coefficient of the plane wave component $e^{i\mathbf{k} \cdot \mathbf{r}}$ as a function of time. According to (18.6) each Fourier component decays exponentially with time with a characteristic decay time $1/\kappa k^2$.

**Problem f:** Show that this implies that in the Fourier expansion (18.2) plane waves with a smaller wavelength decay faster with time than plane waves with larger wavelength. Explain this result physically.

In order to find the Green's function, we need to solve the Fourier integral (18.7). The integrations over the different components $k_i$ of the wavenumber integration all have the same form.

**Problem g:** Show this by giving a proof that the Green's function can be written as:

$$\begin{aligned} G(\mathbf{r}, t) &= \frac{1}{(2\pi)^N} \left( \int e^{-\kappa k_1^2 t + i k_1 x_1} dk_1 \right) \left( \int e^{-\kappa k_2^2 t + i k_2 x_2} dk_2 \right) \\ &\quad \times \cdots \left( \int e^{-\kappa k_N^2 t + i k_N x_N} dk_N \right) \ . \end{aligned} \tag{18.8}$$

You will notice that each of the integrals is of the same form, hence the Green's function can be written as

$$G(x_1, x_2, \cdots, x_N, t) = I(x_1, t) I(x_2, t) \cdots I(x_N, t)$$

with $I(x, t)$ given by

$$I(x, t) = \frac{1}{2\pi} \int_{-\infty}^{\infty} e^{-\kappa k^2 t + i k x} dk \ . \tag{18.9}$$

This means that our problem is solved when the one-dimensional Fourier integral (18.9) is solved. In order to solve this integral it is important to realize that the exponent in the integral is a quadratic function of the integration variable $k$. If the integral were of the form $\int_{-\infty}^{\infty} e^{-\alpha k^2} dk$ the problem would not be difficult because it is known that this integral has the value $\sqrt{\pi/\alpha}$. The problem can be solved by rewriting the integral (18.9) in the form of the integral $\int_{-\infty}^{\infty} e^{-\alpha k^2} dk$.

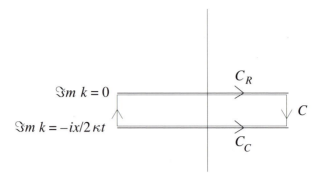

Fig. 18.1.   The contours $C_R$, $C_C$ and $C$ in the complex $k$-plane.

**Problem h:** Complete the square of the exponent in (18.9), i.e. show that

$$-\kappa k^2 t + ikx = -\kappa t \left( k - \frac{ix}{2\kappa t} \right)^2 - \frac{x^2}{4\kappa t} , \qquad (18.10)$$

and use this result to show that $I(x,t)$ can be written as:

$$I(x,t) = \frac{1}{2\pi} e^{-x^2/4\kappa t} \int_{-\infty - ix/2\kappa t}^{\infty - ix/2\kappa t} e^{-\kappa k'^2 t} dk' . \qquad (18.11)$$

With these steps we have achieved our goal of having an integrand of the form $e^{-\alpha k^2}$, but have paid a price. In the integral (18.9) the integration was along the real axis $C_R$, see figure 18.1. In the transformed integral the integration now takes place along the integration path $C_C$ in the complex plane that lies below the real axis, see figure 18.1. However, one can show that when the integration path $C_C$ is replaced by an integration along the real axis the integral has the same value:

$$I(x,t) = \frac{1}{2\pi} e^{-x^2/4\kappa t} \int_{-\infty}^{\infty} e^{-\kappa k^2 t} dk . \qquad (18.12)$$

**Problem i:** When you studied section 16.2 you saw all the material necessary to give a proof that (18.12) is indeed identical to (18.11). Show that this is indeed the case by using that the closed integral along the closed contour $C$ in figure 18.1 vanishes.

**Problem j:** Carry out the integration in (18.12) and derive that

$$I(x,t) = \frac{e^{-x^2/4\kappa t}}{\sqrt{4\pi\kappa t}} , \qquad (18.13)$$

and show that this implies that the Green's function is given by

$$G(\mathbf{r}, t) = \frac{1}{(4\pi\kappa t)^{N/2}} e^{-r^2/4\kappa t} . \qquad (18.14)$$

**Problem k:** This result implies that the Green's function in any dimension has the form of the Gaussian. Show that this Gaussian changes shape with time. Is the Gaussian broadest at early times or at late times? What is the shape of the Green's function in the limit $t \downarrow 0$, i.e. at the time just after the heat force has been applied.

**Problem l:** Sketch the time behavior of the Green's function for a fixed distance $r$. Does the Green's function decay more rapidly as a function of time in three dimensions than in one dimension? Give a physical interpretation of this result.

In one dimension, the Green's function (18.14) is given by

$$G(x,t) = \frac{1}{\sqrt{4\pi\kappa t}} e^{-x^2/4\kappa t} . \tag{18.15}$$

**Problem m:** Show that this is the same function as the solution (10.41) for the heat equation in one dimension when the limits $L \to 0$ and $T_0 L \to 1$ are taken.

These limits to the Gaussian function define the delta function, see section 13.1. This explains why in this limit the Green's function (18.15) and solution (10.41) are identical.

It is a remarkable property of the derivation in this section that the Green's function can be derived with a single derivation for every number of dimensions. It should be noted that this is not the generic case. In many problems, the behavior of the system depends critically of the number of spatial dimensions. We will see in section 18.4 that wave propagation in two dimensions is fundamentally different from wave propagation in one or three dimensions. Another example is chaotic behavior of dynamical systems where the occurrence of chaos is intricately linked to the number of dimensions, see the discussion given by Tabor [86].

## 18.2 The Schrödinger equation with an impulsive source

In this section we study the Green's function for the Schrödinger equation which was introduced in section 7.4:

$$i\hbar \frac{\partial \psi(\mathbf{r}, t)}{\partial t} = -\frac{\hbar^2}{2m} \nabla^2 \psi(\mathbf{r}, t) + V(\mathbf{r}) \psi(\mathbf{r}, t) . \tag{7.14}$$

Solving this equation for a general potential $V(\mathbf{r})$ is a formidable problem, and solutions are known for only very few examples such as the free particle, the harmonic oscillator and the Coulomb potential. We restrict ourselves here to the simplest case of a free particle: this is the case where the potential vanishes ($V(\mathbf{r}) = 0$). The corresponding Green's function satisfies the following partial differential equation:

$$\frac{\hbar}{i}\frac{\partial G(\mathbf{r}, t)}{\partial t} - \frac{\hbar^2}{2m}\nabla^2 G(\mathbf{r}, t) = \delta(\mathbf{r})\delta(t) . \qquad (18.16)$$

Before we compute the Green's function for this problem, let us pause to consider the meaning of this Green's function. First, the Green's function is a solution of Schrödinger's equation for $\mathbf{r} \neq 0$ and $t \neq 0$. This means that $|G|^2$ gives the probability density of a particle (see also section 7.4). However, the right hand side of (18.16) contains a delta-function forcing at time $t = 0$ at location $\mathbf{r} = 0$. This is a source term of $G$ and hence this is a source of the probability of the presence of the particle. One can say that this source term creates the probability for having a particle at the origin at $t = 0$. Of course, this particle will not necessarily remain at the origin, it will move according to the laws of quantum mechanics. This motion is described by (18.16). This means that this equation describes the time evolution of matter waves when matter is injected at $t = 0$ at location $\mathbf{r} = 0$.

**Problem a:** The Green's function $G(\mathbf{r}, t; \mathbf{r}', t')$ gives the wavefunction at location $\mathbf{r}$ and time $t$ for a source of particles at location $\mathbf{r}'$ at time $t'$. Express the Green's function $G(\mathbf{r}, t; \mathbf{r}', t')$ in terms of the solution $G(\mathbf{r}, t)$ of (18.16), and show how you obtained this result. Is this result also valid for the Green's function for the quantum-mechanical harmonic oscillator (where the potential $V(\mathbf{r})$ depends on position)?

In the previous section the Green's function gave the evolution of the temperature field due to a delta-function injection of heat at the origin at time $t = 0$. Similarly, the Green's function of this section describes the time evolution of the probability of a delta-function injection of matter waves at the origin at time $t = 0$. These two Green's functions are not only conceptually very similar, the differential equations (18.1) for the temperature field and (18.16) for the Schrödinger equation are first order differential equations in time and second order differential equations in the space coordinate that have a delta-function excitation on the right hand side. In this section we exploit this similarity and derive the Green's function for Schrödinger's equation from the Green's function for the heat equation derived in the previous section rather than constructing the solution from first principles. This approach is admittedly not very rigorous, but it shows that analogies are useful for making shortcuts.

The principle difference between (18.1) and (18.16) is that the time derivative for Schrödinger's equation is multiplied by $i = \sqrt{-1}$ whereas the heat equation is purely real. We will relate the two equations by introducing the new time variable $\tau$ for the Schrödinger equation that is proportional to the original time: $\tau = \gamma t$.

**Problem b:** How should the proportionality constant $\gamma$ be chosen so that (18.16) transforms to:

$$\frac{\partial G(\mathbf{r},\tau)}{\partial \tau} - \frac{\hbar^2}{2m}\nabla^2 G(\mathbf{r},\tau) = C\delta(\mathbf{r})\delta(\tau)\,. \tag{18.17}$$

The constant $C$ on the right hand side cannot easily be determined from the change of variables $\tau = \gamma t$ because $\gamma$ is not necessarily real and it is not clear how a delta function with a complex argument should be interpreted. For this reason we will not bother to specify $C$.

The key point to note is that this equation is of exactly the same form as the heat equation (18.1), where $\hbar^2/2m$ plays the role of the heat conductivity $\kappa$. The only difference is the constant $C$ on the right hand side of (18.17). However, since the equation is linear, this term only leads to an overall multiplication by $C$.

**Problem c:** Show that the Green's function defined in (18.17) for the Schrödinger equation can be obtained from the Green's function (18.14) for the heat equation by making the following substitutions:

$$\left.\begin{array}{rcl} t & \longrightarrow & it/\hbar\,, \\ \kappa & \longrightarrow & \hbar^2/2m\,, \\ G & \longrightarrow & CG\,. \end{array}\right\} \tag{18.18}$$

It is interesting to note that the 'diffusion constant' $\kappa$ which governs the spreading of the waves with time is proportional to the square of Planck's constant. Classical mechanics follows from quantum mechanics by letting Planck's constant go to zero: $\hbar \to 0$. It follows from (18.18) that in that limit the diffusion constant of the matter waves goes to zero. This reflects the fact that in classical mechanics the probability of the presence of a particle does not spread out with time.

**Problem d:** Use the substitutions (18.18) to show that the Green's function for the Schrödinger equation in $N$ dimensions is given by:

$$G(\mathbf{r},t) = C\frac{1}{(2\pi i\hbar t/m)^{N/2}}e^{imr^2/2\hbar t}\,. \tag{18.19}$$

This Green's function plays a crucial role in the formulation of the *Feynman path integrals* which have been a breakthrough in quantum mechanics as well as in other fields. A very clear description of the Feynman path integrals is given by Feynman and Hibbs [31].

**Problem e:** Sketch the real part of $e^{imr^2/2\hbar t}$ in the Green's function for a fixed time as a function of radius $r$. Does the wavelength of the Green's function increase or decrease with distance?

The Green's function (18.19) actually has an interesting physical meaning which is based on the fact that it describes the propagation of matter waves injected at $t = 0$ at the origin. The Green's function can be written as $G = C \left( 2\pi i \hbar t / m \right)^{-N/2} e^{i\Phi}$, where the phase of the Green's function is given by

$$\Phi = \frac{mr^2}{2\hbar t} \ . \tag{18.20}$$

As you noted in problem e the wavenumber of the waves depends on position. For a plane wave $e^{i\mathbf{k}\cdot\mathbf{r}}$ the phase is given by $\Phi = (\mathbf{k} \cdot \mathbf{r})$ and the wavenumber follows by taking the gradient of this function.

**Problem f:** Show that for a plane wave that

$$\mathbf{k} = \nabla \Phi \ . \tag{18.21}$$

Relation (18.21) has a wider applicability than plane waves. It has been shown by Whitham [99] that for a general phase function $\Phi(\mathbf{r})$ that varies smoothly with $\mathbf{r}$ the local wavenumber $\mathbf{k}(\mathbf{r})$ is defined by (18.21).

**Problem g:** Use this to show that for the Green's function of the Schrödinger equation the local wavenumber is given by

$$\mathbf{k} = \frac{m\mathbf{r}}{\hbar t} \ . \tag{18.22}$$

**Problem h:** Show that this expression is equivalent to:

$$\mathbf{v} = \frac{\hbar \mathbf{k}}{m} \ . \tag{7.20}$$

In problem e you discovered that for a fixed time, the wavelength of the waves decreases when the distance $r$ to the source is increased. This is consistent with (7.20): when a particle has moved further away from the source in a fixed time, its velocity is larger. This corresponds according to (7.20) with a larger wavenumber and hence with a smaller wavelength. This is indeed the behavior that is exhibited by the full wave-function (18.19).

The analysis in this chapter was not very rigorous because the substitution $t \to (i/\hbar)t$ implies that the independent parameter is purely imaginary rather than real. This means that all the arguments used in the previous section for the complex integration should be carefully reexamined. However, a more rigorous analysis shows that (18.19) is indeed the correct Green's function for the Schrödinger equation [31]. The approach taken in this section shows that an educated guess can be very useful in deriving new results. One can in fact argue that many innovations in mathematical physics have been obtained using intuition or analogies rather than formal derivations. Of course, a formal derivation should ultimately substantiate results obtained from a more intuitive approach.

## 18.3 The Helmholtz equation in one, two and three dimensions

The Helmholtz equation plays an important role in mathematical physics because it is closely related to the wave equation. A complete analysis of the Green's function for the wave equation and the Helmholtz equation in different dimensions is given by DeSanto [26]. The Green's function for the wave equation for a medium with constant velocity $c$ satisfies:

$$\nabla^2 G(\mathbf{r}, t; \mathbf{r}_0, t_0) - \frac{1}{c^2} \frac{\partial^2 G(\mathbf{r}, t; \mathbf{r}_0, t_0)}{\partial t^2} = \delta(\mathbf{r} - \mathbf{r}_0)\delta(t - t_0) . \qquad (18.23)$$

As shown in section 18.1 the Green's function depends only on the relative location $\mathbf{r} - \mathbf{r}_0$ and the relative time $t - t_0$ so that without loss of generality we can take the source at the origin ($\mathbf{r}_0 = 0$) and let the source act at time $t_0 = 0$. In addition it follows from symmetry considerations that the Green's function depends only on the relative distance $|\mathbf{r} - \mathbf{r}_0|$ and not on the orientation of the vector $\mathbf{r} - \mathbf{r}_0$. This means that the Green's function then satisfies $G(\mathbf{r}, t; \mathbf{r}_0, t_0) = G(|\mathbf{r} - \mathbf{r}_0|, t - t_0)$ and we need to solve the following equation:

$$\nabla^2 G(r, t) - \frac{1}{c^2} \frac{\partial^2 G(r, t)}{\partial t^2} = \delta(\mathbf{r})\delta(t) . \qquad (18.24)$$

**Problem a:** Under which conditions is this approach justified?

**Problem b:** Use a similar treatment to that in section 18.1 to show that when the Fourier transform (14.43) is used the Green's function satisfies the following equation in the frequency domain:

$$\nabla^2 G(r, \omega) + k^2 G(r, \omega) = \delta(\mathbf{r}) , \qquad (18.25)$$

where the wavenumber $k$ satisfies $k = \omega/c$.

This equation is called the *Helmholtz equation* and is the reformulation of the wave equation in the frequency domain. In the following we suppress the factor $\omega$ in the Green's function but it should be remembered that the Green's function depends on frequency.

Let us first solve (18.25) for one dimension. In that case the Green's function is defined by

$$\frac{d^2G}{dx^2} + k^2G = \delta(x) . \qquad (18.26)$$

**Problem c:** For $x \neq 0$ the right hand side of this equation is equal to zero. Use this to show that for $x \neq 0$ the solution is given by

$$G(x) = A\,e^{ikx} + B\,e^{-ikx} , \qquad (18.27)$$

where $A$ and $B$ are integration constants which need to be determined for the regions $x < 0$ and $x > 0$ separately.

In the Fourier transformation to the time domain, both terms are multiplied by $e^{-i\omega t}$. Using the relation $k = \omega/c$, this means that in the time domain the term $e^{ikx}$ becomes $e^{-i\omega(t-x/c)}$. For increasing time $t$, the phase of the wave remains constant when $x$ increases. This means that the term $e^{ikx}$ corresponds in the time domain to a right-going wave. Similarly, the term $e^{-ikx}$ describes a wave in the time domain that moves to the left. We are looking here for a Green's function that describes waves that move *away* from the source. This means that for $x > 0$ only the term $e^{ikx}$ contributes, so in that region we must have $B = 0$, while for $x < 0$ only the term $e^{-ikx}$ contributes hence we must take $A = 0$. This means that the Green's function is given by

$$G(x) = \begin{cases} B\,e^{-ikx} & \text{for } x < 0 \\ \\ A\,e^{ikx} & \text{for } x > 0 \end{cases} . \qquad (18.28)$$

The constants $A$ and $B$ follow from the requirements that $G(x)$ is continuous at $x = 0$ and that the first derivative is discontinuous at that point.

**Problem d:** Use the theory of section 17.3 to show that the jump in the first derivative at $x = 0$ is given by

$$\left[\frac{dG}{dx}\right]_{x=0-\varepsilon}^{x=0+\varepsilon} = 1 . \qquad (18.29)$$

**Problem e:** Use these results to derive that $A = B = -i/2k$, and use this result to write the Green's function of the Helmholtz equation in one dimension as

$$G^{1D}(x) = \frac{-i}{2k}e^{ik|x|} . \qquad (18.30)$$

Now we solve (18.25) for two and three space dimensions. To do this we consider the case of $N$ dimensions, where $N$ is either 2 or 3. Because the problem is spherically symmetric, we just need to consider a Green's function that depends on radius only.

**Problem f:** Use (9.35) and (9.36) to show that for such a radially symmetric function in two or three dimensions the Laplacian is given by

$$\nabla^2 G(r) = \frac{1}{r^{N-1}} \frac{\partial}{\partial r} \left( r^{N-1} \frac{\partial G}{\partial r} \right) . \qquad (18.31)$$

The differential equation for the Green's function in $N$ dimensions is thus given by

$$\frac{1}{r^{N-1}} \frac{\partial}{\partial r} \left( r^{N-1} \frac{\partial G}{\partial r} \right) + k^2 G(r, \omega) = \delta(\mathbf{r}) . \qquad (18.32)$$

This differential equation is not difficult to solve for two or three space dimensions for locations away from the source ($r \neq 0$). However, we need to consider carefully how the source $\delta(\mathbf{r})$ should be coupled to the solution of the differential equation. This can be achieved by integrating (18.25) over a sphere of radius $R$ centered at the source and letting the radius go to zero.

**Problem g:** Integrate (18.25) over this volume, use Gauss's law and let the radius $R$ go to zero to show that the Green's function satisfies

$$\oint_{S_R} \frac{\partial G}{\partial r} dS = 1 , \qquad (18.33)$$

where the surface integral is over a sphere $S_R$ with radius $R$ in the limit $R \downarrow 0$. Show that this can also be written as

$$\lim_{r \downarrow 0} S_r \frac{\partial G}{\partial r} = 1 , \qquad (18.34)$$

where $S_r$ is the surface of a sphere in $N$ dimensions with radius $r$.

Note that the surface of the sphere in general goes to zero as $r \downarrow 0$ (except in one dimension), which implies that $\partial G/\partial r$ must be infinite in the limit $r \downarrow 0$ in more than one space dimension.

The differential equation (18.32) is a second order differential equation. Such an equation must be supplemented with two boundary conditions. The first boundary condition is given by (18.34), and specifies how the solution is coupled to the source at $\mathbf{r} = 0$. The second boundary condition that we will use reflects the fact that the waves generated by the source will move away from the source. The solutions that we will find will behave for large distances as $e^{\pm ikr}$, but it is not clear whether we should use the upper sign $(+)$ or the lower sign $(-)$.

**Problem h:** Use the Fourier transform (14.42) and the relation $k = \omega/c$ to show that the integrand in the Fourier transformation to the time domain is proportional to $e^{-i\omega(t\mp r/c)}$.* Show that the waves only move away from the source for the upper sign. This means that the second boundary condition dictates that the solution behaves in the limit $r \to \infty$ as $e^{+ikr}$.

The derivative of function $e^{+ikr}$ is given by $ike^{+ikr}$, i.e. the derivative is $ik$ times the original function. When the Green's function behaves for large $r$ as $e^{+ikr}$, then the derivative of the Green's must satisfy the same relation as the derivative of $e^{+ikr}$. This means that the Green's function satisfies for large distance $r$:

$$\frac{\partial G}{\partial r} = ikG .\tag{18.35}$$

This relation specifies that the energy radiates *away* from the source. For this reason (18.35) is called the *radiation boundary condition*.

Now we are at the point where we can actually construct the solution for each dimension. Before we go to two dimensions we first solve the Green's function in three dimensions.

**Problem i:** Make for three dimensions ($N = 3$) the substitution $G(r) = f(r)/r$ and show that (18.32) implies that away from the source the function $f(r)$ satisfies

$$\frac{\partial^2 f}{\partial r^2} + k^2 f = 0 .\tag{18.36}$$

This equation has the solution $Ce^{\pm ikr}$. According to problem h the upper sign should be used and the Green's function is given by $G(r) = Ce^{ikr}/r$. Show that condition (18.34) dictates that $C = -1/4\pi$, so that in three dimensions the Green's function is given by:

$$G^{3D}(r) = \frac{-1}{4\pi}\frac{e^{ikr}}{r} .\tag{18.37}$$

The problem is actually most difficult in two dimensions because in that case the Green's function cannot be expressed in the simplest elementary functions.

---

* The convention of the notation $\mp$ is that the upper sign in $e^{-i\omega(t\mp r/c)}$ corresponds to the upper sign in $e^{\pm ikr}$. Similarly, the lower sign in both expressions corresponds to each other. This means that the solution $e^{-i\omega(t-r/c)}$ corresponds to $e^{+ikr}$ and that the solution $e^{-i\omega(t+r/c)}$ corresponds to $e^{-ikr}$.

| | $J_0(x)$ | $N_0(x)$ |
|---|---|---|
| $x \to 0$ | $1 - \frac{1}{4}x^2 + O(x^4)$ | $\frac{2}{\pi} \ln(x) + O(1)$ |
| $x \gg 1$ | $\sqrt{\frac{2}{\pi x}} \cos\left(x - \frac{\pi}{4}\right) + O(x^{-3/2})$ | $\sqrt{\frac{2}{\pi x}} \sin\left(x - \frac{\pi}{4}\right) + O(x^{-3/2})$ |

Table 18.1.   Leading asymptotic behavior of the Bessel function and Neumann function of order zero.

**Problem j:** Show that in two dimensions ($N = 2$) the differential equation of the Green's function away from the source is given by

$$\frac{\partial^2 G}{\partial r^2} + \frac{1}{r}\frac{\partial G}{\partial r} + k^2 G(r) = 0, \qquad r \neq 0. \qquad (18.38)$$

**Problem k:** This equation cannot be solved in terms of elementary functions. However, there is a close relation between (18.38) and the Bessel equation

$$\frac{d^2 F}{dx^2} + \frac{1}{x}\frac{dF}{dx} + \left(1 - \frac{m^2}{x^2}\right)F = 0. \qquad (18.39)$$

Show that the $G(kr)$ satisfies the Bessel equation for order $m = 0$.

This implies that the Green's function is given by the solution of the zeroth order Bessel equation with argument $kr$. The Bessel equation is a second order differential equation, therefore there are two independent solutions. The solution that is finite everywhere is denoted by $J_m(x)$ and is called the regular Bessel function. The second solution is singular at the point $x = 0$ and is called the Neumann function and denoted by $N_m(x)$. The Green's function is obviously a linear combination of $J_0(kr)$ and $N_0(kr)$. In order to determine how this linear combination is constructed it is crucial to consider the behavior of these functions at the source (i.e. for $x = 0$) and at infinity (i.e. for $x \gg 1$). The required asymptotic behavior can be found in textbooks such as Butkov [18] and Arfken [5] and is summarized in table 18.1.

**Problem l:** Show that neither $J_0(kr)$ nor $N_0(kr)$ behaves for large values of $r$ as $e^{+ikr}$. Show that the linear combination $J_0(kr) + iN_0(kr)$ does behave as $e^{+ikr}$.

The Green's function thus is a linear combination of the regular Bessel function and the Neumann function. This particular combination is called the *first Hankel function of degree zero* and is denoted by $H_0^{(1)}(kr)$. In general the Hankel functions are simply linear combinations of the Bessel function and the Neumann function:

$$\left.\begin{array}{l} H_m^{(1)}(x) \equiv J_m(x) + iN_m(x) \, , \\ H_m^{(2)}(x) \equiv J_m(x) - iN_m(x) \, . \end{array}\right\} \qquad (18.40)$$

**Problem m:** Show that $H_0^{(1)}(kr)$ behaves for large values of $r$ as $e^{+ikr-i\pi/4}/\sqrt{(\pi/2)kr}$ and that in this limit $H_0^{(2)}(kr)$ behaves as $e^{-ikr-i\pi/4}/\sqrt{(\pi/2)kr}$. Use this to argue that the Green's function is given by

$$G(r) = CH_0^{(1)}(kr) \, , \qquad (18.41)$$

where the constant $C$ still needs to be determined.

**Problem n:** This constant follows from the requirement (18.34) at the source. Use (18.40) and the asymptotic value of the Bessel function and the Neumann function given in table 18.1 to derive the asymptotic behavior of the Green's function near the source and use this to show that $C = -i/4$.

This result implies that in two dimensions the Green's function of the Helmholtz equation is given by

$$G^{2D}(r) = \frac{-i}{4} H_0^{(1)}(kr) \, . \qquad (18.42)$$

Summarizing these results and reverting to the more general case of a source at location $\mathbf{r}_0$ it follows that in one, two and three dimensions the Green's functions of the Helmholtz equation are given by

$$\left.\begin{array}{l} G^{1D}(x, x_0) = \dfrac{-i}{2k} e^{ik|x-x_0|} \, , \\[2mm] G^{2D}(\mathbf{r}, \mathbf{r}_0) = \dfrac{-i}{4} H_0^{(1)}(k \, |\mathbf{r} - \mathbf{r}_0|) \, , \\[2mm] G^{3D}(\mathbf{r}, \mathbf{r}_0) = \dfrac{-1}{4\pi} \dfrac{e^{ik|\mathbf{r}-\mathbf{r}_0|}}{|\mathbf{r} - \mathbf{r}_0|} \, . \end{array}\right\} \qquad (18.43)$$

Note that in two and three dimensions the Green's function is singular at the source $\mathbf{r}_0$.

**Problem o:** Show that these singularities are *integrable*, i.e. show that when the Green's function is integrated over a sphere with finite radius around the source the result is finite.

There is a physical reason why the Green's function in two and three dimensions has an integrable singularity. Suppose one has a source that is not a point source but that the source is constant within a sphere with radius $R$ centered around the origin. The response $p$ to this source is given by $p(\mathbf{r}) = \int_{r'<R} G(\mathbf{r}, \mathbf{r}')dV'$, where the integration over the variable $\mathbf{r}'$ is over a sphere with radius $R$. It follows from this expression that the response at the origin is given by

$$p(\mathbf{r}=0) = \int_{r'<R} G(\mathbf{r}=0, \mathbf{r}')dV' . \tag{18.44}$$

Since the excitation of this field is finite everywhere, the response $p(\mathbf{r}=0)$ should be finite as well. This implies that the integral (18.44) should be finite as well, which is a different way of stating that the singularity of the Green's function must be integrable.

## 18.4 The wave equation in one, two and three dimensions

In this section we consider the Green's function for the wave equation in one, two and three dimensions. This means that we consider solutions to the wave equation with an impulsive source at location $\mathbf{r}_0$ at time $t_0$:

$$\nabla^2 G(\mathbf{r}, t; \mathbf{r}_0, t_0) - \frac{1}{c^2} \frac{\partial^2 G(\mathbf{r}, t; \mathbf{r}_0, t_0)}{\partial t^2} = \delta(\mathbf{r} - \mathbf{r}_0)\delta(t - t_0) . \tag{18.23}$$

It was shown in the previous section that this Green's function depends only on the relative distance $|\mathbf{r} - \mathbf{r}_0|$ and the relative time $t - t_0$. For the case of a source at the origin ($\mathbf{r}_0 = 0$) acting at time zero ($t_0 = 0$) the time domain solution follows by applying a Fourier transform to the solution $G(\mathbf{r}, \omega)$ of the previous section. This Fourier transform is simplest in three dimensions, hence we will start with this case.

**Problem a:** Apply the Fourier transform (14.42) to the three-dimensional Green's function (18.37) and use the relation $k = \omega/c$ and the properties of the delta function to show that the Green's function is given in the time domain by

$$G^{3D}(\mathbf{r}, t) = -\frac{1}{4\pi r}\delta\left(t - \frac{r}{c}\right) . \tag{18.45}$$

**Problem b:** Now consider the wave equation with a general source term $S(\mathbf{r},t)$:

$$\nabla^2 p(\mathbf{r}, t) - \frac{1}{c^2}\frac{\partial^2 p(\mathbf{r}, t_0)}{\partial t^2} = S(\mathbf{r}, t) . \tag{18.46}$$

Use the Green's function (18.45) to show that a solution of this equation is given by

$$
p(\mathbf{r}, t) = -\frac{1}{4\pi} \int \frac{S\left(\mathbf{r}', t - \frac{|\mathbf{r} - \mathbf{r}'|}{c}\right)}{|\mathbf{r} - \mathbf{r}'|} dV' . \tag{18.47}
$$

Note that since $|\mathbf{r} - \mathbf{r}'|$ is always positive, the response $p(\mathbf{r}, t)$ depends only on the source function at *earlier* times. The solution therefore has a causal behavior and the Green's function (18.45) is called the *retarded Green's function*. However, in several applications one does not want to use a Green's function that depends on excitation at earlier times. An example is reflection seismology, in which one records the wave field at the surface, and from these observations one wants to reconstruct the wave field at *earlier* times while it was being reflected off layers inside the Earth. (See the treatment in section 7.3 and the work of Schneider [78].) A Green's function with waves that propagate towards the source and are then annihilated by the source can be obtained by replacing the radiation condition (18.35) by $\partial G/\partial r = -ikG$. The only difference is the minus sign on the right hand side, which is equivalent to replacing $k$ by $-k$.

**Problem c:** Apply the Fourier transform (14.42) to the three-dimensional Green's function (18.37) with $k$ replaced by $-k$ and show that the resulting Green's function in the time is given by

$$
G^{3D, \, advanced}(\mathbf{r}, t) = -\frac{1}{4\pi r} \delta\left(t + \frac{r}{c}\right) , \tag{18.48}
$$

and that the following function is a solution of the wave equation (18.46):

$$
p(\mathbf{r}, t) = -\frac{1}{4\pi} \int \frac{S\left(t + \frac{|\mathbf{r} - \mathbf{r}'|}{c}\right)}{|\mathbf{r} - \mathbf{r}'|} dV' . \tag{18.49}
$$

Note that in this representation the wave field is expressed in terms of the source function at *later* times. For this reason the Green's function (18.48) is called the *advanced Green's function*. The fact that the wave equation has both a retarded and an advanced solution is mathematically due to the wave equation (18.46) being invariant for time-reversal. This means that when one replaces $t$ by $-t$ the equation does not change. Physically this means that the wave equation does not know the 'direction of time'. In practice one most often works with the retarded Green's function, but keep in mind that in some applications, such as exploration seismology, the advanced Green's functions are crucial. In the remaining part of this

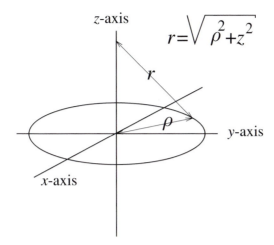

Fig. 18.2.    Definition of the variables $r$ and $\rho$.

section we will focus exclusively on the retarded Green's functions that represent causal solutions.

In order to obtain the Green's function for two dimensions in the time domain one could apply a Fourier transform to the solution (18.42). This involves taking the Fourier transform of a Hankel function, and it is not obvious how this Fourier integral should be solved (although it can be solved). Here we follow an alternative route by recognizing that the Green's function in two dimensions is identical to the solution of the wave equation in three dimensions when the source is not a point source but a cylindrical source. In other words, we obtain the two-dimensional Green's function by considering the wave field in three dimensions that is generated by a source which is distributed homogeneously along the $z$-axis. In order to separate the distance to the origin from the distance to the $z$-axis the variables $r$ and $\rho$ are used, see figure 18.2.

**Problem d:** Show that

$$G^{2D}(\rho, t) = \int_{-\infty}^{\infty} G^{3D}(r, t)dz \; . \tag{18.50}$$

**Problem e:** Use the Green's function (18.45) and the relation

$$r = \sqrt{\rho^2 + z^2}$$

to show that

$$G^{2D}(\rho, t) = -\frac{1}{2\pi} \int_{0}^{\infty} \frac{\delta\left(t - \dfrac{\sqrt{\rho^2 + z^2}}{c}\right)}{\sqrt{\rho^2 + z^2}} dz \; . \tag{18.51}$$

Note that the integration interval has been changed from $(-\infty, \infty)$ to $(0, \infty)$, and show how this can be achieved.

The distance $r$ in three dimensions no longer appears in this expression.

**Problem f:** The integral (18.51) can be solved by introducing the new integration variable $u \equiv \sqrt{\rho^2 + z^2}$, instead of the old integration variable $z$. Show that with this new variable the integral (18.51) can be written as

$$G^{2D}(\rho, t) = -\frac{1}{2\pi} \int_\rho^\infty \frac{\delta\left(t - \dfrac{u}{c}\right)}{\sqrt{u^2 - \rho^2}} \, du : \qquad (18.52)$$

pay attention to the limits of integration!

**Problem g:** Use the property $\delta(ax) = \delta(x)/|a|$ to rewrite this integral and evaluate the resulting integral separately for $t < \rho/c$ and $t > \rho/c$. Finally denote the distance $\rho$ in the two-dimensional $(x, y)$-plane by $r$ to show that

$$G^{2D}(r, t) = \begin{cases} 0 & \text{for} \quad t < r/c \\[2ex] -\dfrac{1}{2\pi} \dfrac{1}{\sqrt{t^2 - r^2/c^2}} & \text{for} \quad t > r/c \end{cases} . \qquad (18.53)$$

This Green's function and the Green's function for the three-dimensional case are shown in figure 18.3. There is a fundamental difference between the Green's function for two dimensions and the Green's function (18.45) for three dimensions. In three dimensions the Green's function is a delta-function $\delta(t - r/c)$ modulated by the geometrical spreading $-1/4\pi r$. This means that the response to a delta-function source has the same shape as the input function $\delta(t)$ that excites the wave field. An impulsive input leads to an impulsive output with a time delay given by $r/c$ and the solution is only nonzero at the wave front $t = r/c$. However, (18.53) shows that an impulsive input in two dimensions leads to a response that is not impulsive. The response has an infinite duration and decays with time as $1/\sqrt{t^2 - r^2/c^2}$: the solution is not only nonzero *at* the wave front $t = r/c$, it is nonzero everywhere *within* this wave front. This means that in two dimensions an impulsive input leads to a sound response that is of infinite duration. One can therefore say that:

*Any word spoken in two dimensions will reverberate forever (albeit weakly).*

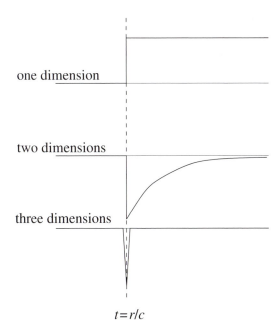

one dimension

two dimensions

three dimensions

$t = r/c$

Fig. 18.3.   The Green's function of the wave equation in one, two and three dimensions as a function of time.

The approach we have taken to compute the Green's function in two dimensions is interesting in that we solved the problem first in a higher dimension and retrieved the solution by integrating over one space dimension. Note that for this trick it is not necessary that this higher-dimensional space indeed exists! (Although in this case it does.) Remember that we took this approach because we did not want to evaluate the Fourier transform of a Hankel function. We can also turn this around: the Green's function (18.53) can be used to determine the Fourier transform of the Hankel function.

**Problem h:** Show that the Fourier transform of the Hankel function is given by:

$$\int_{-\infty}^{\infty} H_0^{(1)}(x) e^{-iqx} dx = \begin{cases} 0 & \text{for } q < 1 \\[2mm] \dfrac{2}{i\pi} \dfrac{1}{\sqrt{q^2 - 1}} & \text{for } q > 1 \end{cases} . \quad (18.54)$$

Hint: take the Fourier transform $G^{2D}(\mathbf{r}, \mathbf{r}_0)$ in (18.43) in order to obtain the Green's function for two dimensions in the time domain and compare the result with the corresponding expression (18.53). Make a suitable change of variables to arrive at (18.54).

Let us continue with the Green's function of the wave equation in one dimension in the time domain.

**Problem i:** Use the Green's function for one dimension of the last section
to show that in the time domain

$$G^{1D}(x,t) = -\frac{ic}{4\pi} \int_{-\infty}^{\infty} \frac{1}{\omega} e^{-i\omega(t-|x|/c)} d\omega . \tag{18.55}$$

This integral resembles the integral used for the calculation of the Green's
function in three dimensions. The only difference is the term $1/\omega$ in the
integrand, and because of this term we cannot immediately evaluate the
integral. However, the $1/\omega$ term can be removed by differentiating (18.55)
with respect to time, and the remaining integral can then be evaluated
analytically.

**Problem j:** Show that

$$\frac{\partial G^{1D}(x,t)}{\partial t} = \frac{c}{2} \delta \left( t - \frac{|x|}{c} \right) . \tag{18.56}$$

**Problem k:** This expression can be integrated but one condition is
needed to specify the integration constant that appears. We will
use here that at $t = -\infty$ the Green's function vanishes. Show that
with this condition the Green's function is given by:

$$G^{1D}(x,t) = \begin{cases} 0 & \text{for} \quad t < |x|/c \\ c/2 & \text{for} \quad t > |x|/c \end{cases} . \tag{18.57}$$

Just as in two dimensions the solution is nonzero everywhere *within* the
expanding wave front and not only on the wave front $|x| = ct$ as in three
dimensions. However, there is an important difference: in two dimensions
the solution changes for all times with time whereas in one dimension the
solution is constant except for $t = |x|/c$. The human ear is only sensitive
to pressure variations, it is insensitive to a static pressure. Therefore a
one-dimensional human will only hear a sound at $t = |x|/c$ but not at
later times.

In order to appreciate the difference in sound propagation in one, two
and three space dimensions, the Green's functions for the different dimensions are shown in figure 18.3. Note the dramatic change in the response
for different numbers of dimensions. This change in the properties of the
Green's function with change in dimension has been used somewhat jok-
ingly by Morley [57] to give 'a simple proof that the world is three dimensional.' When you worked through the sections 18.1 and 18.2 you learned
that for both the heat equation and the Schrödinger equation the solution does not depend fundamentally on the number of dimensions. This
is in stark contrast with the solutions of the wave equation which depend
critically on the number of dimensions.

# 19

---

# Normal modes

Many physical systems have the property that they can carry out oscillations only at certain specific frequencies. As a child (and hopefully also as an adult) you will have discovered that a swing in a playground will move only with a very specific natural period, and that the force that pushes the swing is only effective when the period of the force matches the period of the swing. The patterns of motion at which a system oscillates are called the *normal modes* of the system. A swing has one normal mode, but you have seen in section 12.6 that a simple model of a tri-atomic molecule has three normal modes. An example of a normal mode of a system is given in figure 19.1 which shows the pattern of oscillation of a metal plate which is driven by an oscillator at a fixed frequency. The screw in the middle of the plate shows the point at which the force on the plate is applied. Sand is sprinkled on the plate. When the frequency of the external force is equal to the frequency of a normal mode of the plate, the motion of the plate is given by the motion that corresponds to that specific normal mode. Such a pattern of oscillation has nodal lines where the motion vanishes. These nodal lines are visible because the sand on the plate collects at these lines.

In this chapter, the normal modes of a variety of systems are analyzed. Normal modes play an important role in a variety of applications because the eigenfrequencies of normal modes provide important information of physical systems. Examples are the normal modes of the Earth which provide information about the internal structure of our planet, and the spectral lines of light emitted by atoms which have led to the advent of quantum mechanics and its description of the internal structure of atoms. In addition, normal modes are used in this chapter to introduce some properties of special functions, such as Bessel functions and Legendre functions. This is achieved by analyzing the normal modes of a system in one, two and three dimensions in the sections 19.1–19.3.

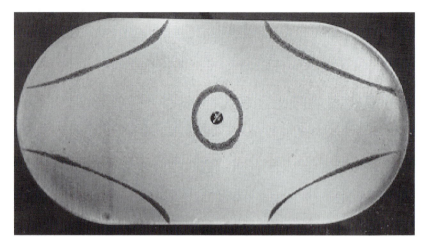

Fig. 19.1.   Sand on a metal plate which is driven by an oscillator at a frequency that corresponds to one of the eigenfrequencies of the plate. This figure was prepared by John Scales at the Colorado School of Mines.

## 19.1  The normal modes of a string

In this and the following two sections we assume that the motion of the system is governed by the Helmholtz equation

$$\nabla^2 u + k^2 u = 0 \ . \tag{19.1}$$

In this expression the wavenumber $k$ is related to the angular frequency $\omega$ by the relation

$$k = \frac{\omega}{c} \ . \tag{19.2}$$

For simplicity we assume the system to be homogeneous, which means that the velocity $c$ is constant. This in turn implies that the wavenumber $k$ is constant. In the sections 19.1–19.3 we consider a body of radius $R$. Since a circle or a sphere of radius $R$ has a diameter $2R$ we will consider here a string of length $2R$ in order to be able to make meaningful comparisons. It is assumed that the endpoints of the string are fixed so that the boundary conditions are:

$$u(0) = u(2R) = 0 \ . \tag{19.3}$$

**Problem a:** Show that the solutions of (19.1) that satisfy the boundary conditions (19.3) are given by $\sin(k_n x)$ with the wavenumber $k_n$ given by

$$k_n = \frac{n\pi}{2R} \ , \tag{19.4}$$

where $n$ is an integer.

For a number of purposes it is useful to normalize the modes: this means that we require that the modes $u_n(x)$ satisfy the condition $\int_0^{2R} u_n^2(x)dx = 1$.

**Problem b:** Show that the normalized modes are given by

$$u_n(x) = \frac{1}{\sqrt{R}} \sin(k_n x) \,. \qquad (19.5)$$

**Problem c:** Sketch the modes for several values of $n$ as a function of the distance $x$.

**Problem d:** The modes $u_n(x)$ are orthogonal, which means that the inner product $\int_0^{2R} u_n(x)u_m(x)dx$ vanishes when $n \neq m$. Give a proof of this property to derive that

$$\int_0^{2R} u_n(x)u_m(x)dx = \delta_{nm} \,. \qquad (19.6)$$

We conclude from this section that the modes of a string are oscillatory functions with a wavenumber that can only have discrete well-defined values $k_n$. According to (19.2) this means that the string can only vibrate at discrete frequencies which are given by

$$\omega_n = \frac{n\pi c}{2R} \,. \qquad (19.7)$$

This property will be familiar to you because you probably know that a guitar string vibrates only at very specific frequencies which determine the pitch of the sound that you hear. The results of this section imply that each string oscillates not only at one particular frequency, but at many discrete frequencies. The oscillation with the lowest frequency is given by (19.7) with $n = 1$: this is called the *fundamental mode* or *ground-tone*. This is what the ear perceives as the pitch of the tone. The oscillations corresponding to larger values of $n$ are called the *higher modes* or *overtones*. The particular mix of overtones determines the timbre of the signal. If the higher modes are strongly excited the ear perceives this sound as metallic, whereas just the fundamental mode is perceived as a smooth sound. The reader who is interested in the theory of musical instruments should consult ref. [73].

The discrete modes are not a peculiarity of the string. Most systems that support waves and that are of a finite extent support modes. For example, in figure 14.1 the spectrum of the sound of a soprano saxophone is shown. This spectrum is characterized by well-defined peaks that correspond to the modes of the air-waves in the instrument. Mechanical systems in general have discrete modes, these modes can be destructive when they are excited at their resonance frequency. The matter waves in atoms are organized in modes as well, and this is ultimately the reason why atoms in an excited state emit light only at very specific frequencies, called spectral lines.

## 19.2 The normal modes of drum

In the previous section we looked at the modes of a one-dimensional system. Here we derive the modes of a two-dimensional system which is a model of a drum. We consider a two-dimensional membrane that satisfies the Helmholtz equation (19.1). The membrane is circular and has radius $R$. At the edge $r = R$, the membrane cannot move, this means that in cylindrical coordinates the boundary condition for the waves $u(r, \varphi)$ is given by:

$$u(R, \varphi) = 0 . \tag{19.8}$$

In order to find the modes of the drum we will use *separation of variables*, which means that we seek solutions that can be written as a product of a function that depends only on $r$ and a function that depends only $\varphi$:

$$u(r, \varphi) = F(r)G(\varphi) . \tag{19.9}$$

**Problem a:** Insert this solution in the Helmholtz equation, use the expression of the Laplacian in cylindrical coordinates, and show that the resulting equation can be written as

$$\underbrace{\left[ \frac{1}{F(r)} r \frac{\partial}{\partial r} \left( r \frac{\partial F}{\partial r} \right) + k^2 r^2 \right]}_{(A)} = - \underbrace{\frac{1}{G(\varphi)} \frac{\partial^2 G}{\partial \varphi^2}}_{(B)} . \tag{19.10}$$

**Problem b:** The terms labelled (A) depend on the variable $r$ only whereas the terms labelled (B) depend only on the variable $\varphi$. These terms can only be equal for all values of $r$ and $\varphi$ when they depend neither on $r$ nor on $\varphi$, i.e. when they are a constant. Use this to show that $F(r)$ and $G(\varphi)$ satisfy the following differential equations:

$$\frac{d^2 F}{dr^2} + \frac{1}{r} \frac{dF}{dr} + \left( k^2 - \frac{\mu}{r^2} \right) F = 0 , \tag{19.11}$$

$$\frac{d^2 G}{d\varphi^2} + \mu G = 0 , \tag{19.12}$$

where $\mu$ is a constant that is not yet known.

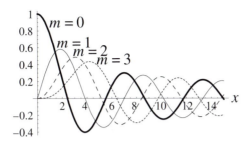

Fig. 19.2. The Bessel functions $J_m(x)$ for several orders $m$.

These differential equations need to be supplemented with boundary conditions. The boundary conditions for $F(r)$ follow from the requirements that this function is finite everywhere and that the displacement vanishes at the edge of the drum:

$$F(r) \text{ is finite everywhere} , \qquad F(R) = 0 . \qquad (19.13)$$

The boundary condition for $G(\varphi)$ follows from the requirement that if we rotate the drum through $360°$, every point on the drum returns to its original position. This means that the modes satisfy the requirement that $u(r, \varphi) = u(r, \varphi + 2\pi)$. This implies that $G(\varphi)$ satisfies the *periodic boundary condition*:

$$G(\varphi) = G(\varphi + 2\pi) . \qquad (19.14)$$

**Problem c:** The general solution of (19.12) is given by

$$G(\varphi) = e^{\pm i\sqrt{\mu}\varphi} .$$

Show that the boundary condition (19.14) implies that $\mu = m^2$, with $m$ an integer.

This means that the dependence of the modes on the angle $\varphi$ is given by:

$$G(\varphi) = e^{im\varphi} . \qquad (19.15)$$

The value $\mu = m^2$ can be inserted in (19.11). The resulting equation then bears a close resemblance to the Bessel equation:

$$\frac{d^2 J_m}{dx^2} + \frac{1}{x}\frac{dJ_m}{dx} + \left(1 - \frac{m^2}{x^2}\right) J_m = 0 . \qquad (19.16)$$

This equation has two independent solutions: the Bessel function $J_m(x)$ which is finite everywhere and the Neumann function $N_m(x)$ which is singular at $x = 0$. The Bessel functions $J_m(x)$ for several orders $m$ are shown in figure 19.2. All the Bessel functions are oscillatory functions which decay with increasing values of the argument $x$. Note that there is a phase shift of a quarter cycle between the successive orders $m$. Near $x = 0$ the lowest order Bessel functions behaves as $J_0(x) \sim 1$, while the higher order Bessel functions behave as $J_1(x) \sim x$ and $J_2(x) \sim x^2$. This is due to the fact that the Bessel function $J_m(x)$ behaves as $x^m$ for small values of $x$.

**Problem d:** Show that the general solution of (19.11) can be written as:

$$F(r) = A J_m(kr) + B N_m(kr) \,, \tag{19.17}$$

with $A$ and $B$ integration constants.

**Problem e:** Use the boundary conditions of $F(r)$ to show that $B = 0$ and that the wavenumber $k$ must take a value such that $J_m(kR) = 0$.

This last condition for the wavenumber is analogous to the condition (19.4) for the one-dimensional string. For both the string and the drum the wavenumber can only take discrete values: these values are dictated by the condition that the displacement vanishes at the outer boundary of the string or the drum. For the drum there are for every value of the angular degree $m$ infinitely many wavenumbers that satisfy the requirement $J_m(kR) = 0$. These wavenumbers are labelled with a subscript $n$, but since these wavenumbers are different for each value of the angular order $m$, the allowed wavenumbers carry two indices and are denoted by $k_n^{(m)}$. They satisfy the condition

$$J_m(k_n^{(m)} R) = 0 \,. \tag{19.18}$$

The zeroes (or roots) of the Bessel function $J_m(x)$ are not known in closed analytical form. However, numerical tables exists for the roots of Bessel functions, see for example, table 9.4 of Abramowitz and Stegun [1]. Take a look at this reference which contains a bewildering collection of formulas, graphs and tables of mathematical functions. The lowest order zeroes of the Bessel functions $J_0(x)$, $J_1(x), \ldots, J_5(x)$ are shown in table 19.1.

Using the results in this section it follows that the modes of the drum are given by

$$u_{nm}(r, \varphi) = J_m(k_n^{(m)} r) e^{im\varphi} \,. \tag{19.19}$$

**Problem f:** Let us first consider the $\varphi$-dependence of these modes. Show that when one follows the mode $u_{nm}(r, \varphi)$ along a complete circle around the origin one encounters exactly $m$ oscillations of that mode.

Table 19.1. The roots of the Bessel function $J_m(x)$, these are the values of $x$ for which $J_m(x) = 0$.

| $n$ | $m = 0$ | $m = 1$ | $m = 2$ | $m = 3$ | $m = 4$ | $m = 5$ |
|---|---|---|---|---|---|---|
| 1 | 2.40482 | 3.83171 | 5.13562 | 6.38016 | 7.58834 | 8.77148 |
| 2 | 5.52007 | 7.01559 | 8.41724 | 9.76102 | 11.06471 | 12.33860 |
| 3 | 8.65372 | 10.17347 | 11.61984 | 13.01520 | 14.37254 | 15.70017 |
| 4 | 11.79153 | 13.32369 | 14.79595 | 16.22347 | 17.61597 | 18.98013 |
| 5 | 14.93091 | 16.47063 | 17.95982 | 19.4092 | 20.82693 | 22.21780 |
| 6 | 18.07106 | 19.61586 | 21.11700 | 22.58273 | 24.01902 | 25.43034 |
| 7 | 21.21163 | 22.76008 | 24.27011 | 25.74817 | 27.19909 | 28.62662 |

**Problem g:** Find the eigenfrequencies of the five modes of the drum with the lowest frequencies and make a sketch of the associated standing waves of the drum. Use (19.18) and table 19.1 to determine what the values of $n$ and $m$ are for these five modes.

**Problem h:** Compute the separation between the different zero-crossings for a fixed value of $m$. To which number does this separation converge for the zero-crossings at large values of $x$?

The shape of the Bessel function is more difficult to see than the properties of the functions $e^{im\varphi}$. As shown in section 9.7 of Butkov [18], these functions satisfy a large number of properties which include recursion relations and series expansions. However, at this point the following facts are most important:

- The Bessel functions $J_m(x)$ are oscillatory functions which decay with distance: in a sense they behave as decaying standing waves. We will return to this issue in section 19.5.

- The Bessel functions satisfy an orthogonality relation similar to the orthogonality relation (19.6) for the modes of the string. This orthogonality relation is treated in more detail in section 19.4.

## 19.3 The normal modes of a sphere

In this section we consider the normal modes of a spherical surface with radius $R$. We only consider the modes that are associated with the waves that propagate along the surface, hence we do not consider wave motion in the *interior* of the sphere. The modes are assumed to satisfy the wave equation (19.1). Since the waves propagate on the spherical surface, they are only a function of the angles $\theta$ and $\varphi$ that are used in spherical coordinates: $u = u(\theta, \varphi)$. Using the Laplacian expressed in spherical coordinates, wave equation (19.1) is then given by

$$\frac{1}{R^2} \left[ \frac{1}{\sin\theta} \frac{\partial}{\partial\theta} \left( \sin\theta \frac{\partial u}{\partial\theta} \right) + \frac{1}{\sin^2\theta} \frac{\partial^2 u}{\partial\varphi^2} \right] + k^2 u = 0 . \tag{19.20}$$

Again, we seek a solution by applying separation of variables by writing the solution in a form similar to (19.9):

$$u(\theta, \varphi) = F(\theta)G(\varphi) . \tag{19.21}$$

**Problem a:** Insert this into (19.20) and apply separation of variables to show that $F(\theta)$ satisfies the following differential equation:

$$\sin\theta \frac{d}{d\theta} \left( \sin\theta \frac{dF}{d\theta} \right) + \left( k^2 R^2 \sin^2\theta - \mu \right) F = 0 , \tag{19.22}$$

and that $G(\varphi)$ satisfies (19.12), where the unknown constant $\mu$ does not depend on $\theta$ or $\varphi$.

To make further progress we have to apply boundary conditions. Just as with the drum in section 19.2, the system is invariant when a rotation over $2\pi$ is applied: $u(\theta, \varphi) = u(\theta, \varphi + 2\pi)$. This means that $G(\varphi)$ satisfies the same differential equation (19.12) as for the case of the drum and the same periodic boundary condition (19.14). The solution is therefore given by $G(\varphi) = e^{im\varphi}$ and the separation constant satisfies $\mu = m^2$, with $m$ an integer. Using this, the differential equation for $F(\theta)$ can be written as:

$$\frac{1}{\sin\theta} \frac{d}{d\theta} \left( \sin\theta \frac{dF}{d\theta} \right) + \left( k^2 R^2 - \frac{m^2}{\sin^2\theta} \right) F = 0 . \tag{19.23}$$

Before we continue let us compare this equation with (19.11) for the modes of the drum which we can rewrite as

$$\frac{1}{r} \frac{d}{dr} \left( r \frac{dF}{dr} \right) + \left( k^2 - \frac{m^2}{r^2} \right) F = 0 . \tag{19.11}$$

Note that these equations are identical when we compare $r$ in (19.11) with $\sin\theta$ in (19.23). There is a good reason for this. Suppose that we have a source in the middle of the drum. Then the variable $r$ measures the distance from a point on the drum to the source. This can be compared with the case of waves on a spherical surface that are excited by a source at the north pole. In that case, $\sin\theta$ is a measure of the distance from a point to the source point. The only difference is that $\sin\theta$ enters the equation rather than the true angular distance $\theta$. This is a consequence of the fact that the surface is curved, and this curvature leaves an imprint on the differential equation that the modes satisfy.

**Problem b:** The differential equation (19.11) was reduced in section 19.2 to the Bessel equation by changing to a new variable $x = kr$. Define a new variable

$$x \equiv \cos \theta \qquad (19.24)$$

and show that the differential equation for $F$ is given by

$$\frac{d}{dx}\left[\left(1 - x^2\right)\frac{dF}{dx}\right] + \left(k^2 R^2 - \frac{m^2}{1 - x^2}\right) F = 0 . \qquad (19.25)$$

The solution of this differential equation is given by the associated Legendre functions $P_l^m(x)$. These functions are described in great detail in section 9.8 of Butkov [18]. In fact, just like the Bessel equation, the differential equation (19.25) has a solution that is regular as well as a solution $Q_l^m(x)$ that is singular at the point $x = 1$ where $\theta = 0$. However, since the modes are finite everywhere, they are given by the regular solution $P_l^m(x)$ only.

The wavenumber $k$ is related to frequency by the relation $k = \omega/c$. At this point it is not clear what $k$ is, hence the eigenfrequencies of the spherical surface are not yet known. It is shown in section 9.8 of Butkov [18] that:

- The associated Legendre functions are only finite when the wavenumber satisfies

$$k^2 R^2 = l(l + 1) , \qquad (19.26)$$

where $l$ is a positive integer. Using this in (19.25) implies that the associated Legendre functions satisfy the following differential equation:

$$\frac{1}{\sin \theta}\frac{d}{d\theta}\left[\sin \theta \frac{dP_l^m(\cos \theta)}{d\theta}\right] + \left[l(l+1) - \frac{m^2}{\sin^2 \theta}\right] P_l^m(\cos \theta) = 0 .$$
$$(19.27)$$

Seen as a function of $x$ ($= \cos \theta$) this is equivalent to the following differential equation

$$\frac{d}{dx}\left[\left(1 - x^2\right)\frac{dP_l^m(x)}{dx}\right] + \left[l(l+1) - \frac{m^2}{1 - x^2}\right] P_l^m(x) = 0 . \quad (19.28)$$

- The integer $l$ must be larger than or equal to the absolute value of the angular order $m$.

**Problem c:** Show that the last condition can also be written as:

$$-l \leq m \leq l . \qquad (19.29)$$

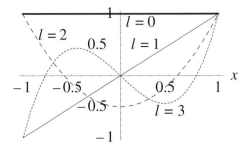

Fig. 19.3.    The Legendre polynomials $P_l^0(x)$ for several degrees $l$.

**Problem d:** Derive that the eigenfrequencies of the modes are given by

$$\omega_l = \sqrt{l(l+1)}\,\frac{c}{R} \; . \qquad\qquad (19.30)$$

It is interesting to compare this result with the eigenfrequencies (19.7) of the string. The eigenfrequencies of the string all have the same spacing in frequency, but the eigenfrequencies of the spherical surface are not spaced at the same interval. In musical jargon one would say that the overtones of a string are harmonious: this means that the eigenfrequencies of the overtones are multiples of the eigenfrequency of the ground tone. In contrast, the overtones of a spherical surface are not harmonious.

**Problem e:** Show that for large values of $l$ the eigenfrequencies of the spherical surface have an almost equal spacing.

**Problem f:** The eigenfrequency $\omega_l$ only depends on the order $l$ and not on the degree $m$. For each value of $l$, the angular degree $m$ can according to (19.29) take the values $-l, -l+1, \ldots, l-1, l$. Show that this implies that for every value of $l$, there are $(2l+1)$ modes with the same eigenfrequency.

When different modes have the same eigenfrequency one speaks of *degenerate modes*.

The Legendre functions $P_l^0(x)$ for several degrees $l$ are shown in figure 19.3 as a function of the variable $x$. The value $x = 1$ corresponds to $\cos\theta = 1$ or $\theta = 0$: this is the north pole of the spherical coordinate system. The value $x = -1$ corresponds to the south pole of the spherical coordinate system. The number of oscillations of these functions increases with the degree $l$. All Legendre functions $P_l^0(x)$ have the same value at the north pole: $P_l^0(x = 1) = 1$.

The results we obtained imply that the modes on a spherical surface are given by $P_l^m(\cos\theta)e^{im\varphi}$. We used here that the variable $x$ is related to the angle $\theta$ through (19.24). The modes of the spherical surface are called *spherical harmonics*. These eigenfunctions for $m \geq 0$ are given by:

$$Y_{lm}(\theta, \varphi) = (-1)^m \sqrt{\frac{2l+1}{4\pi} \frac{(l-m)!}{(l+m)!}} P_l^m(\cos\theta) e^{im\varphi} \qquad m \geq 0. \quad (19.31)$$

For $m < 0$ the spherical harmonics are defined by the relation

$$Y_{lm}(\theta, \varphi) = (-1)^m Y_{l,-m}(\theta, \varphi). \qquad (19.32)$$

You may wonder where the square-root in front of the associated Legendre function comes from. One can show that with this numerical factor the spherical harmonics are normalized when integrated over the sphere:

$$\iint |Y_{lm}|^2 \, d\Omega = 1, \qquad (19.33)$$

where $\iint \cdots d\Omega$ denotes an integration over the unit sphere. You should be aware of the fact that different authors use different definitions of the spherical harmonics. For example, one could also define the spherical harmonics as $\tilde{Y}_{lm}(\theta, \varphi) = P_l^m(\cos\theta) e^{im\varphi}$ because the functions also account for the normal modes of a spherical surface.

**Problem g:** Show that the modes defined in this way satisfy $\iint |\tilde{Y}_{lm}|^2 \, d\Omega = 4\pi/(2l+1) \times (l+m)!/(l-m)!$.

This means that the modes defined in this way are not normalized when integrated over the sphere. There is no reason why one cannot work with this convention, as long as one accounts for the fact that in this definition the modes are not normalized. Throughout this book we will use the definition (19.31) for the spherical harmonics. In doing so we follow the normalization that is used by Edmonds [29].

The real parts of the lowest order spherical harmonics are shown in figure 19.4, in which these functions are projected on the Earth's surface. Only those spherical harmonics with $m \geq 0$ are shown, the spherical harmonics for $m < 0$ follow from (19.32). Just as with the Bessel functions, the associated Legendre functions satisfy recursion relations and a large number of other properties which are described in detail in section 9.8 of Butkov [18]. The most important properties of the spherical harmonics $Y_{lm}(\theta, \varphi)$ are:

- These functions display $m$ oscillations when the angle $\varphi$ increases by $2\pi$. In other words, there are $m$ oscillations along one circle of constant latitude.

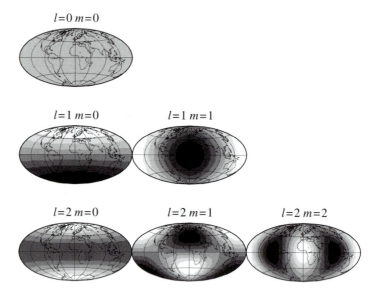

Fig. 19.4.   The real parts of the lowest order spherical harmonics. The value of
the order $l$ and degree $m$ is shown above each panel. Negative values are shown
in white, positive values in black. Each panel shows the surface of the complete
sphere using a Hammer projection. The world map is shown to provide spatial
orientation. The thin solid lines correspond to lines of constant latitude and
longitude on the sphere. Figure courtesy of Jeannot Trampert.

- The associated Legendre functions $P_l^m(\cos\theta)$ behave like Bessel
  functions in the sense that they behave like standing waves with
  an amplitude that decays from the pole. We will return to this issue
  in section 19.6.

- The number of oscillations between the north pole of the sphere and
  the south pole of the sphere increases with the difference $(l - m)$.

- The spherical harmonics are orthogonal for a suitably chosen inner
  product; this orthogonality relation is derived in section 19.4.

- The spherical harmonics are the eigenfunctions of the Laplacian on
  the sphere.

**Problem h:** Give a proof of this last property by showing that

$$\nabla_1^2 Y_{lm}(\theta, \varphi) = -l\,(l+1)\,Y_{lm}(\theta, \varphi)\,, \tag{19.34}$$

where the Laplacian on the unit sphere is given by

$$\nabla_1^2 = \frac{1}{\sin\theta}\frac{\partial}{\partial\theta}\left(\sin\theta\frac{\partial}{\partial\theta}\right) + \frac{1}{\sin^2\theta}\frac{\partial^2}{\partial\varphi^2}\,. \tag{19.35}$$

This property is extremely useful in many applications, because the action of the Laplacian on a sphere can be replaced by the much simpler multiplication by the constant $-l\,(l+1)$ when spherical harmonics are concerned.

## 19.4 Normal modes and orthogonality relations

The normal modes of a physical system often satisfy orthogonality relations when a suitably chosen inner product for the eigenfunctions is used. In this section this is illustrated by studying once again the normal modes of the Helmholtz equation (19.1) for different geometries. In this section we derive first the general orthogonality relation for these normal modes. This is then applied to the normal modes of the previous sections to derive the orthogonality relations for Bessel functions and associated Legendre functions.

Let us consider two normal modes of the Helmholtz equation (19.1), and let these modes be called $u_p$ and $u_q$. At this point we leave it open whether the modes are defined on a line, on a surface of arbitrary shape or in a volume. The integration over the region of space in which the modes are defined is denoted as $\int \cdots d^N x$, where $N$ is the dimension of this space. The wavenumbers of these modes, which are effectively the corresponding eigenvalues of the Helmholtz equation, are defined by $k_p$ and $k_q$, respectively. In other words, the modes satisfy the equations:

$$\nabla^2 u_p + k_p^2 u_p = 0 \,, \tag{19.36}$$

$$\nabla^2 u_q + k_q^2 u_q = 0 \,. \tag{19.37}$$

The subscript $p$ may stand for a single mode index such as in the index $n$ for the wavenumber $k_n$ for the modes of a string, or it may stand for a number of indices such as the indices $nm$ that label the eigenfunctions (19.19) of a circular drum.

**Problem a:** Multiply (19.36) by $u_q^*$, take the complex conjugate of (19.37) and multiply the result by $u_p$. Subtract the resulting equations and integrate this over the region of space for which the modes are defined to show that

$$\int \left( u_q^* \nabla^2 u_p - u_p \nabla^2 u_q^* \right) d^N x + \left( k_p^2 - k_q^{*2} \right) \int u_q^* u_p d^N x = 0 \,. \tag{19.38}$$

**Problem b:** Use the theorem of Gauss to derive that

$$\int u_q^* \nabla^2 u_p d^N x = \oint u_q^* \nabla u_p \cdot d\mathbf{S} - \int \left( \nabla u_q^* \cdot \nabla u_p \right) d^N x \,, \tag{19.39}$$

where the integral $\oint \cdots d\mathbf{S}$ is over the surface that bounds the body. If you have trouble deriving this, you can consult (7.10) where a similar result was used for the derivation of the representation theorem for acoustic waves.

**Problem c:** Use the last result to show that

$$\oint \left( u_q^* \nabla u_p - u_p \nabla u_q^* \right) \cdot d\mathbf{S} + \left( k_p^2 - k_q^{*2} \right) \int u_q^* u_p d^N x = 0 . \quad (19.40)$$

**Problem d:** The result is now expressed in the first term as an integral over the boundary of the body. The second term contains a volume integral and this term leads to the orthogonality relation of the modes. Let us assume that on this boundary the modes satisfy one of the three boundary conditions: (i) $u = 0$, (ii) $\hat{\mathbf{n}} \cdot \nabla u = 0$ (where $\hat{\mathbf{n}}$ is the unit vector perpendicular to the surface) or (iii) $\hat{\mathbf{n}} \cdot \nabla u = \alpha u$ (where $\alpha$ is a constant). Show that for all of these boundary conditions the surface integral in (19.40) vanishes.

The last result implies that when the modes satisfy one of these boundary conditions

$$\left( k_p^2 - k_q^{*2} \right) \int u_q^* u_p d^N x = 0 . \quad (19.41)$$

Let us first consider the case in which the modes are equal, i.e. in which $p = q$. In that case the integral reduces to $\int |u_p|^2 d^N x$ which is guaranteed to be positive. Equation (19.41) then implies that $k_p^2 = k_p^{*2}$, so that the wavenumbers $k_p$ must be real: $k_p = k_p^*$. For this reason the complex conjugate of the wavenumbers can be dropped and (19.41) can be written as:

$$\left( k_p^2 - k_q^2 \right) \int u_q^* u_p d^N x = 0 . \quad (19.42)$$

Now consider the case of two different modes for which the wavenumbers $k_p$ and $k_q$ are different. In that case the term $\left( k_p^2 - k_q^2 \right)$ is nonzero, hence in order to satisfy (19.42) the modes must satisfy

$$\int u_q^* u_p d^N x = 0 \qquad \text{for} \qquad k_p \neq k_q . \quad (19.43)$$

This finally gives the *orthogonality relation* of the modes in the sense that it states that the modes are orthogonal for the following inner product: $\langle f \cdot g \rangle \equiv \int f^* g \, d^N x$. Note that the inner product for which the modes are orthogonal follows from the Helmholtz equation (19.1) which defines the modes.

Let us now consider this orthogonality relation for the modes of the string, the drum and the spherical surface of the previous sections. For the string, the orthogonality relation was derived in problem d of section 19.1 and you can see that equation (19.6) is identical to the general orthogonality relation (19.43). For the circular drum the modes are given by (19.19).

**Problem e:** Use (19.19) for the modes of the circular drum to show that the orthogonality relation (19.43) for this case can be written as:

$$\int_0^R \int_0^{2\pi} J_{m_1}(k_{n_1}^{(m_1)}r) J_{m_2}(k_{n_2}^{(m_2)}r) e^{i(m_1-m_2)\varphi} d\varphi \, r dr = 0$$

$$\text{for} \quad k_{n_1}^{(m_1)} \neq k_{n_2}^{(m_2)} . \quad (19.44)$$

Explain where the factor $r$ comes from in the integration.

**Problem f:** This integral can be separated into an integral over $\varphi$ and an integral over $r$. The $\varphi$-integral is given by $\int_0^{2\pi} e^{i(m_1-m_2)\varphi}d\varphi$. Show that this integral vanishes when $m_1 \neq m_2$:

$$\int_0^{2\pi} e^{i(m_1-m_2)\varphi} d\varphi = 0 \qquad \text{for} \qquad m_1 \neq m_2 . \qquad (19.45)$$

Note that you obtained this relation earlier in (16.9) in the derivation of the residue theorem.

Expression (19.45) implies that the modes $u_{n_1 m_1}(r, \varphi)$ and $u_{n_2 m_2}(r, \varphi)$ are orthogonal when $m_1 \neq m_2$ because the $\varphi$-integral in (19.44) vanishes when $m_1 \neq m_2$. Let us now consider why the different modes of the drum are orthogonal when $m_1$ and $m_2$ are equal to the same integer $m$. In that case (19.44) implies that

$$\int_0^R J_m(k_{n_1}^{(m)}r) J_m(k_{n_2}^{(m)}r) \, r \, dr = 0 \qquad \text{for} \quad n_1 \neq n_2 . \qquad (19.46)$$

Note that we have used here that $k_{n_1}^{(m)} \neq k_{n_2}^{(m)}$ when $n_1 \neq n_2$. This integral defines an orthogonality relation for Bessel functions. Note that both Bessel functions in this relation are of the same degree $m$ but that the wavenumbers in the argument of the Bessel functions differ. Note the resemblance between this expression and the orthogonality relation (19.6) of the modes of the string which can be written as

$$\int_0^{2R} \sin(k_n x) \, \sin(k_m x) \, dx = 0 \qquad \text{for} \quad n \neq m . \qquad (19.47)$$

The presence of the term $r$ in the integral (19.46) comes from the fact that the modes of the drum are orthogonal for the integration over the total area of the drum. In cylindrical coordinates this leads to a factor $r$ in the integration.

**Problem g:** Take you favorite book on mathematical physics and find
   an alternative derivation of the orthogonality relation (19.46) of the
   Bessel functions of the same degree $m$.

Note finally that the modes $u_{n_1 m_1}(r, \varphi)$ and $u_{n_2 m_2}(r, \varphi)$ are orthogonal
when $m_1 \neq m_2$ because the $\varphi$-integral satisfies (19.45), whereas the modes
are orthogonal when $n_1 \neq n_2$ but with the same order $m$ because the $r$-
integral (19.46) vanishes in that case. This implies that the eigenfunctions
of the drum defined in (19.19) satisfy the following orthogonality relation:

$$\int_0^R \int_0^{2\pi} u^*_{n_1 m_1}(r, \varphi) u_{n_2 m_2}(r, \varphi) \, d\varphi \, rdr = C\delta_{n_1 n_2}\delta_{m_1 m_2} , \qquad (19.48)$$

where $\delta_{ij}$ is the Kronecker delta and $C$ is a constant that depends on $n_1$
and $m_1$.

   A similar analysis can be applied to the spherical harmonics $Y_{lm}(\theta, \varphi)$
which are the eigenfunctions of the Helmholtz equation on a spherical surface. You may wonder in that case what the boundary conditions of these
eigenfunctions are because in the step from (19.40) to (19.41) the boundary conditions of the modes have been used. A closed surface, however,
has no boundary. This means that the surface integral in (19.40) vanishes.
This in turn means that the orthogonality relation (19.43) holds despite
the fact that the spherical harmonics do not satisfy one of the boundary
conditions that was used in problem d. Let us now consider the inner prod-
uct of two spherical harmonics on the sphere: $\iint Y^*_{l_1 m_1}(\theta, \varphi) Y_{l_2 m_2}(\theta, \varphi) d\Omega$.

**Problem h:** Show that the $\varphi$-integral in the integration over the sphere
   is of the form $\int_0^{2\pi} e^{i(m_2 - m_1)} d\varphi$ and that this integral is equal to
   $2\pi\delta_{m_1 m_2}$.

This implies that the spherical harmonics are orthogonal when $m_1 \neq m_2$
because of the $\varphi$-integration. We will now continue with the case in which
$m_1 = m_2$, and denote this common value with the single index $m$.

**Problem i:** Use the general orthogonality relation (19.43) to derive that
   the associated Legendre functions satisfy the following orthogonality
   relation:

$$\int_0^\pi P_{l_1}^m(\cos\theta) P_{l_2}^m(\cos\theta) \sin\theta d\theta = 0 \qquad \text{when } l_1 \neq l_2 . \qquad (19.49)$$

   Note the common value of the degree $m$ in the two associated Legendre functions. Also show explicitly that the condition $k_{l_1} \neq k_{l_2}$ is
   equivalent to the condition $l_1 \neq l_2$.

**Problem j:** Use a substitution of variables to show that this orthogonality relation can also be written as

$$\int_{-1}^{1} P_{l_1}^m(x) P_{l_2}^m(x) dx = 0 \qquad \text{when } l_1 \neq l_2 . \tag{19.50}$$

**Problem k:** Find an alternative derivation of this orthogonality relation in the literature.

The result you obtained in problem h implies that the spherical harmonics are orthogonal when $m_1 \neq m_2$ because of the $\varphi$-integration, whereas problem i implies that the spherical harmonics are orthogonal when $l_1 \neq l_2$ because of the $\theta$-integration. This means that the spherical harmonics satisfy the following orthogonality relation:

$$\iint Y_{l_1 m_1}^*(\theta, \varphi) Y_{l_2 m_2}(\theta, \varphi) d\Omega = \delta_{l_1 l_2} \delta_{m_1 m_2} . \tag{19.51}$$

The numerical constant multiplying the delta functions is equal to 1. This is a consequence of the square-root term in (19.31) that pre-multiplies the associated Legendre functions. You should be aware of the fact that when a different convention is used for the normalization of the spherical harmonics a normalization factor appears on the right hand side of the orthogonality relation (19.51) of the spherical harmonics.

## 19.5 Bessel functions behave as decaying cosines

As we have seen in section 19.2 the modes of the circular drum are given by $J_m(kr)e^{im\varphi}$, where the Bessel function satisfies the differential equation (19.16) and where $k$ is a wavenumber chosen in such a way that the displacement at the edge of the drum vanishes. We show in this section that the waves that propagate through the drum have an approximately constant wavelength, but that their amplitude decays with the distance to the center of the drum. The starting point of the analysis is the Bessel equation

$$\frac{d^2 J_m}{dx^2} + \frac{1}{x} \frac{dJ_m}{dx} + \left(1 - \frac{m^2}{x^2}\right) J_m = 0 . \tag{19.16}$$

If the terms $(1/x)dJ_m/dx$ and $m^2/x^2$ were absent in (19.16) the Bessel equation would reduce to the differential equation $d^2 F/dx^2 + F = 0$ whose solutions are given by a superposition of $\cos x$ and $\sin x$. We can therefore expect the Bessel functions to display an oscillatory behavior when $x$ is large.

It follows directly from (19.16) that the term $m^2/x^2$ is relatively small for large values of $x$, specifically when $x \gg m$. However, it is not obvious under which conditions the term $(1/x)dJ_m/dx$ is relatively small. Fortunately this term can be transformed away.

**Problem a:** Write $J_m(x) = x^\alpha g_m(x)$, insert this in the Bessel equation (19.16), and show that the term with the first derivative vanishes when $\alpha = -1/2$ and that the resulting differential equation for $g_m(x)$ is given by

$$\frac{d^2 g_m}{dx^2} + \left(1 - \frac{m^2 - 1/4}{x^2}\right) g_m = 0 . \tag{19.52}$$

Up to this point we have made no approximations. Although we have transformed the first derivative term out of the Bessel equation, we still cannot solve (19.52). However, when $x \gg m$ the term proportional to $1/x^2$ in this expression is relatively small. This means that for large values of $x$ the function $g_m(x)$ satisfies the approximate differential equation $d^2 g_m/dx^2 + g_m \approx 0$.

**Problem b:** Show that the solution of this equation is given by $g_m(x) \approx A \cos(x + \varphi)$, where $A$ and $\varphi$ are constants. Also show that this implies that the Bessel function is approximately given by:

$$J_m(x) \approx A \frac{\cos(x + \varphi)}{\sqrt{x}} . \tag{19.53}$$

This approximation is obtained from a local analysis of the Bessel equation. Since all values of the constants $A$ and $\varphi$ lead to a solution that approximately satisfies the differential equation (19.52), it is not possible to retrieve the precise values of these constants from the analysis in this section. An analysis based on the asymptotic evaluation of the integral representation of the Bessel function [11] shows that:

$$J_m(x) = \sqrt{\frac{2}{\pi x}} \cos\left[x - (2m+1)\frac{\pi}{4}\right] + O(x^{-3/2}) . \tag{19.54}$$

**Problem c:** As a check on the accuracy of this asymptotic expression let us compare the zeroes of this approximation with the zeroes of the Bessel functions as given in table 19.1. In problem h of section 19.2 you found that the separation of the zero-crossings tends to $\pi$ for large values of $x$. Explain this using the approximate expression (19.54). How large must $x$ be for the different values of the degree $m$ so that the error in the spacing of the zero-crossings is less than 0.01?

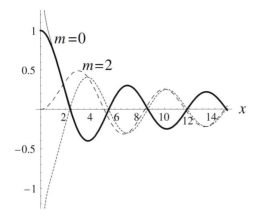

Fig. 19.5. The Bessel functions $J_0(x)$ (thick solid line) and $J_2(x)$ (dashed line) and their approximations (thin solid line and dotted line respectively). Note that for $m = 0$ the approximation is better for smaller values of $x$ than it is for $m = 2$.

The asymptotic expression (19.54) can be compared in figure 19.5 with the Bessel functions $J_m(x)$ for the degrees $m = 0$ and $m = 2$, respectively. For large values of $x$, the approximation (19.54) is very good. Note that for $m = 0$ the approximation is better for smaller values of $x$ than it is for $m = 2$. This is related to the fact that the approximation (19.54) is valid under the condition $(m^2 - 1/4)/x^2 \ll 1$. This requirement is satisfied when $x \gg m$, hence the approximation (19.54) is for a given value of $x$ better for the smaller degrees $m$ than for larger degrees $m$.

Physically, (19.54) states that Bessel functions behave like standing waves with a constant wavelength and which decay with distance as $1/\sqrt{kr}$. (Here it is used that the modes are given by the Bessel functions with argument $x = kr$.) How can we explain this decay of the amplitude with distance? First let us note that (19.54) expresses the Bessel function in a cosine, hence this is a representation of the Bessel function as a standing wave. However, using the relation $\cos x = \left(e^{ix} + e^{-ix}\right)/2$ the Bessel function can be written as two travelling waves that depend on the distance $r$ as $e^{\pm ikr}/\sqrt{kr}$ and that interfere to give the standing wave pattern of the Bessel function. Now let us consider a propagating wave $A(r)e^{ikr}$ in two dimensions; in this expression $A(r)$ is an amplitude that is at this point unknown. The energy of the wave varies with the square of the wave field, and thus depends on $|A(r)|^2$. The energy current therefore also varies as $|A(r)|^2$. Consider an outgoing wave as shown in figure 19.6. The total energy flux through a ring of radius $r$ is given by the energy current times the circumference of the ring, which means that the flux is equal to $2\pi r\,|A(r)|^2$. Since energy is conserved, this total energy flux is the same for all values of $r$, which means that $2\pi r\,|A(r)|^2 = \text{constant}$.

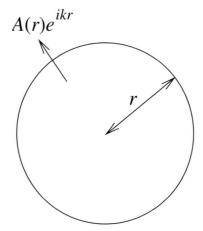

Fig. 19.6.   An expanding wavefront with radius $r$.

**Problem d:** Show that this implies that $A(r) \sim 1/\sqrt{r}$ .

This is the same dependence on distance as the $1/\sqrt{x}$ decay of the approximation (19.54) of the Bessel function. This means that the decay of the Bessel function with distance is dictated by the requirement of energy conservation.

### 19.6 Legendre functions behave as decaying cosines

The technique used in the previous section for the approximation of the Bessel function can also be applied to spherical harmonics. We show in this section that the spherical harmonics behave asymptotically as standing waves on a sphere with an amplitude decay that is determined by the condition that energy is conserved. The spherical harmonics are proportional to the associated Legendre functions with argument $\cos\theta$. The starting point of our analysis therefore is the differential equation for $P_l^m(\cos\theta)$ that was derived in section 19.3:

$$\frac{1}{\sin\theta}\frac{d}{d\theta}\left[\sin\theta\frac{dP_l^m(\cos\theta)}{d\theta}\right] + \left[l\,(l+1) - \frac{m^2}{\sin^2\theta}\right]P_l^m(\cos\theta) = 0 \,. \quad (19.27)$$

Let us assume that we have a source at the north pole, where $\theta = 0$. Far away from the source, the term $m^2/\sin^2\theta$ in the last term on the left hand side is much smaller than the constant $l\,(l+1)$.

**Problem a:** Show that the words 'far away from the source' stand for the requirement

$$\sin\theta \gg \frac{m}{\sqrt{l\,(l+1)}} \,, \quad (19.55)$$

and show that this implies that the approximation that we derive breaks down near the north pole as well as near the south pole of the employed system of spherical coordinates. In addition, the asymptotic expressions that we derive are most accurate for large values of the angular order $l$ and small values of the degree $m$.

**Problem b:** Just as in the previous section we can transform the first derivative in the differential equation (19.27) away: here this can be achieved by writing $P_l^m(\cos\theta) = (\sin\theta)^\alpha \, g_l^m(\theta)$. Insert this substitution into the differential equation (19.27) and show that the first derivative $dg_l^m/d\theta$ disappears when $\alpha = -1/2$ and that the resulting differential equation for $g_l^m(\theta)$ is given by:

$$\frac{d^2 g_l^m}{d\theta^2} + \left[\left(l + \frac{1}{2}\right)^2 - \frac{m^2 - 1/4}{\sin^2\theta}\right] g_l^m(\theta) = 0 . \tag{19.56}$$

It is interesting to note the resemblance of this equation to the corresponding expression (19.52) in the analysis of the Bessel function.

**Problem c:** If the term $(m^2 - 1/4)/\sin^2\theta$ were absent, this equation would be simple to solve. Show that this term is small compared to the constant $(l + \frac{1}{2})^2$ when the requirement (19.55) is satisfied.

**Problem d:** Show that under this condition the associated Legendre functions satisfy the following approximation:

$$P_l^m(\cos\theta) \approx A \frac{\cos\left[\left(l + \frac{1}{2}\right)\theta + \gamma\right]}{\sqrt{\sin\theta}} , \tag{19.57}$$

where $A$ and $\gamma$ are constants.

Just as in the previous section the constants $A$ and $\gamma$ cannot be obtained from this analysis because (19.57) satisfies the approximate differential equation for *any* values of these constant. As shown in (2.5.58) of ref. [29] the asymptotic relation of the associated Legendre functions is given by:

$$P_l^m(\cos\theta) \approx (-l)^m \sqrt{\frac{2}{\pi l \sin\theta}} \cos\left[\left(l + \frac{1}{2}\right)\theta - (2m + 1)\frac{\pi}{4}\right] + O(l^{-3/2}) . \tag{19.58}$$

Just like the Bessel functions the spherical harmonics behave like a standing wave given by a cosine that is multiplied by a factor $1/\sqrt{\sin\theta}$ which modulates the amplitude.

**Problem e:** Use the same reasoning as you used in problem d of section 19.5 to explain that this amplitude decrease follows from the requirement of energy conservation. In doing so you may find figure 19.7 helpful.

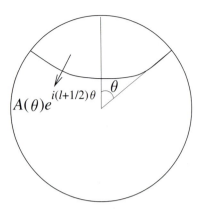

Fig. 19.7.   An expanding wavefront on a spherical surface at a distance $\theta$ from the source.

**Problem f:** Deduce from (19.58) that the wavelength of the associated Legendre functions measured in radians is given by $2\pi/(l + \frac{1}{2})$.

This last result can be used to find the number of oscillations in the spherical harmonics when one moves around the globe once. For simplicity we consider here the case of a spherical harmonic $Y_l^0(\theta, \varphi)$ for degree $m = 0$. When one goes from the north pole to the south pole, the angle $\theta$ increases from 0 to $\pi$. The number of oscillations which fit in this interval is given by $\pi/wavelength$, and according to problem f this number is equal to $\pi/[2\pi/(l + \frac{1}{2})] = (l + \frac{1}{2})/2$. This is the number of wavelengths that fit on half the globe. When one returns from the south pole to the north pole one encounters another $(l + \frac{1}{2})$ oscillations. This means that the total number of waves that fit around the globe is given by $(l + \frac{1}{2})$. It may surprise you that the number of oscillations that one encounters making one loop around the globe is not an integer. One would expect that the requirement of constructive interference would dictate that an integer number of wavelengths should 'fit' in this interval. The reason why the total number of oscillations is $(l + \frac{1}{2})$ rather than the integer $l$ is that near the north and south poles the asymptotic approximation (19.58) breaks down; this follows from the requirement (19.55).

The fact that $(l + \frac{1}{2})$ rather than $l$ oscillations fit on the globe has a profound effect in quantum mechanics. In the first attempts to explain the line spectra of light emitted by atoms, Bohr postulated that an integer number of waves has to fit on a sphere: this can be expressed as $\oint k \, ds = 2\pi n$, where $k$ is the local wave-number. This condition could not explain the observed spectra of light emitted by atoms. However, the arguments in this section imply that the number of wavelengths that fit on a sphere should be given by the requirement

$$\oint kds = 2\pi \left(n + \frac{1}{2}\right). \tag{19.59}$$

This is the *Bohr–Sommerfeld quantization rule*, and was the earliest result in quantum mechanics which provided an explanation of the line spectra of light emitted by atoms. More details on this issue and the cause of the factor $\frac{1}{2}$ in the quantization rule can be found in the refs. [86] and [16]. The effect of this factor on the Earth's normal modes is discussed in a pictorial way by Dahlen and Henson [24].

The asymptotic expression (19.58) can give a useful insight into the relation between modes and travelling waves on a sphere. Let us first return to the modes on the string, which according to (19.5) are given by $\sin(k_n x)$. For simplicity, we will leave out normalization constants in the arguments. The wave motion associated with this mode is given by the real part of $\sin(k_n x)e^{-i\omega_n t}$, with $\omega_n = k_n/c$. These modes therefore denote a standing wave. However, using the decomposition $\sin(k_n x) = (e^{ik_n x} - e^{-ik_n x})/2i$, the mode can in the time domain also be seen as a superposition of two waves $e^{i(k_n x - \omega_n t)}$ and $e^{-i(k_n x + \omega_n t)}$. These are two travelling waves which move in opposite directions.

**Problem g:** Suppose we excite a string at the left hand side at $x = 0$. We know we can account for the motion of the string as a superposition of standing waves $\sin(k_n x)$. However, we can consider these modes to exist also as a superposition of waves $e^{\pm i k_n x}$ which move in opposite directions. The wave $e^{ik_n x}$ moves away from the source at $x = 0$. However the wave $e^{-ik_n x}$ moves *towards* the source at $x = 0$. Give a physical explanation of why in the string travelling waves also move *towards* the source.

On a sphere the situation is completely analogous. The modes can be written according to (19.58) as standing waves

$$\cos\left[\left(l + \frac{1}{2}\right)\theta - (2m + 1)\frac{\pi}{4}\right]/\sqrt{\sin\theta}$$

on the sphere. However, using the relation $\cos x = \left(e^{ix} + e^{-ix}\right)/2$ the modes can also be seen as a superposition of travelling waves $e^{i(l+\frac{1}{2})\theta}/\sqrt{\sin\theta}$ and $e^{-i(l+\frac{1}{2})\theta}/\sqrt{\sin\theta}$ on the sphere.

**Problem h:** Explain why the first wave travels away from the north pole while the second wave travels towards the north pole.

**Problem i:** Suppose that the waves are excited by a source at the north pole. According to the last problem the motion of the sphere can alternatively be seen as a superposition of standing waves or of travelling waves. The travelling wave $e^{i\left(l+\frac{1}{2}\right)\theta}/\sqrt{\sin\theta}$ moves away from the source. Explain physically why there is also a travelling wave $e^{-i\left(l+\frac{1}{2}\right)\theta}/\sqrt{\sin\theta}$ moving *towards* the source.

These results imply that the oscillations of the Earth can be seen as either a superposition of normal modes, or a superposition of waves that travel along the Earth's surface in opposite directions. The waves that travel along the Earth's surface are called *surface waves*. The relation between normal modes and surface waves is treated in more detail by Dahlen [23] and by Snieder and Nolet [84].

## 19.7 Normal modes and the Green's function

In section 12.6 we analyzed the normal modes of a system of three coupled masses. This system had three normal modes, and each mode could be characterized by a vector $\hat{\mathbf{v}}^{(n)}$ with the displacement of the three masses and by an eigenfrequency $\omega_n$. The response of the system to a force $\mathbf{F}$ acting on the three masses with time dependence $e^{-i\omega t}$ was derived to be:

$$\mathbf{x} = \frac{1}{m}\sum_{n=1}^{3}\frac{\hat{\mathbf{v}}^{(n)}\left(\hat{\mathbf{v}}^{(n)}\cdot\mathbf{F}\right)}{\left(\omega_n^2 - \omega^2\right)} . \tag{12.67}$$

This means that the Green's function of this system is given by the following dyad:

$$\mathbf{G} = \frac{1}{2\pi m}\sum_{n=1}^{3}\frac{\hat{\mathbf{v}}^{(n)}\hat{\mathbf{v}}^{(n)T}}{\left(\omega_n^2 - \omega^2\right)} . \tag{19.60}$$

The factor $1/2\pi$ is due to the fact that a delta-function force $f(t) = \delta(t)$ in the time domain corresponds with the Fourier transform (14.43) to $F(\omega) = 1/2\pi$ in the frequency domain. In this section we derive the Green's function for a general oscillating system that can be continuous. An important example is the Earth, which is a body that has well-defined normal modes and where the displacement is a continuous function of the space coordinates.

We consider a system that satisfies the following equation of motion:

$$\rho\ddot{u} + Hu = F . \tag{19.61}$$

The field $u$ can be either a scalar field or a vector field. The operator $H$ at this point is very general, the only requirement that we impose is that this operator is Hermitian, which means that we require that

$$(f \cdot Hg) = (Hf \cdot g) , \tag{19.62}$$

where the inner product is defined as $(f \cdot h) \equiv \int f^* g \, dV$. In the frequency domain, the equation of motion is given by

$$-\rho \omega^2 u + H u = F(\omega) . \qquad (19.63)$$

Let the normal modes of the system be denoted by $u^{(n)}$; the normal modes describe the oscillations of the system in the absence of any external force. The normal modes therefore satisfy the following expression

$$H u^{(n)} = \rho \omega_n^2 u^{(n)} , \qquad (19.64)$$

where $\omega_n$ is the eigenfrequency of this mode.

**Problem a:** Take the inner product of this expression with a mode $u^{(m)}$, and use the fact that $H$ is Hermitian to derive that

$$\left( \omega_n^2 - \omega_m^{*2} \right) \left( u^{(m)} \cdot \rho u^{(n)} \right) = 0 . \qquad (19.65)$$

Note the resemblance of this expression to (19.41) for the modes of a system that obeys the Helmholtz equation.

**Problem b:** Just like in section 19.4 one can show that the eigenfrequencies are real by setting $m = n$, and one can derive that different modes are orthogonal with respect to the following inner product:

$$\left( u^{(m)} \cdot \rho u^{(n)} \right) = \delta_{nm} \qquad \text{for} \quad \omega_m \neq \omega_n . \qquad (19.66)$$

Use (19.65) to give a proof of this orthogonality relation.

Note the presence of the density term $\rho$ in this inner product.

In the analysis of this section it was crucial for the operator $H$ to be Hermitian. This property of $H$ has two important implications: (1) the eigenvalues of $H$ are real and (2) the eigenfunctions are orthogonal. It is for this reason that it is crucial to establish whether the operator $H$ is Hermitian or not. In general, the operator of a dynamical system is Hermitian when the system is invariant for time-reversal. This means that the equations are invariant when one lets the clock run backward, or mathematically when one replaces $t$ by $-t$. Dissipation in general breaks the symmetry for time-reversal. It was shown in detail by Dahlen and Tromp [25] that attenuation in the Earth makes the eigenfrequencies of the Earth complex and that the normal modes of an attenuating Earth do not satisfy the orthogonality relation (19.66).

Let us now return to the inhomogeneous problem (19.63) where an external force $F(\omega)$ is present. Assuming that the normal modes form a

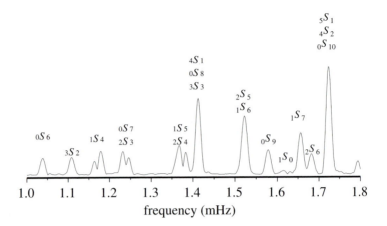

Fig. 19.8. Amplitude spectrum of the vertical component of the ground motion at a seismic station in Tucson, Arizona after the 9 June 1994 Bolivian earthquake. The numbers $_nS_l$ denote the different normal modes of the Earth. Figure courtesy of Arwen Deuss.

complete set, the response to this force can be written as a sum of normal modes:

$$u = \sum_n c_n u^{(n)} , \tag{19.67}$$

where the $c_n$ are unknown coefficients.

**Problem c:** Find these coefficients by inserting (19.67) into the equation of motion (19.63) and by taking the inner product of the result with a mode $u^{(m)}$ to derive that

$$c_m = \frac{\left(u^{(m)} \cdot F\right)}{\omega_m^2 - \omega^2} . \tag{19.68}$$

This means that the response of the system can be written as:

$$u = \sum_n \frac{u^{(n)} \left(u^{(n)} \cdot F\right)}{\omega_n^2 - \omega^2} . \tag{19.69}$$

Note the resemblance of this expression to (12.67) for a system of three masses. The main difference is that the derivation in this section is also valid for continuous vibrating systems such as the Earth.

It is instructive to rewrite this expression taking the dependence of the space coordinates explicitly into account:

$$u(\mathbf{r}) = \sum_n \frac{u^{(n)}(\mathbf{r}) \int u^{*(n)}(\mathbf{r}')F(\mathbf{r}')dV'}{\omega_n^2 - \omega^2} . \tag{19.70}$$

It follows from this expression that the Green's function is given by

$$G(\mathbf{r}, \mathbf{r}', \omega) = \frac{1}{2\pi} \sum_n \frac{u^{(n)}(\mathbf{r}) u^{*(n)}(\mathbf{r}')}{\omega_n^2 - \omega^2} . \qquad (19.71)$$

When the mode is a vector, one should take the transpose of the mode $u^{*(n)}(\mathbf{r}')$. Note the similarity between this expression for the Green's function of a continuous medium and the Green's function (19.60) for a discrete system. In this sense, the Earth behaves in the same way as a tri-atomic molecule. For both systems, the dyadic representation of the Green's function provides a compact way to account for the response of the system to external forces.

Note that the response is strongest when the frequency $\omega$ of the external force is close to one of the eigenfrequencies $\omega_n$ of the system. This implies for example for the Earth that modes with a frequency close to the frequency of the external forcing are most strongly excited. If we jump up and down with a frequency of 1 Hz, we excite the Earth's fundamental mode with a period of about 1 hour only very weakly. In addition, a mode is most effectively excited when the inner product of the forcing $F(\mathbf{r}')$ in (19.70) is maximal. This means that a mode is most strongly excited when the spatial distribution of the force equals the displacement $u^{(n)}(\mathbf{r}')$ of the mode.

**Problem d:** Show that a mode is not excited when the force acts only at one of the nodal lines of that mode.

The normal modes of the Earth leave a clear imprint of the motion of the Earth after a strong earthquake. The ground motion after the 9 June 1994 earthquake in Bolivia was recorded in Tucson, Arizona. Figure 19.8 shows the amplitude spectrum of the vertical component of the ground motion after the earthquake. Note that the frequency is given in units of millihertz because the Earth oscillates slowly. The peaks in the amplitude spectrum correspond to the normal modes of the Earth. These peaks are described by the function $1/(\omega_n^2 - \omega^2)$ in (19.71). In reality the detailed structure of these resonances is also affected by the attenuation in the Earth and by a number of factors that perturb the Earth's normal modes.

As a next step we consider the Green's function in the time domain. This function follows by applying the Fourier transform (14.42) to the Green's function (19.71).

**Problem e:** Show that this gives:

$$G(\mathbf{r}, \mathbf{r}', t) = \frac{1}{2\pi} \sum_n u^{(n)}(\mathbf{r}) u^{*(n)}(\mathbf{r}') \int_{-\infty}^{\infty} \frac{e^{-i\omega t}}{\omega_n^2 - \omega^2} d\omega . \qquad (19.72)$$

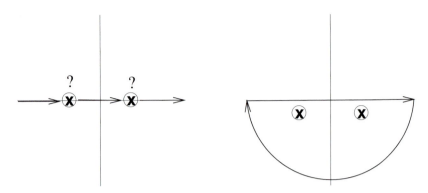

Fig. 19.9. The location of the poles and the integration path in the complex $\omega$-plane. The poles are indicated with a cross. The left hand panel shows the original situation where the poles are located on the integration path at location $\pm\omega_n$. The right hand panel shows the location of the poles when a slight anelastic damping is present.

The integrand is singular at the frequencies $\omega = \pm\omega_n$ of the normal modes. These singularities are located on the integration path, as shown in the left hand panel of figure 19.9. At the singularity at $\omega = \omega_n$ the integrand behaves as $1/[2\omega_n(\omega - \omega_n)]$. The contribution of these singularities is poorly defined because the integral $\int 1/(\omega - \omega_n) \, d\omega$ is not defined.

   This situation is comparable to the treatment in section 16.4 of the response of a particle in syrup to an external forcing. When this particle was subjected to a damping $\beta$, the integrand in the Fourier transform to the time domain had a singularity in the lower half-plane. This gave a causal response; as shown in (16.35) the response was only different from zero at times *later* than the time at which the forcing was applied. This suggests that we can obtain a well-defined causal response of the Green's function (19.72) when we introduce a slight damping. This damping breaks the invariance of the problem for time-reversal, and is responsible for a causal response. At the end of the calculation we can let the damping parameter go to zero. Damping can be introduced by giving the eigenfrequencies of the normal modes a small negative imaginary component: $\pm\omega_n \to \pm\omega_n - i\eta$, where $\eta$ is a small positive number.

**Problem f:** The time dependence of the oscillation of a normal mode is given by $e^{-i\omega_n t}$. Show that with this replacement the modes decay with a decay time that is given by

$$\tau = 1/\eta . \tag{19.73}$$

This last property means that when we ultimately set $\eta = 0$ the decay time becomes infinite: in other words, the modes are not attenuated in that limit.

With the replacement $\pm\omega_n \rightarrow \pm\omega_n - i\eta$ the poles that are associated with the normal modes are located in the lower $\omega$-plane: this situation is shown in figure 19.9. Now that the singularities are moved from the integration path, the theory of complex integration can be used to evaluate the resulting integral.

**Problem g:** Use the theory of contour integration as treated in chapter 16 to derive that the Green's function is given in the time domain by:

$$G(\mathbf{r}, \mathbf{r}', t) = \begin{cases} 0 & \text{for } t < 0 \\ \displaystyle\sum_n \frac{u^{(n)}(\mathbf{r})u^{*(n)}(\mathbf{r}')}{\omega_n} \sin\omega_n t & \text{for } t > 0 \end{cases} \qquad (19.74)$$

Hint: use the same steps as in the derivation of the function (16.35) and let the damping parameter $\eta$ go to zero at the end of the integration.

This result gives a causal response because the Green's function is only nonzero at times $t > 0$, which is later than the time $t = 0$ when the delta-function forcing is nonzero. The total response is given as a sum over all the modes. Each mode leads to a time signal $\sin(\omega_n t)$ in the modal sum: this is a periodic oscillation with the frequency $\omega_n$ of the mode. The singularities in the integrand of the Green's function (19.72) at the pole positions $\omega = \pm\omega_n$ are thus associated in the time domain with a harmonic oscillation with angular frequency $\omega_n$. Note that the Green's function is continuous at the time $t = 0$ of excitation. It is interesting to compare this Green's function with the Green's function (17.21) for the girl on the swing. The only differences are that in (19.74) modes are present in the Green's function and that a summation over the modes is carried out. This difference in due the fact that the harmonic oscillator has only one mode and that this mode of oscillation does not have a spatial extent.

**Problem h:** Use the Green's function (19.74) to derive that the response of the system to a force $F(\mathbf{r}, t)$ is given by:

$$u(\mathbf{r}, t) = \sum_n \frac{1}{\omega_n} u^{(n)}(\mathbf{r}) \int \int_{-\infty}^{t} u^{*(n)}(\mathbf{r}') \sin\omega_n (t - t') F(\mathbf{r}', t') \, dt' dV'.$$

$$(19.75)$$

Justify the integration limit in the $t'$-integration.

The results in this section imply that the total Green's function of a system is known once the normal modes are known. The total response

can then be obtained by summing the contribution of each normal mode
to the total response. This technique is called *normal-mode summation*,
and is often used to obtain the low-frequency response of the Earth to an
excitation [25] [27]. However, in the seismological literature one usually
treats a source signal that is given by a step function at $t = 0$ rather
than a delta function because this is a more accurate description of the
slip on a fault during an earthquake [3]. This leads to a time dependence
$(1 - \cos(\omega_n t))$ rather than the time dependence $\sin(\omega_n t)$ in the response
(19.74) to a delta-function excitation.

### 19.8 Guided waves in a low-velocity channel

In this section we consider a system that strictly speaking does not have
normal modes, but that can support solutions that behave like travelling
waves in one direction and as modes in another direction. The waves in
such a system propagate as *guided waves*. Consider a system in two dimensions ($x$ and $z$), where the velocity depends only on the $z$-coordinate.
We assume that the wave-field satisfies the Helmholtz equation (19.1) in
the frequency domain:

$$\nabla^2 u + \frac{\omega^2}{c^2(z)} u = 0 . \tag{19.76}$$

In this section we consider a simple model of a layer of thickness $H$ that
extends from $z = 0$ to $z = H$ in which the velocity is given by $c_1$. This
layer is embedded in a medium with a constant velocity $c_0$. The geometry
of the problem is shown in figure 19.10. Since the system is invariant in
the $x$-direction, the problem can be simplified by a Fourier transform over
the $x$-coordinate:

$$u(x, z) = \int_{-\infty}^{\infty} U(k, z) e^{ikx} dk . \tag{19.77}$$

**Problem a:** Show that $U(k, z)$ satisfies the following ordinary differential equation:

$$\frac{d^2 U}{dz^2} + \left[ \frac{\omega^2}{c(z)^2} - k^2 \right] U = 0 . \tag{19.78}$$

It is important to note at this point that the frequency $\omega$ is a fixed constant, and that according to (19.77) the variable $k$ is an integration variable that assumes all values in the integration (19.77). For this reason one
should not at this point use the relation $k = \omega/c(z)$.

Now consider the special case of the model shown in figure 19.10. We
require that the waves outside the layer move away from the layer.

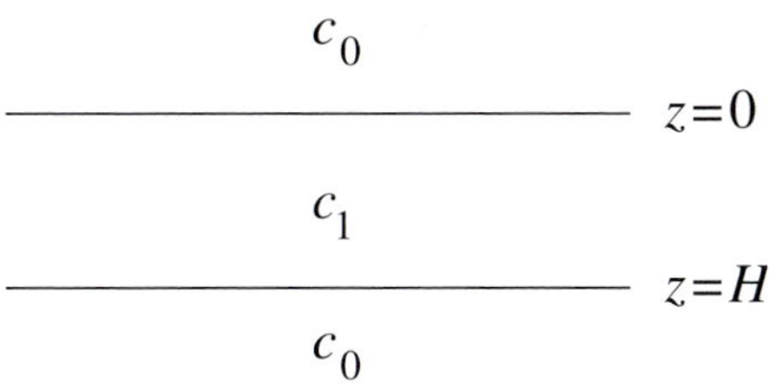

$$c_0$$
$$\overline{\hspace{5cm}} \quad z=0$$
$$c_1$$
$$\overline{\hspace{5cm}} \quad z=H$$
$$c_0$$

Fig. 19.10. Geometry of the model of a single layer sandwiched between two homogeneous half-spaces.

**Problem b:** Show that this implies that the solution for $z < 0$ is given by $Ae^{-ik_0 z}$ and the solution for $z > H$ is given by $Be^{+ik_0 z}$ where $A$ and $B$ are unknown integration constants and where $k_0$ is given by

$$k_0 = \sqrt{\frac{\omega^2}{c_0^2} - k^2}\ . \tag{19.79}$$

**Problem c:** Show that within the layer the wave field is given by $C\cos(k_1 z) + D\sin(k_1 z)$ with $C$ and $D$ integration constants and $k_1$ given by

$$k_1 = \sqrt{\frac{\omega^2}{c_1^2} - k^2}\ . \tag{19.80}$$

The solution in the three regions of space therefore takes the following form:

$$U(k,z) = \begin{cases} Ae^{-ik_0 z} & \text{for} & z < 0 \\ C\cos(k_1 z) + D\sin(k_1 z) & \text{for} & 0 < z < H\ . \\ Be^{+ik_0 z} & \text{for} & z > H \end{cases} \tag{19.81}$$

We now have the general form of the solution within the layer and the two half-spaces on either side of the layer. Boundary conditions are needed to find the integration constants $A$, $B$, $C$ and $D$. For this system both $U$ and $dU/dz$ are continuous at $z = 0$ and $z = H$.

**Problem d:** Use the results of problem b and problem c to show that these requirements impose the following constraints on the integration constants:

$$\left.\begin{array}{l} A - C = 0\ , \\ ik_0 A + k_1 D = 0\ , \\ -Be^{ik_0 H} + C\cos k_1 H + D\sin k_1 H = 0\ , \\ ik_0 Be^{ik_0 H} + k_1 C\cos k_1 H - k_1 D\sin k_1 H = 0\ . \end{array}\right\} \tag{19.82}$$

This is a linear system of four equations for the four unknowns $A$, $B$, $C$ and $D$. Note that this is a homogeneous system of equations, because the right hand sides vanish. Such a homogeneous system of equations only has nonzero solutions when the determinant of the system of equations vanishes.

**Problem e:** Show that this requirement leads to the following condition:

$$\tan(k_1 H) = \frac{-2ik_0 k_1}{k_1^2 + k_o^2} . \qquad (19.83)$$

This equation is implicitly an equation for the wave-number $k$, because according to (19.79) and (19.80) both $k_0$ and $k_1$ are functions of the wavenumber $k$. Equation (19.83) implies that the system can only support waves when the wavenumber $k$ is such that expression (19.83) is satisfied. The system, strictly speaking, does not have normal modes, because the waves propagate in the $x$-direction. However, in the $z$-direction the waves only 'fit' in the layer for very specific values of the wavenumber $k$. These waves are called 'guided waves' because they propagate along the layer with a well-defined phase velocity that follows from the relation $c(\omega) = \omega/k$. Be careful not to confuse this phase velocity $c(\omega)$ with the velocities $c_1$ and $c_0$ in the layer and the half-spaces outside the layer. At this point we do not know yet what the phase velocities of the guided waves are.

The phase velocity follows from expression (19.83) because this expression is implicitly an equation for the wavenumber $k$. At this point we consider the case of a low-velocity layer, i.e. we assume that $c_1 < c_0$. In this case $1/c_0 < 1/c_1$. We look for guided waves with a wavenumber in the following interval: $\omega/c_0 < k < \omega/c_1$.

**Problem f:** Show that in that case $k_1$ is real and $k_0$ is purely imaginary. Write $k_0 = i\kappa_0$ and show that

$$\kappa_0 = \sqrt{k^2 - \frac{\omega^2}{c_0^2}} . \qquad (19.84)$$

**Problem g:** Show that the solution decays exponentially away from the low-velocity channel both in the half-space $z < 0$ and the half-space $z > H$.

The fact that the waves decay exponentially with the distance to the plate means that the guided waves are trapped near the low-velocity layer. Waves that decay exponentially are called *evanescent waves*.

**Problem h:** Use (19.83) to show that the wavenumber of the guided
waves satisfies the following relation:

$$\tan \sqrt{\frac{\omega^2}{c_1^2} - k^2} H = \frac{2\sqrt{k^2 - \frac{\omega^2}{c_0^2}}\sqrt{\frac{\omega^2}{c_1^2} - k^2}}{\omega^2 \left( \frac{1}{c_1^2} - \frac{1}{c_0^2} \right)}. \tag{19.85}$$

For a fixed value of $\omega$ this expression constitutes a constraint on the
wavenumber $k$ of the guided waves. Unfortunately, it is not possible to
solve this equation for $k$ in closed form. Such an equation is called a
*transcendental equation*.

**Problem j:** Make a sketch of both the left hand side and the right hand
side of (19.85) as a function of $k$. Show that the two curves have a
finite number of intersection points.

These intersection points correspond to the $k$-values of the guided waves.
The corresponding phase velocity $c = \omega/k$ in general depends on the
frequency $\omega$. This means that these guided waves are *dispersive,* which
means that the different frequency components travel with a different
phase velocity. It is for this reason that (19.83) is called the *dispersion
relation.*

Dispersive waves occur in many different situations. When electromagnetic waves propagate between plates or in a layered structure, guided
waves result [42]. The atmosphere, and most importantly the ionosphere,
is an excellent waveguide for electromagnetic waves [36]. This leads to a
large variety of electromagnetic guided waves in the upper atmosphere
with exotic names such as 'pearls', 'whistlers', 'tweaks', 'hydromagnetic
howling' and 'serpentine emissions'; colorful names associated with the
sounds these phenomena would make if they were audible, or with the
patterns they generate in frequency–time diagrams. These perturbations
are excited for example by the electromagnetic fields generated by lightning. Guided waves play a crucial role in telecommunication, because light
propagates through optical fibers as guided waves [48]. The fact that these
waves are guided prohibits the light from propagating out of the fiber, and
this allows for the transmission of light signals over extremely large distances.

In the Earth the wave velocity increases rapidly with depth. Elastic
waves can be guided near the Earth's surface and the different modes
are called 'Rayleigh waves' and 'Love waves' [3]. These surface waves in
the Earth are a prime tool for mapping the shear velocity within the
Earth [82].

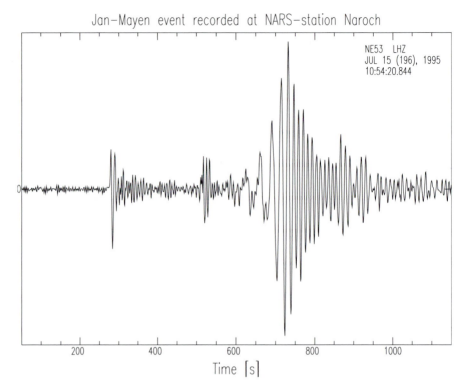

Fig. 19.11. Vertical component of the ground motion at a seismic station in Naroch (Belarus) after an earthquake at Jan-Mayen Island. This station is part of the Network of Autonomously Recording Seismographs (NARS) which is operated by Utrecht University.

Since the surface waves in the Earth are trapped near the Earth's surface, they effectively propagate in two dimensions rather than in three dimensions. The surface waves therefore suffer less from geometrical spreading than the body waves that propagate through the interior of the Earth. For this reason, it is the surface waves that do most damage after an earthquake. This is illustrated in figure 19.11 which shows the vertical displacement at a seismic station in Naroch (Belarus) after an earthquake at Jan-Mayen Island. Around $t = 300$ s and $t = 520$ s impulsive waves arrive: these are the body waves that travel through the interior of the Earth. The wave with the largest amplitude that arrives between $t = 650$ s and $t = 900$ s is the surface wave that is guided along the Earth's surface. Note that the waves that arrive around $t = 700$ s have a lower frequency content than the waves that arrive later, around $t = 850$ s. This is due to the fact that the group velocity of the low-frequency components of the surface wave is higher than the group velocity of the high-frequency components. Hence it is ultimately the dispersion of the Rayleigh waves that causes the change in the apparent frequency of the surface wave arrival.

## 19.9 Leaky modes

The guided waves in the previous section decay exponentially with the distance to the low-velocity layer. Intuitively, the fact that the waves are confined to a region near a low-velocity layer can be understood as follows. Waves are refracted from regions of high velocity to a region of low velocity. This means that the waves that stray out of the low-velocity channel are refracted back in the channel. Effectively this traps the waves in the vicinity of the channel. This explanation suggests that for a high-velocity channel the waves are refracted *away* from the channel. The resulting wave pattern then corresponds to waves that preferentially move away from the high-velocity layer. For this reason we consider in this section the waves that propagate through the system shown in figure 19.10 but now we consider the case of a high-velocity layer where $c_1 > c_0$.

In this case, $1/c_1 < 1/c_0$, and we consider waves with a wavenumber that is confined to the following interval: $\omega/c_1 < k < \omega/c_0$.

**Problem a:** Show that in this case the wavenumber $k_1$ is imaginary and that it can be written as $k_1 = i\kappa_1$, with

$$\kappa_1 = \sqrt{k^2 - \frac{\omega^2}{c_1^2}}, \qquad (19.86)$$

and show that the dispersion relation (19.83) is given by:

$$\tan(i\kappa_1 H) = \frac{-2k_0\kappa_1}{\kappa_1^2 - k_o^2}. \qquad (19.87)$$

**Problem b:** Use the relation $\cos x = \left(e^{ix} + e^{-ix}\right)/2$ and the related expression for $\sin x$ to rewrite the dispersion relation (19.87) in the following form:

$$i\tanh(\kappa_1 H) = \frac{-2k_0\kappa_1}{\kappa_1^2 - k_o^2}. \qquad (19.88)$$

In this expression all quantities are real when $k$ is real. The factor $i$ on the left hand side implies that this equation cannot be satisfied for real values of $k$. The only way in which the dispersion relation (19.88) can be satisfied is if $k$ is complex. What does it mean if the wavenumber is complex? Suppose that the dispersion relation is satisfied for a complex wavenumber $k = k_r + ik_i$, with $k_r$ and $k_i$ the real and imaginary parts. In the time domain a solution behaves for a fixed frequency as $U(k, z)e^{i(kx-\omega t)}$. This means that for complex values of the wavenumber the solution behaves as $U(k, z)e^{-k_i x}e^{i(k_r x-\omega t)}$. This is a wave that propagates in the $x$-direction with phase velocity $c = \omega/k_r$ and that decays exponentially with the propagation distance $x$.

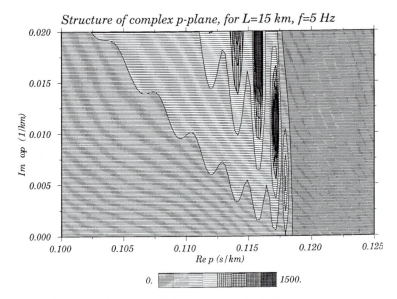

Fig. 19.12. Contour diagram of the function $|F(k)|$ for a high-velocity layer with velocity $c_1 = 8.4$ km/s and a thickness $H = 15$ km which is embedded between two half-spaces with velocity $c_0 = 8$ km/s, for waves with a frequency of 5 Hz. The horizontal axis is given by $k_r/\omega$ and the vertical axis by $k_i$.

The exponential decay of the wave with the propagation distance $x$ is due to the fact that the wave energy refracts out of the high-velocity layer. A different way of understanding this exponential decay is to consider the character of the wave-field outside the layer.

**Problem c:** Show that in the two half-spaces outside the high-velocity layer the waves propagate away from the layer. Hint: analyze the wavenumber $k_0$ in the half-spaces and consider the corresponding solution in these regions.

This means that wave energy is continuously radiated away from the high-velocity layer. The exponential decay of the mode with propagation distance $x$ is thus due to the fact that wave energy continuously leaks out of the layer. For this reason one speaks of *leaky modes* [95]. In the Earth a well-observed leaky mode is the S-PL wave. This is a mode in which a transverse propagating wave in the mantle is coupled to a wave that is trapped in the Earth's crust.

In general there is no simple way to find the complex wavenumber $k$ for which the dispersion relation (19.88) is satisfied. However, the presence of leaky modes can be seen in figure 19.12 where the absolute value of the following function is shown in the complex $k$-plane:

$$F(k) \equiv 1/\left[ i \tanh(\kappa_1 H) + \frac{2k_0\kappa_1}{\kappa_1^2 - k_o^2} \right] . \tag{19.89}$$

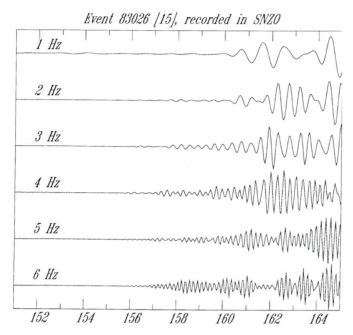

Fig. 19.13. Seismic waves recorded in Wellington after an earthquake in the Tonga-Kermadec subduction zone, from ref. [35]. The different traces correspond to the waves band-pass filtered with a center frequency indicated at each trace. The horizontal axis gives the time since the earthquake occured in units of seconds.

**Problem d:** Show that this function is infinite for the $k$-values that correspond to a leaky mode.

The function $F(k)$ in figure 19.12 is computed for a high-velocity layer with a thickness of 15 km and a velocity of 8.4 km/s that is embedded between two half-spaces with a velocity of 8 km/s. The frequency of the wave is 5 Hz. In this figure, the horizontal axis is given by $\Re e\,(p) = k_r/\omega$ while the vertical axis is given by $\Im m\,(\omega p) = k_i$. The quantity $p$ is called the slowness and is defined as $\Re e\,(p) = k_r/\omega = 1/c(\omega)$. The leaky modes show up in figure 19.12 as a number of localized singularities of the function $F(k)$.

**Problem e:** What is the propagation distance over which the amplitude of the mode with the lowest phase velocity decays with a factor $1/e$?

Leaky modes have been used by Gubbins and Snieder [35] to analyze waves which have propagated along a subduction zone. (A subduction zone is a plate in the Earth that slides downward in the mantle.) By a fortuitous geometry, compressive waves that are excited by earthquakes in

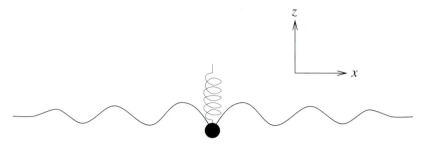

Fig. 19.14.   Geometry of an oscillating mass that is coupled to a spring.

the Tonga-Kermadec region travel to a seismic station in Wellington (New Zealand) and propagate for a large distance through the Tonga-Kermadec subduction zone. At the station in Wellington, a high-frequency wave arrives before the main compressive wave. This can be seen in figure 19.13 where such a seismogram is shown band-pass filtered at different frequencies. It can clearly be seen that waves with a frequency around 6 Hz arrive before waves with a frequency around 1 Hz. This observation can be explained by the propagation of a leaky mode through the subduction zone. The physical reason why the high-frequency components arrive before the lower-frequency components is that the high-frequency waves 'fit' in the high-velocity layer in the subducting plate, whereas the lower-frequency components do not fit in the high-velocity layer and are more influenced by the slower material outside the high-velocity layer. Of course, the energy leaks out of the high-velocity layer so that this arrival is very weak. From the data it could be inferred that in the subduction zone a high-velocity layer with a thickness between 6 and 10 km is present [35].

## 19.10  Radiation damping

Up to this point we have considered systems that are freely oscillating. When such systems are of finite extent, such a system displays undamped free oscillations. In the previous section leaky modes were introduced. In such a system, energy is radiated away, which leads to an exponential decay of waves that propagate through the system. In a similar way, a system that has normal modes when it is isolated from its surroundings can display damped oscillations when it is coupled to the external world.

   As a simple prototype of such a system, consider a mass $m$ that can move in the $z$-direction which is coupled to a spring with spring constant $\kappa$. The mass is attached to a string which is under a tension $T$ and which has a mass $\rho$ per unit length. The system is shown in figure 19.14. The total force acting on the mass is the sum of the force $-\kappa z$ exerted by the spring and the force $F_s$ that is generated by the string:

$$m\ddot{z} + \kappa z = F_s , \qquad (19.90)$$

where $z$ denotes the vertical displacement of the mass. The motion of the waves that propagate in the string is given by the wave equation:

$$u_{xx} - \frac{1}{c^2} u_{tt} = 0 , \qquad (19.91)$$

where $u$ is the displacement of the string in the vertical direction and $c$ is given by

$$c = \sqrt{\frac{T}{\rho}} . \qquad (19.92)$$

Let us first consider the case in which no external force is present and the mass is not coupled to the spring.

**Problem a:** Show that in that case the equation of motion is given by

$$\ddot{z} + \omega_0^2 z = 0 , \qquad (19.93)$$

with $\omega_0$ given by

$$\omega_0 = \sqrt{\frac{\kappa}{m}} . \qquad (19.94)$$

One can say that the mass that is not coupled to the string has one free oscillation with angular frequency $\omega_0$. The fact that the system has one free oscillation is a consequence of the fact that this mass can move only in the vertical direction, hence it has only one degree of freedom.

Before we couple the mass to the string let us first analyze the wave motion in the string in the absence of the mass.

**Problem b:** Show that any function $f(t - x/c)$ satisfies the wave equation (19.91). Show that this function describes a wave that moves in the positive $x$-direction with velocity $c$.

**Problem c:** Show that any function $g(t + x/c)$ satisfies the wave equation (19.91). Show that this function describes a wave that moves in the negative $x$-direction with velocity $c$.

The general solution is a superposition of the rightward and leftward moving waves:

$$u(x, t) = f\left(t - \frac{x}{c}\right) + g\left(t + \frac{x}{c}\right) . \qquad (19.95)$$

This general solution is called the d'Alembert solution.

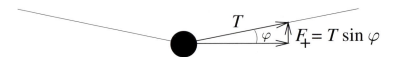

Fig. 19.15.   Sketch of the force exerted by the spring on the mass.

Now we want to describe the motion of the coupled system. Let us first assume that the mass oscillates with a prescribed displacement $z(t)$ and find the waves that this displacement generates in the string. We consider here the situation in which there are no waves moving towards the mass. This means that to the right of the mass the waves can only move rightward and to the left of the mass there are only leftward moving waves.

**Problem d:** Show that this radiation condition implies that the waves in the string are given by:

$$u(x,t) = \begin{cases} f\left(t - \dfrac{x}{c}\right) & \text{for} \quad x > 0 \\[3mm] g\left(t + \dfrac{x}{c}\right) & \text{for} \quad x < 0 \end{cases} \qquad (19.96)$$

**Problem e:** At $x = 0$ the displacement of the mass is the same as the displacement of the string. Show that this implies that $f(t) = g(t) = z(t)$, so that

$$u(x,t) = \begin{cases} z\left(t - \dfrac{x}{c}\right) & \text{for} \quad x > 0 \\[3mm] z\left(t + \dfrac{x}{c}\right) & \text{for} \quad x < 0 \end{cases} \qquad (19.97)$$

Now we have solved the problem of finding the wave motion in the string given the motion of the mass. To complete our description of the system we also need to specify how the motion of the string affects the mass. In other words, we need to find the force $F_s$ in (19.90) given the motion of the string. This force can be derived from figure 19.15. The vertical component $F_+$ of the force acting on the mass from the right hand side of the string is given by $F_+ = T \sin \varphi$, where $T$ is the tension in the string. When the motion in the spring is sufficiently weak we can approximate: $F_+ = T \sin \varphi \approx T \varphi \approx T \tan \varphi \approx T u_x(x = 0^+, t)$. In the last identity we used that the derivative $u_x(x = 0^+, t)$ gives the slope of the string on the right of the point $x = 0$.

**Problem f:** Use a similar reasoning to determine the force acting on the mass from the left hand part of the spring and show that the net force acting on the spring is given by

$$F_s(t) = T\left(u_x(x = 0^+, t) - u_x(x = 0^-, t)\right) , \qquad (19.98)$$

where $u_x(x = 0^-, t)$ is the $x$-derivative of the displacement in the string just to the left of the mass.

**Problem g:** Show that this expression implies that the net force that acts on the mass is equal to the *kink* in the spring at the location of the mass.

You may not feel comfortable with the fact the we used the approximation of a small angle $\varphi$ in the derivation of (19.98). However, keep in mind that the wave equation (19.91) is derived using the same approximation and that this wave equation therefore is only valid for small displacements of the string.

At this point we have assembled all the ingredients for solving the coupled problem.

**Problem h:** Use (19.97) and (19.98) to derive that the force exerted by the spring on the mass is given by

$$F_s(t) = -\frac{2T}{c}\,\dot{z} , \qquad (19.99)$$

and that the motion of the mass is therefore given by:

$$\ddot{z} + \frac{2T}{mc}\dot{z} + \omega_0^2 z = 0 . \qquad (19.100)$$

It is interesting to compare this expression for the motion of the mass that is coupled to the string with the equation of motion (19.93) for the mass that is not coupled to the string. The string leads to a term $(2T/mc)\dot{z}$ in the equation of motion that damps the motion of the mass. How can we explain this damping physically? When the mass moves, the string moves with it at location $x = 0$. Any motion in the string at that point excites waves propagating in the string. This means that the string radiates wave energy away from the mass whenever the mass moves. Since energy is conserved, the energy that is radiated in the string must be withdrawn from the energy of the moving mass. This means that the mass loses energy whenever it moves; this effect is described by the damping term in equation (19.100). This damping process is called *radiation damping*, because it is the radiation of waves that damps the motion of the mass.

The system described in this section is extremely simple. However, it does contain the essential physics of radiation damping. Many systems in physics that display normal modes are not quite isolated from their surroundings. Interactions of the systems with their surroundings often lead to the radiation of energy, and hence to a damping of the oscillation of the system.

One example of such a system is an atom in an excited state. In the absence of external influences such an atom will not emit any light and will not decay. However, when such an atom can interact with electromagnetic fields, it can emit a photon and subsequently decay.

A second example is a charged particle that moves in a synchrotron. In the absence of external fields, such a particle will continue forever in a circular orbit without any change in its speed. In reality, a charged particle is coupled to electromagnetic fields. This has the effect that a charged particle that is accelerated emits electromagnetic radiation, called synchrotron radiation [42]. The radiated energy corresponds to an energy loss of the particle, so that the particle slows down. This is actually the reason why accelerators such as those used at CERN and Fermilab are so large. The acceleration of a particle in a circular orbit with radius $r$ at the given velocity $v$ is given by $v^2/r$. This means that for a fixed velocity $v$ the larger the radius of the orbit is, the smaller the acceleration is, and the weaker the energy loss due to the emission of synchrotron radiation is. This is why one needs huge machines to accelerate tiny particles to an extreme energy.

**Problem i:** The modes in the plate in figure 19.1 are also damped because of radiation damping. What form of radiation is emitted by this oscillating plate?

# 20

## Potential theory

Potential fields play an important role in physics and geophysics because they describe the behavior of gravitational and electric fields as well as a number of other fields. Conversely, measurements of potential fields provide important information about the internal structure of bodies. For example, measurements of the electric potential at the Earth's surface when a current is sent into the Earth give information about the electrical conductivity while measurements of the Earth's gravity field or geoid provide information about the mass distribution within the Earth.

An example of this can be seen in figure 20.1 in which the gravity anomaly over the northern part of the Yucatan peninsula in Mexico is shown [38]. The coast is visible as a thin white line. Note the clear ring structure that is visible in the gravity signal. These rings have led to the discovery of the Chicxulub crater which was caused by the massive impact of a meteorite. Note that the diameter of the impact crater is about 150 km! This crater is presently hidden by thick layers of sediments: at the surface the only apparent imprint of this crater is the presence of underground water-filled caves called 'cenotes' at the outer edge of the crater. It was the measurement of the gravity field that made it possible to find this massive impact crater.

The equation that the gravitational or electrical potential satisfies depends critically on the Laplacian of the potential. As shown in section 5.5 the gravitational field has the mass-density as its source:

$$(\nabla \cdot \mathbf{g}) = -4\pi G\rho \,. \tag{5.18}$$

The gravity field $\mathbf{g}$ is (minus) the gradient of the gravitational potential: $\mathbf{g} = -\nabla V$. This means that the gravitational potential satisfies the following partial differential equation:

$$\nabla^2 V(\mathbf{r}) = 4\pi G\rho \,. \tag{20.1}$$

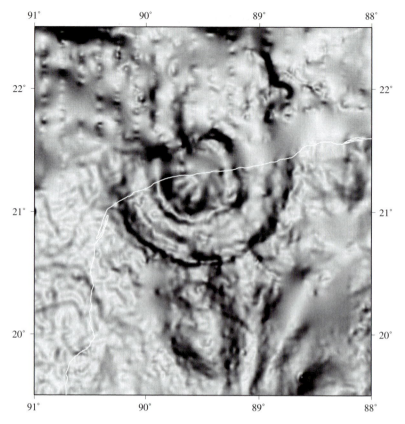

Fig. 20.1.   Gravity field over the Chicxulub impact crater on the northern coast
of Yucatan (Mexico). The coastline is shown by a white line. The numbers along
the vertical and horizontal axes refer to the latitude and longitude respectively.
The magnitude of the horizontal gradient of the Bouguer gravity anomaly is
shown, details can be found in ref. [38]. Courtesy of M. Pilkington and A.R.
Hildebrand.

This equation is called Poisson's equation and is the prototype of the
equations that occur in potential field theory. Note that the mathematical
structures of the equations of the gravitational field and the electrical field
are identical (compare (5.13) and (5.18)), therefore the results derived in
this chapter for the gravitational field can be used directly for the electrical
field as well by replacing the mass-density by the charge-density and by
making the following replacement:

$$\underbrace{4\pi G}_{\text{gravity}} \Leftrightarrow \underbrace{-1/\varepsilon_0}_{\text{electrostatics}} \quad . \tag{20.2}$$

The theory of potential fields is treated in great detail by Blakeley [14].

## 20.1 The Green's function of the gravitational potential

Poisson's equation (20.1) can be solved using a Green's function technique. In essence the derivation of the Green's function yields the well-known result that the gravitational potential for a point-mass $m$ is given by $-Gm/r$. The use of Green's functions was introduced in great detail in chapter 17. The Green's function $G(\mathbf{r}, \mathbf{r}')$ that describes the gravitational potential at location $\mathbf{r}$ generated by a point mass at location $\mathbf{r}'$ satisfies the following differential equation:

$$\nabla^2 G(\mathbf{r}, \mathbf{r}') = \delta\left(\mathbf{r} - \mathbf{r}'\right) . \tag{20.3}$$

Take care not to confuse the Green's function $G(\mathbf{r}, \mathbf{r}')$ with the gravitational constant $G$.

**Problem a:** Show that the solution of (20.1) is:

$$V(\mathbf{r}) = 4\pi G \int G(\mathbf{r}, \mathbf{r}')\rho(\mathbf{r}')dV' . \tag{20.4}$$

**Problem b:** The differential equation (20.3) has translational invariance, and is invariant for rotations. Show that this implies that $G(\mathbf{r}, \mathbf{r}') = G(|\mathbf{r} - \mathbf{r}'|)$. Show by placing the point-mass at the origin by setting $\mathbf{r}' = 0$ that $G(r)$ satisfies

$$\nabla^2 G(r) = \delta\left(\mathbf{r}\right) . \tag{20.5}$$

**Problem c:** Use the expression for the Laplacian in spherical coordinates to show that for $r > 0$ (20.5) is given by

$$\frac{1}{r^2}\frac{\partial}{\partial r}\left(r^2\frac{\partial G(r)}{\partial r}\right) = 0 . \tag{20.6}$$

**Problem d:** Integrate this equation with respect to $r$ to derive that the solution is given by $G(r) = A/r + B$, where $A$ and $B$ are integration constants.

The constant $B$ in the potential does not contribute to the forces that are associated with this potential because $\nabla B = 0$. For this reason the arbitrary constant $B$ can be taken to be equal to zero. The potential is therefore given by

$$G(r) = -\frac{A}{r} . \tag{20.7}$$

**Problem e:** The constant $A$ can be found by integrating (20.5) over a sphere of radius $R$ centered around the origin. Show that Gauss's theorem implies that $\int \nabla^2 G(r)dV = \oint \nabla G \cdot d\mathbf{S}$, use (20.7) in the right hand side of this expression and show that this gives $A = 1/4\pi$. Note that this result is independent of the radius $R$ that you have used.

**Problem f:** Show that the Green's function is given by:

$$G(\mathbf{r}, \mathbf{r}') = -\frac{1}{4\pi}\frac{1}{|\mathbf{r} - \mathbf{r}'|} \ . \tag{20.8}$$

With (20.4) this implies that the gravitational potential is given by:

$$V(\mathbf{r}) = -G \int \frac{\rho(\mathbf{r}')}{|\mathbf{r} - \mathbf{r}'|}dV' \ . \tag{20.9}$$

This general expression is useful for a variety of different purposes, and we will make extensive use of it. By taking the gradient of this expression one obtains the gravitational acceleration $\mathbf{g}$. This acceleration was also derived in (7.5) for the special case of a spherically symmetric mass distribution. Surprisingly it is a nontrivial calculation to derive (7.5) by taking the gradient of (20.9).

## 20.2 Upward continuation in a flat geometry

Suppose that one has a body with variable mass in two dimensions and that the mass-density is only nonzero in the half-space $z < 0$. In this section we determine the gravitational potential $V$ above the half-space when the potential is specified at the plane $z = 0$ that forms the upper boundary of this body. The geometry of this problem is sketched in figure 20.2. This problem is of relevance for the interpretation of gravity measurements taken above the Earth's surface using aircraft or satellites because the first step in this interpretation is to relate the values of the potential at the Earth's surface to the measurements taken above the surface. This process is called *upward continuation*.

Mathematically the problem can be stated this way. Suppose one is given the function $V(x, z = 0)$, what is the function $V(x, z)$? When we know that there is no mass above the surface it follows from (20.1) that the potential satisfies:

$$\nabla^2 V(\mathbf{r}) = 0 \qquad \text{for} \qquad z > 0 \ . \tag{20.10}$$

It is instructive to solve this problem by making a Fourier expansion of the potential in the variable $x$:

$$V(x, z) = \int_{-\infty}^{\infty} v(k, z)e^{ikx}dk \ . \tag{20.11}$$

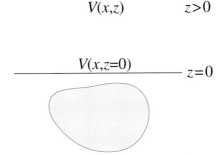

$V(x,z)$          $z>0$

$V(x,z=0)$          $z=0$

Fig. 20.2. Geometry of the upward continuation problem. A mass anomaly (shaded) leaves an imprint on the potential at $z = 0$. The upward continuation problem states how the potential at the surface $z = 0$ is related to the potential $V(x, z)$ at greater height.

**Problem a:** Show that for $z = 0$ the Fourier coefficients can be expressed in terms of the known value of the potential at the edge of the half-space:

$$v(k, z = 0) = \frac{1}{2\pi} \int_{-\infty}^{\infty} V(x, z = 0)e^{-ikx} dx . \qquad (20.12)$$

**Problem b:** Use Poisson's equation (20.10) and the Fourier expansion (20.11) to derive that the Fourier components of the potential satisfy for $z > 0$ the following differential equation:

$$\frac{\partial^2 v(k, z)}{\partial z^2} - k^2 v(k, z) = 0 . \qquad (20.13)$$

**Problem c:** Show that the general solution of this differential equation can be written as $v(k, z) = A(k)e^{+|k|z} + B(k)e^{-|k|z}$ and explain why absolute values of the wavenumber $k$ in the exponent can be taken.

Since the potential must remain finite at great height ($z \to \infty$) the coefficient $A(k)$ must be equal to zero. Setting $z = 0$ shows that $B(k) = v(k, z = 0)$, so that the potential is given by:

$$V(x, z) = \int_{-\infty}^{\infty} v(k, z = 0)e^{ikx}e^{-|k|z} dk . \qquad (20.14)$$

This expression is interesting because it states that the different Fourier components of the potential decay as $e^{-|k|z}$ with height.

**Problem d:** Explain that this implies that the short-wavelength components in the potential field decay must faster with height than the long-wavelength components.

The decrease of the Fourier components with the distance $z$ to the surface is problematic when one wants to infer the mass-density in the body from measurements of the potential or from gravity at a great height above the surface, because the influence of mass perturbations on the gravitational field decays rapidly with height. The measurement of the short-wavelength component of the potential at a great height therefore carries virtually no information about the small-scale details of the density distribution within the Earth. This is the reason why gravity measurements from space are preferably carried out using satellites in low orbits rather than in high orbits. Similarly, for gravity surveys at sea, a gravity meter has been developed that is towed far below the sea surface [106]. The idea is that by towing the gravity meter closer to the sea bed, the gravity signal generated at the sub-surface for short wavelengths suffers less from the exponential decay due to upward continuation.

**Problem e:** Take the gradient of (20.14) to find the vertical component of the gravity field. Use the resulting expression to show that the gravity field $\mathbf{g}$ is less sensitive to the exponential decay due to upward continuation than the potential $V$.

This last result is the reason why satellites in low orbits are used to measure the Earth's gravitational potential and satellites in high orbits are used to measure gravity. In fact, a space-borne gradiometer [74] is presently being developed. This instrument measures the *gradient* of the gravity vector by monitoring the differential motion between two masses in the satellite. Taking the gradient of the gravity leads to another factor of $k$ in the Fourier expansion so that the effects of upward continuation are further reduced.

   We now explicitly express the potential at height $z$ to the potential at the surface $z = 0$.

**Problem f:** Insert (20.12) into (20.14) to show that the upward continuation of the potential is given by:

$$V(x, z) = \int_{-\infty}^{\infty} H(x - x', z)V(x', z = 0)dx' , \qquad (20.15)$$

with

$$H(x, z) = \frac{1}{2\pi} \int_{-\infty}^{\infty} e^{-|k|z} e^{ikx} dk . \qquad (20.16)$$

Note that (20.15) has exactly the same structure as (14.57) for a time-independent linear filter. The only difference is that the variable $x$ now plays the role of the variable $t$ in (14.57). This means that we can consider upward continuation as a linear filtering operation. The convolutional filter $H(x, z)$ maps the potential from the surface $z = 0$ onto the potential at height $z$.

**Problem g:** Show that this filter is given by:

$$H(x, z) = \frac{1}{\pi} \frac{z}{z^2 + x^2} \,. \qquad (20.17)$$

**Problem h:** Sketch this filter as a function of $x$ for a large value of $z$ and a small value of $z$.

**Problem i:** Equation (20.15) implies that at the surface $z = 0$ this filter is given by $H(x, z = 0) = \delta(x)$, with $\delta(x)$ the Dirac delta function. Convince yourself of this by showing that $H(x, z)$ becomes more and more peaked round $x = 0$ when $z \to 0$ and by proving that for all values of $z$ the filter function satisfies $\int_{-\infty}^{\infty} H(x, z)dx = 1$.

## 20.3 Upward continuation in a flat geometry in three dimensions

The analysis in the previous section is valid for a flat geometry in two dimensions. However, the theory can readily be extended to three dimensions by including another horizontal coordinate $y$ in the derivation.

**Problem a:** Show that the theory in the previous section up to equation (20.16) can be generalized by carrying out a Fourier transformation over both $x$ and $y$. Show in particular that in three dimensions:

$$V(x, y, z) = \int\!\!\!\int_{-\infty}^{\infty} H^{3D}(x - x', y - y', z)V(x', y', z = 0)dx'dy' \,, \quad (20.18)$$

with

$$H^{3D}(x, y, z) = \frac{1}{(2\pi)^2} \int\!\!\!\int_{-\infty}^{\infty} e^{-\sqrt{k_x^2 + k_y^2}\, z} e^{i(k_x x + k_y y)} dk_x dk_y \,. \quad (20.19)$$

The only difference with the case in two dimensions is that integral (20.19) leads to a different upward continuation function than integral (20.16) for the two-dimensional case. The integral can be solved by switching the $k$-integral to cylindrical coordinates. The product $k_x x + k_y y$ can be written as $kr \cos \varphi$, where $k$ and $r$ are the length of the **k**-vector and the position vector in the horizontal plane.

**Problem b:** Use this to show that $H^{3D}$ can be written as

$$H^{3D}(x, y, z) = \frac{1}{(2\pi)^2} \int_0^\infty \int_0^{2\pi} k e^{-kz} e^{ikr \cos \varphi} d\varphi dk .\qquad (20.20)$$

**Problem c:** The Bessel function has the following integral representation:

$$J_0(x) = \frac{1}{2\pi} \int_0^{2\pi} e^{ix \cos \theta} d\theta .\qquad (20.21)$$

Use this result to write the upward continuation filter as

$$H^{3D}(x, y, z) = \frac{1}{2\pi} \int_0^\infty e^{-kz} J_0(kr) k dk .\qquad (20.22)$$

It appears that we have only made the problem more complex, because the integral of the Bessel function is not trivial. Fortunately, books and tables exist with a bewildering collection of integrals. For example, in equation (6.621.1) of Gradshteyn and Ryzhik [34] you can find an expression for the following integral: $\int_0^\infty e^{-\alpha x} J_\nu(\beta x) x^{\mu-1} dx$.

**Problem d:** What are the values of $\alpha$, $\nu$, $\beta$ and $\mu$ if we want to use this integral to solve the integration in (20.22)?

Take a look at the form of the integral in Gradshteyn and Ryzhik [34]. You will probably be discouraged by what you find because the result is expressed in hypergeometric functions, which means that you now have the new problem of finding out what these functions are. There is, however, a way out because (6.611.1) of ref. [34] gives the following integral:

$$\int_0^\infty e^{-\alpha x} J_\nu(\beta x) dx = \frac{\beta^{-\nu} \left\{ \sqrt{\alpha^2 + \beta^2} - \alpha \right\}^\nu}{\sqrt{\alpha^2 + \beta^2}} .\qquad (20.23)$$

This is not quite the integral that we want because it does not contain a term that corresponds to the factor $k$ in (20.22). However, we can introduce such a factor by differentiating (20.23) with respect to $\alpha$.

**Problem e:** Do this to show that the upward continuation operator is given by

$$H^{3D}(x, y, z) = \frac{1}{2\pi} \frac{z}{(x^2 + y^2 + z^2)^{3/2}} .\qquad (20.24)$$

You can make the problem simpler by first inserting the appropriate value of $\nu$ in (20.23).

**Problem f:** Compare the upward continuation operator (20.24) for three dimensions with the corresponding operator for two dimensions in (20.17). Which of these operators decays more rapidly as a function of the horizontal distance? Can you explain this difference physically?

**Problem g:** In section 20.2 you showed that the integral of the upward continuation operator over the horizontal distance is equal to 1. Show that the same holds in three dimensions, i.e. show that $\iint_{-\infty}^{\infty} H^{3D}(x, y, z)dxdy = 1$. The integration simplifies by using cylindrical coordinates.

Comparing the upward continuation operators in different dimensions one finds that these operators are different functions of the horizontal distance in the space domain. However, a comparison of (20.16) with (20.19) shows that in the wavenumber domain the upward continuation operators in two and three dimensions have the same dependence on wavenumber. The same is actually true for the Green's function of the wave equation in one, two or three dimensions. As you can see in section 18.4 these Green's functions are very different in the space domain, but one can show that in the wavenumber domain they are given in each dimension by the same expression.

## 20.4 The gravity field of the Earth

In this section we obtain an expression for the gravitational potential outside the Earth for an arbitrary distribution of the mass density $\rho(\mathbf{r})$ within the Earth. This could be done be using the Green's function that is appropriate for Poisson's equation (20.1). As an alternative we solve the problem here by expanding both the mass-density and the potential in spherical harmonics and by using a Green's function technique for every component in the spherical harmonics expansion separately.

When using spherical harmonics, the natural coordinate system is a system of spherical coordinates. For every value of the radius $r$ both the density and the potential can be expanded in spherical harmonics:

$$\rho(r, \theta, \varphi) = \sum_{l=0}^{\infty} \sum_{m=-l}^{l} \rho_{lm}(r)Y_{lm}(\theta, \varphi) \tag{20.25}$$

and

$$V(r, \theta, \varphi) = \sum_{l=0}^{\infty} \sum_{m=-l}^{l} V_{lm}(r)Y_{lm}(\theta, \varphi) . \tag{20.26}$$

**Problem a:** Show that the expansion coefficients for the density are given by:

$$\rho_{lm}(r) = \int Y_{lm}^*(\theta, \varphi)\rho(r, \theta, \varphi)d\Omega , \qquad (20.27)$$

where $\int (\cdots) d\Omega$ denotes an integration over the unit sphere.

Equation (20.1) for the gravitational potential contains the Laplacian. The Laplacian in spherical coordinates can be decomposed as

$$\nabla^2 = \frac{1}{r^2}\frac{\partial}{\partial r}\left(r^2\frac{\partial}{\partial r}\right) + \frac{1}{r^2}\nabla_1^2 , \qquad (20.28)$$

with $\nabla_1^2$ the Laplacian on the unit sphere:

$$\nabla_1^2 = \frac{1}{\sin\theta}\frac{\partial}{\partial\theta}\left(\sin\theta\frac{\partial}{\partial\theta}\right) + \frac{1}{\sin^2\theta}\frac{\partial^2}{\partial\varphi^2} . \qquad (19.35)$$

The reason why an expansion in spherical harmonics is used for the density and the potential is that the spherical harmonics are the eigenfunctions of the operator $\nabla_1^2$ (see (19.34) or p. 379 of Butkov [18]):

$$\nabla_1^2 Y_{lm} = -l(l+1) Y_{lm} . \qquad (19.34)$$

**Problem b:** Insert (20.25) and (20.26) into the Laplace equation, and use (19.34) and (19.35) for the Laplacian of the spherical harmonics to show that the expansion coefficients $V_{lm}(r)$ of the potential satisfy the following differential equation:

$$\frac{1}{r^2}\frac{\partial}{\partial r}\left[r^2\frac{\partial V_{lm}(r)}{\partial r}\right] - \frac{l(l+1)}{r^2}V_{lm}(r) = 4\pi G\rho_{lm}(r) . \qquad (20.29)$$

What we have gained by making the expansion in spherical harmonics is that (20.29) is an ordinary differential equation in the variable $r$ whereas the original equation (20.1) is a partial differential equation in the variables $r$, $\theta$ and $\varphi$. The differential equation (20.29) can be solved using the Green's function technique described in section 17.3. Let us first consider a mass $\delta(r - r')$ located at a radius $r'$. The response to this mass is the Green's function $G_l$ that satisfies the following differential equation:

$$\frac{1}{r^2}\frac{\partial}{\partial r}\left[r^2\frac{\partial G_l(r, r')}{\partial r}\right] - \frac{l(l+1)}{r^2}G_l(r, r') = \delta(r - r') . \qquad (20.30)$$

Note that this equation depends on the angular order $l$ but not on the angular degree $m$. For this reason the Green's function $G_l(r, r')$ depends on $l$ but not on $m$.

**Problem c:** The Green's function can be found by first solving the differential equation (20.30) for $r \neq r'$. Show that when $r \neq r'$ the general solution of the differential equation (20.30) can be written as $G_l = Ar^l + Br^{-(l+1)}$, where the constants $A$ and $B$ do not depend on $r$.

**Problem d:** In the regions $r < r'$ and $r > r'$ the constants $A$ and $B$ in general have different values. Show that the requirement that the potential is everywhere finite implies that $B = 0$ for $r < r'$ and that $A = 0$ for $r > r'$. The solution can therefore be written as:

$$G_l(r, r') = \begin{cases} Ar^l & \text{for} \quad r < r' \\ Br^{-(l+1)} & \text{for} \quad r > r' \end{cases} . \tag{20.31}$$

The integration constants follow in the same way as in the analysis of section 17.3. One constraint on the integration constants follows from the requirement than the Green's function is continuous in the point $r = r'$. The other constraint follows by multiplying (20.30) by $r^2$ and integrating the resulting equation over $r$ from $r' - \varepsilon$ to $r' + \varepsilon$.

**Problem e:** Show by taking the limit $\varepsilon \to 0$ that this leads to the requirement

$$\left[ r^2 \frac{\partial G_l(r, r')}{\partial r} \right]_{r=r'-\varepsilon}^{r=r'+\varepsilon} = r'^2 . \tag{20.32}$$

**Problem f:** Use this condition with the continuity of $G_l$ to find the coefficients $A$ and $B$ and show that the Green's function is given by:

$$G_l(r, r') = \begin{cases} -\dfrac{1}{(2l+1)} \dfrac{r^l}{r'^{(l-1)}} & \text{for} \quad r < r' \\ -\dfrac{1}{(2l+1)} \dfrac{r'^{(l+2)}}{r^{(l+1)}} & \text{for} \quad r > r' \end{cases} . \tag{20.33}$$

**Problem g:** Use this result to derive that the solution of (20.29) is:

$$\begin{aligned} V_{lm}(r) &= -\frac{4\pi G}{(2l+1)} \frac{1}{r^{l+1}} \int_0^r \rho_{lm}(r') r'^{(l+2)} dr' \\ &\quad - \frac{4\pi G}{(2l+1)} r^l \int_r^\infty \rho_{lm}(r') \frac{1}{r'^{(l-2)}} dr' . \end{aligned} \tag{20.34}$$

Hint: split the integration over $r'$ into the interval $0 < r' < r$ and the interval $r' > r$.

**Problem h:** Let us now consider the potential outside the Earth. The radius of the Earth is denoted by the symbol $a$. Use the above expression to show that the potential outside the Earth is given by

$$V(r, \theta, \varphi) = -\sum_{l=0}^{\infty} \sum_{m=-l}^{l} \frac{4\pi G}{(2l+1)} \frac{1}{r^{l+1}} \int_0^a \rho_{lm}(r') r'^{(l+2)} dr' \, Y_{lm}(\theta, \varphi) \, .$$

$$(20.35)$$

**Problem i:** Eliminate $\rho_{lm}$ using the result of problem a and show that the potential is finally given by:

$$V(r, \theta, \varphi) = -\sum_{l=0}^{\infty} \sum_{m=-l}^{l} \frac{4\pi G}{(2l+1)} \frac{1}{r^{l+1}}$$

$$\times \int_0^a \rho(r', \theta', \varphi') r'^l Y_{lm}^*(\theta', \varphi') dV' \, Y_{lm}(\theta, \varphi) \, . \quad (20.36)$$

Note that the integration in this expression is over the volume of the Earth rather than over the distance $r'$ to the Earth's center.

Let us reflect on the relation between this result and the derivation of upward continuation in a Cartesian geometry of the section 20.2. Equation (20.36) can be compared with (20.14). In (20.14) the potential is written as an integration over wavenumber and the potential is expanded in basis functions $e^{ikx}$, whereas in (20.36) the potential is written as a summation over the degree $l$ and order $m$ and the potential is expanded in basis functions $Y_{lm}$. In both expressions the potential is written as a sum over basis functions with increasingly shorter wavelength as the summation index $l$ or the integration variable $k$ increases. The decay of the potential with height is in both geometries faster for a potential with rapid horizontal variations than for a potential with smooth horizontal variations. In both cases the potential decreases when the height $z$ (or the radius $r$) increases. In a flat geometry the potential decreases as $e^{-|k|z}$ whereas in a spherical geometry the potential decreases as $r^{-(l+1)}$. This difference in the reduction of the potential with distance is due to the difference in the geometry in the two problems.

Expressions (20.9) and (20.36) both express the gravitational potential due a density distribution $\rho(\mathbf{r})$, therefore these expressions must be identical.

**Problem j:** Use the equivalence of these expressions to derive that for $r' < r$ the following identity holds:

$$\frac{1}{|\mathbf{r} - \mathbf{r}'|} = \sum_{l=0}^{\infty} \sum_{m=-l}^{l} \frac{4\pi}{(2l+1)} Y_{lm}(\theta, \varphi) Y_{lm}^*(\theta', \varphi') \frac{r'^l}{r^{l+1}} \, . \quad (20.37)$$

The derivation in this section could also have been made using (20.37) as a starting point because this expression can be derived by using the *generating function* of Legendre polynomials and by using the addition theorem to obtain the $m$-summation [5] [42]. However, these concepts are not needed in the treatment in this section which is based only on the expansion of functions in spherical harmonics and on the use of Green's functions.

As a last exercise let us consider the special case of a spherically symmetric mass distribution: $\rho = \rho(r)$. For such a mass distribution the potential is given by

$$V(\mathbf{r}) = -\frac{GM}{r} \, ,$$
(20.38)

where $M$ is the total mass of the body. The gradient of this potential is indeed equal to the gravitational acceleration given in (7.5) for a spherically symmetric mass $M$.

**Problem i:** Derive the potential (20.38) from (20.36) by considering the special case that the mass-density depends only on the radius.

## 20.5 Dipoles, quadrupoles and general relativity

We have seen in the section 7.2 that a spherically symmetric mass leads to a gravitational field $\mathbf{g}(\mathbf{r}) = -GM\hat{\mathbf{r}}/r^2$, which corresponds to a gravitational potential $V(\mathbf{r}) = -GM/r$. Similarly, the electric potential due to a spherically symmetric charge distribution is given by $V(\mathbf{r}) = q/4\pi\varepsilon_0 r$, where $q$ is the total charge. In this section we investigate what happens if we place a positive charge and a negative charge close together. Since there is no negative mass, we treat for the moment the electrical potential, but we will see in section 20.6 that the results also have a bearing on the gravitational potential. The theory developed here is not only important in electrostatics, it also accounts for the measurable effect of the ellipsoidal shape of the Earth on its gravitational field. This application is treated in more detail in section 20.7.

Consider a charge distribution that consists of a positive charge $+q$ placed at position $\mathbf{a}/2$ and a negative charge $-q$ is placed at position $-\mathbf{a}/2$ as shown in figure 20.3.

**Problem a:** The total charge of this system is zero. What would you expect the electric potential to be at positions that are very far from the charges compared to the distance between the charges?

**Problem b:** The potential follows by adding the potentials for the two point charges. Show that the electric potential generated by these

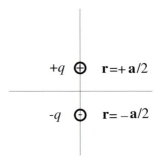

Fig. 20.3.   Two opposite charges that constitute an electric dipole.

two charges is given by

$$4\pi\varepsilon_0 V(\mathbf{r}) = \frac{q}{|\mathbf{r} - \mathbf{a}/2|} - \frac{q}{|\mathbf{r} + \mathbf{a}/2|} \ . \tag{20.39}$$

**Problem c:** Ultimately we will place the charges very close to the origin by taking the limit $a \to 0$. We can therefore restrict our attention to the special case that $a \ll r$. Use a first order Taylor expansion to show that up to order $a$:

$$\frac{1}{|\mathbf{r} - \mathbf{a}/2|} = \frac{1}{r} + \frac{1}{2r^3} (\mathbf{r} \cdot \mathbf{a}) \ . \tag{20.40}$$

Hint: use that $|\mathbf{r} - \mathbf{a}/2|^{-1} = [(\mathbf{r} - \mathbf{a}/2) \cdot (\mathbf{r} - \mathbf{a}/2)]^{-1/2}$.

**Problem d:** Insert this into (20.39) and derive that the electric potential is given by:

$$4\pi\varepsilon_0 V(\mathbf{r}) = - \frac{q (\mathbf{r} \cdot \mathbf{a})}{r^3} \ . \tag{20.41}$$

Now suppose we bring the charges in figure 20.3 closer and closer together, and suppose we let the charge $q$ increase so that the product $\mathbf{p} = q\mathbf{a}$ is constant, then the electric potential is given by:

$$4\pi\varepsilon_0 V(\mathbf{r}) = - \frac{(\hat{\mathbf{r}} \cdot \mathbf{p})}{r^2} \ , \tag{20.42}$$

where we have used that $\mathbf{r} = r\hat{\mathbf{r}}$. The vector $\mathbf{p}$ is called the *dipole vector*. We will see in the next section how the dipole vector can be defined for arbitrary charge or mass distributions.

In problem a you might have guessed that the electric potential would go to zero at great distance. Of course the potential due to the combined charges goes to zero much faster than the potential due to a single charge only: the electric potential of a dipole vanishes as $1/r^2$ compared to the

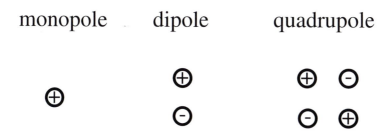

Fig. 20.4.   The definition of the monopole, dipole and quadrupole in terms of electric charges

$1/r$ decay of the potential for a single charge. Many physical systems, such as neutral atoms, consist of neutral combinations of positive and negative charges. The lesson we learn from (20.41) is that such a neutral combination of charges may generate a nonzero electric field and that such a system will in general interact with other electromagnetic systems. For example, atoms interact to leading order with the radiation field (light) through their *dipole moment* [76]. In chemistry, the dipole moment of molecules plays a crucial role in the distinction between polar and apolar substances. Water would not have its many wonderful properties if it did not have a dipole moment.

Let us now consider the electric field generated by an electric dipole.

**Problem e:** Take the gradient of (20.42) to show that this field is given by

$$\mathbf{E}(\mathbf{r}) = \frac{1}{4\pi\varepsilon_0 r^3}\left[\mathbf{p} - 3\hat{\mathbf{r}}\left(\hat{\mathbf{r}}\cdot\mathbf{p}\right)\right].\tag{20.43}$$

Hint: either use the expression of the gradient in spherical coordinates or take the gradient in Cartesian coordinates and use (5.7).

The electric field generated by an electric dipole has the same form as the magnetic field generated by a magnetic dipole as shown in (5.3). The mathematical reason for this is that the magnetic field satisfies (5.14) which states that $(\nabla \cdot \mathbf{B}) = 0$, while the electric field in free space satisfies according to (5.13) the field equation: $(\nabla \cdot \mathbf{E}) = 0$. However, there is an important difference. The electric field is generated by electric charges, and this field satisfies the equation $(\nabla \cdot \mathbf{E}) = \rho(\mathbf{r})/\varepsilon_0$. In the example in this section we created a dipole field by taking two opposite charges and putting them closer and closer together. However, the magnetic field satisfies $(\nabla \cdot \mathbf{B}) = 0$ *everywhere*. The reason for this is that the magnetic equivalent of the electric charge, the *magnetic monopole*, has not been discovered in nature.

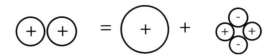

Fig. 20.5.   The decomposition of a double star in a gravitational monopole and a gravitational quadrupole.

The fact that magnetic monopoles have not been observed in nature seems puzzling, because we have seen that the magnetic dipole field has the same form as the electric dipole field which was constructed by putting two opposite electric charges close together. The reason for the analogy between an electric and a magnetic dipole field is not that the magnetic dipole can be seen as a combination of positive and negative magnetic 'charges' placed close together. In the context of classical electromagnetism, the magnetic dipole field is generated by a current that runs in a small circular loop. On a microscopic scale a magnetic dipole is generated by the *spin* of particles. This means that the electric and magnetic dipole fields have fundamentally different origins.

The starting point of the derivation in this section is the electric field of a single point charge. Such as single point charge is called a monopole, see figure 20.4. The field of this charge decays as $1/r^2$. If we put two opposite charges close together we can create a dipole, see figure 20.4. Its electric field is derived in (20.43); this field decays as $1/r^3$. We can also put two opposite dipoles together as shown in figure 20.4. The resulting charge distribution is called a quadrupole. To leading order the electric fields of the dipoles that constitute the quadrupole cancel, and we will see in the next section that the electric potential for a quadrupole decays as $1/r^3$ so that the electric field decays with distance as $1/r^4$.

You may wonder whether the concept of a dipole or quadrupole can also be used for the gravity field because for the electric field these concepts are based on the presence of both positive and negative charges whereas we know that only positive mass occurs in nature. However, there is nothing to keep us from computing the gravitational field for a negative mass, and this is actually quite useful. As an example, let us consider a double star that consists of two heavy stars which rotate around their joint center of gravity. The first order field is the monopole field that is generated by the joint mass of the stars. However, as shown in figure 20.5 the mass of the two stars can be seen as approximately the sum of a monopole and a quadrupole consisting of two positive and two negative charges. Since the stars rotate, the gravitational quadrupole rotates as well, and this is the reason why rotating double stars are seen as a prime source for the generation of *gravitational waves* [62]. However, gravitational waves that spread in space with time cannot be described by the classic expression (20.1) for the gravitational potential.

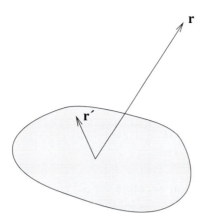

Fig. 20.6. Definition of the integration variable $\mathbf{r}'$ within the mass and the observation point $\mathbf{r}$ outside the mass.

**Problem g:** Can you explain why (20.1) cannot account for propagating gravitational waves?

A proper description of gravitational waves depends on the general theory of relativity [62]. Huge detectors for gravitational waves are being developed [7], because these waves can be used to investigate the theory of general relativity as well as the astronomical objects that generate gravitational waves.

## 20.6 The multipole expansion

Now that we have learned that the concepts of the monopole, dipole and quadrupole are relevant for both the electric field and the gravity field we continue the analysis with the gravitational field. In this section we derive the *multipole expansion* in which the total field is written as a superposition of a monopole field, a dipole field, a quadrupole field, an octupole field, etc.

Consider the situation shown in figure 20.6 in which a finite body has a mass-density $\rho(\mathbf{r}')$. The gravitational potential generated by this mass is given by:

$$V(\mathbf{r}) = -G \int \frac{\rho(\mathbf{r}')}{|\mathbf{r} - \mathbf{r}'|} dV' . \qquad (20.9)$$

We consider the potential at a distance that is much larger than the size of the body. Since the integration variable $\mathbf{r}'$ is limited by the size of the body, a 'large distance' means in this context that $r \gg r'$. We therefore make a Taylor expansion of the term $1/|\mathbf{r} - \mathbf{r}'|$ in the small parameter $(r'/r)$ which is much smaller than unity.

**Problem a:** Show that

$$|\mathbf{r} - \mathbf{r}'| = \sqrt{r^2 - 2(\mathbf{r} \cdot \mathbf{r}') + r'^2} \,. \tag{20.44}$$

**Problem b:** Use a Taylor expansion in the small parameter $r'/r$ to show that:

$$\frac{1}{|\mathbf{r} - \mathbf{r}'|} = \frac{1}{r} \left\{ 1 + \frac{1}{r} (\hat{\mathbf{r}} \cdot \mathbf{r}') + \frac{1}{2r^2} \left[ 3(\hat{\mathbf{r}} \cdot \mathbf{r}')^2 - r'^2 \right] + O\left(\frac{r'}{r}\right)^3 \right\}. \tag{20.45}$$

Be careful that you account for all the terms of order $r'$ correctly. Also be aware of the distinction between the position vector $\mathbf{r}$ and the unit vector $\hat{\mathbf{r}}$.

From this point on we will ignore the terms of order $(r'/r)^3$.

**Problem c:** Insert the expansion (20.45) into (20.9) and show that the gravitational potential can be written as a sum of different contributions:

$$V(\mathbf{r}) = V_{mon}(\mathbf{r}) + V_{dip}(\mathbf{r}) + V_{qua}(\mathbf{r}) + \cdots, \tag{20.46}$$

with

$$V_{mon}(\mathbf{r}) = -\frac{G}{r} \int \rho(\mathbf{r}') dV', \tag{20.47}$$

$$V_{dip}(\mathbf{r}) = -\frac{G}{r^2} \int \rho(\mathbf{r}') (\hat{\mathbf{r}} \cdot \mathbf{r}') \, dV', \tag{20.48}$$

$$V_{qua}(\mathbf{r}) = -\frac{G}{2r^3} \int \rho(\mathbf{r}') \left[ 3(\hat{\mathbf{r}} \cdot \mathbf{r}')^2 - r'^2 \right] dV'. \tag{20.49}$$

It thus follows that the gravitational potential can be written as the sum of terms that decay with increasing powers of $r^{-n}$. Let us analyze these terms in turn. The term $V_{mon}(\mathbf{r})$ in (20.47) is the simplest since the volume integral of the mass-density is simply the total mass of the body: $\int \rho(\mathbf{r}') dV' = M$. This means that this term is given by

$$V_{mon}(\mathbf{r}) = -\frac{GM}{r}. \tag{20.50}$$

This is the potential generated by a point mass $M$. To leading order, the gravitational field is the same as if all the mass of the body were concentrated in the origin. The mass distribution within the body does not affect this part of the gravitational field at all. Because the resulting field is the same as for a point mass, this field is called the *monopole field*.

For the analysis of the term $V_{dip}(\mathbf{r})$ in (20.48) it is useful to define the center of gravity $\mathbf{r}_g$ of the body:

$$\mathbf{r}_g \equiv \frac{\int \rho(\mathbf{r}')\mathbf{r}'dV'}{\int \rho(\mathbf{r}')dV'} . \tag{20.51}$$

This is simply a weighted average of the position vector with the mass-density as weight functions. Note that the word 'weight' here has a double meaning!

**Problem d:** Show that $V_{dip}(\mathbf{r})$ is given by:

$$V_{dip}(\mathbf{r}) = -\frac{GM}{r^2}(\hat{\mathbf{r}} \cdot \mathbf{r}_g) . \tag{20.52}$$

Note that this potential has exactly the same form as the potential (20.42) for an electric dipole. For this reason $V_{dip}(\mathbf{r})$ is called the dipole field.

**Problem e:** Compared to the monopole term, the dipole term decays as $1/r^2$ rather than $1/r$. The monopole term does not depend on $\hat{\mathbf{r}}$, the direction of observation. Show that the dipole term varies with the direction of observation as $\cos\theta$ and show how the angle $\theta$ must be defined.

**Problem f:** You may be puzzled by the fact that the gravitational potential contains a dipole term, despite the fact that there is no negative mass. Draw a figure similar to figure 20.5 to show that a displaced mass can be written as an undisplaced mass plus a mass dipole.

Of course, one is completely free in the choice of the origin of the coordinate system. If one chooses the origin to be at the center of mass of the body, then $\mathbf{r}_g = 0$ and the dipole term vanishes.

We now analyze the term $V_{qua}(\mathbf{r})$ in (20.49). It can be seen from this expression that this term decays with distance as $1/r^3$. For this reason, this term is called the quadrupole field. The dependence of the quadrupole field on the direction is more complex than for the monopole field and the dipole field. In the determination of the directional dependence of the quadrupole term it is useful to use the double contraction between two tensors. Tensors are treated in chapter 21. For the moment you only need to know that the double contraction is defined as:

$$(\mathbf{A} : \mathbf{B}) \equiv \sum_{i,j} A_{ij}B_{ij} . \tag{20.53}$$

The double contraction generalizes the concept of the inner product of two vectors $(\mathbf{a} \cdot \mathbf{b}) = \sum_i a_i b_i$ to matrices or tensors of rank two. A double contraction occurs for example in the following identity

$$1 = (\hat{\mathbf{r}} \cdot \hat{\mathbf{r}}) = (\hat{\mathbf{r}} \cdot \mathbf{I}\hat{\mathbf{r}}) = (\hat{\mathbf{r}}\hat{\mathbf{r}} : \mathbf{I}) , \tag{20.54}$$

where $\mathbf{I}$ is the identity operator. Note that the term $\hat{\mathbf{r}}\hat{\mathbf{r}}$ is a dyad. If you are unfamiliar with the concept of a dyad you may want to look at section 12.1 before continuing with this chapter.

**Problem g:** Use these results to show that $V_{qua}(\mathbf{r})$ can be written as:

$$V_{qua}(\mathbf{r}) = -\frac{G}{2r^3}\left(\hat{\mathbf{r}}\hat{\mathbf{r}} : \mathbf{T}\right) , \qquad (20.55)$$

where $\mathbf{T}$ is the *inertia tensor* defined as

$$\mathbf{T} = \int \rho(\mathbf{r})\left(3\mathbf{r}\mathbf{r} - \mathbf{I}r^2\right) dV . \qquad (20.56)$$

Note that we have renamed the integration variable $\mathbf{r}'$ in the quadrupole moment tensor as $\mathbf{r}$.

**Problem h:** Show that in explicit matrix notation $\mathbf{T}$ is given by:

$$\mathbf{T} = \int \rho(\mathbf{r}) \begin{pmatrix} 2x^2 - y^2 - z^2 & 3xy & 3xz \\ 3xy & 2y^2 - x^2 - z^2 & 3yz \\ 3xz & 3yz & 2z^2 - x^2 - y^2 \end{pmatrix} dV . \qquad (20.57)$$

Note the resemblance between (20.55) and (20.52). For the dipole field the directional dependence is described by the single contraction $(\hat{\mathbf{r}} \cdot \mathbf{r}_g)$ whereas for the quadrupole field directional dependence is now given by the double contraction $(\hat{\mathbf{r}}\hat{\mathbf{r}} : \mathbf{T})$. This double contraction leads to a greater angular dependence of the quadrupole term than for the monopole term and the dipole term.

To find the angular dependence, we use that the inertia tensor $\mathbf{T}$ is a real symmetric $3 \times 3$ matrix. This matrix therefore has three orthogonal eigenvectors $\hat{\mathbf{v}}^{(i)}$ with corresponding eigenvalues $\lambda_i$. Using expression (12.45) this implies that the quadrupole moment tensor can be written as:

$$\mathbf{T} = \sum_{i=1}^{3} \lambda_i \hat{\mathbf{v}}^{(i)}\hat{\mathbf{v}}^{(i)} . \qquad (20.58)$$

**Problem i:** Use this result to show that the quadrupole term can be written as

$$V_{qua}(\mathbf{r}) = -\frac{G}{2r^3} \sum_{i=1}^{3} \lambda_i \cos^2\Psi_i , \qquad (20.59)$$

where the $\Psi_i$ denote the angles between the eigenvectors $\hat{\mathbf{v}}^{(i)}$ and the observation direction $\hat{\mathbf{r}}$; see figure 20.7 for the definition of these angles.

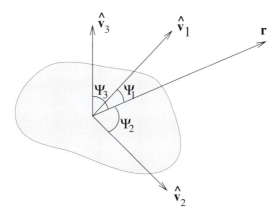

Fig. 20.7. Definition of the angles $\Psi_i$.

The directional dependence of (20.59) varies as $\cos^2\Psi_i = (\cos2\Psi_i + 1)/2$; this implies that the quadrupole field varies through two periods when $\Psi_i$ increases from 0 to $2\pi$. This contrasts with the monopole field, which does not depend on the direction at all, as well as with the dipole field that varies according to problem e as $\cos\theta$. There is actually a close connection between the different terms in the multipole expansion and spherical harmonics. This can be seen by comparing the multipole terms (20.47)–(20.49) with (20.36) for the gravitational potential. In (20.36), the different terms decay with distance as $r^{-(l+1)}$ and have an angular dependence $Y_{lm}(\theta, \varphi)$. Similarly, the multipole terms decay as $r^{-1}$, $r^{-2}$ and $r^{-3}$, respectively, and depend on the direction as $\cos 0$, $\cos\theta$ and $\cos 2\Psi$ respectively.

## 20.7 The quadrupole field of the Earth

Let us now investigate what the multipole expansion implies for the gravity field of the Earth. The monopole term is by far the dominant term. It explains why an apple falls from a tree, why the Moon orbits the Earth and most other manifestations of gravity that we observe in daily life. The dipole term has in this context no physical meaning whatsoever. This can be seen from (20.52) which states that the dipole term only depends on the distance from the Earth's center of gravity to the origin of the coordinate system. Since we are completely free in choosing the origin, the dipole term can be made to vanish by choosing the origin of the coordinate system as the Earth's center of gravity. It is through the quadrupole field that some of the subtleties of the Earth's gravity field become manifest.

**Problem a:** The quadrupole field vanishes when the mass distribution in the Earth is spherically symmetric. Show this by computing the inertia tensor $\mathbf{T}$ when $\rho = \rho(r)$.

The dominant departure of the shape of the Earth from spherical is the flattening of the Earth due to its rotation. If that is the case, then by symmetry one eigenvector of $\mathbf{T}$ must be aligned with the Earth's axis of rotation, and the two other eigenvectors must be perpendicular to the axis of rotation. By symmetry these other eigenvectors must correspond to equal eigenvalues. When we choose a coordinate system with the $z$-axis along the Earth's axis of rotation the eigenvectors are therefore given by the unit vectors $\hat{\mathbf{z}}$, $\hat{\mathbf{x}}$ and $\hat{\mathbf{y}}$ with eigenvalues $\lambda_z$, $\lambda_x$ and $\lambda_y$, respectively. The last two eigenvalues identical because of the rotational symmetry around the Earth's axis of rotation: hence $\lambda_y = \lambda_x$.

Let us first determine the eigenvalues. Once the eigenvalues are known, the quadrupole moment tensor follows from (20.59). The eigenvalues could be found in the standard way by solving $\det(\mathbf{T} - \lambda\mathbf{I}) = 0$, but this is unnecessarily difficult. Once we know the eigenvectors the eigenvalues can easily be found from (20.58).

**Problem b:** Take twice the inner product of (20.58) with the eigenvector $\hat{\mathbf{v}}^{(j)}$ to show that

$$\lambda_j = \hat{\mathbf{v}}^{(j)} \cdot \mathbf{T} \cdot \hat{\mathbf{v}}^{(j)} \; . \tag{20.60}$$

**Problem c:** Use this with (20.57) to show that the eigenvalues are given by:

$$\left.\begin{array}{l} \lambda_x = \displaystyle\int \rho(\mathbf{r}) \left(2x^2 - y^2 - z^2\right) dV \; , \\[2mm] \lambda_y = \displaystyle\int \rho(\mathbf{r}) \left(2y^2 - x^2 - z^2\right) dV \; , \\[2mm] \lambda_z = \displaystyle\int \rho(\mathbf{r}) \left(2z^2 - x^2 - y^2\right) dV \; . \end{array}\right\} \tag{20.61}$$

It is useful to relate these eigenvalues to the Earth's *moments of inertia*. The moment of inertia of the Earth around the $z$-axis is defined as:

$$C \equiv \int \rho(\mathbf{r}) \left(x^2 + y^2\right) dV \; , \tag{20.62}$$

whereas the moment of inertia around the $x$-axis is defined as

$$A \equiv \int \rho(\mathbf{r}) \left(y^2 + z^2\right) dV \; . \tag{20.63}$$

By symmetry the moment of inertia around the $y$-axis is given by the same moment $A$. These moments describe the rotational inertia around the coordinate axes as shown in figure 20.8. The eigenvalues in (20.61) can be related to these moments of inertia. Because of the assumed axisymmetric density distribution, the integral of $y^2$ in (20.61) is equal to the integral of $x^2$. The eigenvalue $\lambda_x$ in (20.61) is therefore given by: $\lambda_x = \int \rho(\mathbf{r}) \left(x^2 - z^2\right) dV = \int \rho(\mathbf{r}) \left(x^2 + y^2 - y^2 - z^2\right) dV = C - A$.

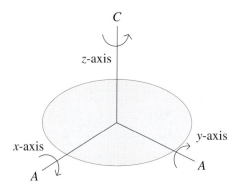

Fig. 20.8. Definition of the moments of inertia $A$ and $C$ for an Earth with cylindrical symmetry around the rotation axis.

**Problem d:** Apply a similar treatment to the other eigenvalues to show that:

$$\lambda_x = \lambda_y = C - A , \qquad \lambda_z = -2(C - A) . \qquad (20.64)$$

**Problem e:** Use these eigenvalues in (20.58) together with (20.55) and expression (3.7) for the unit vector $\hat{\mathbf{r}}$ to show that the quadrupole field is given by:

$$V_{qua}(\mathbf{r}) = \frac{G}{2r^3} (C - A) \left( 3 \cos^2 \theta - 1 \right) . \qquad (20.65)$$

The Legendre polynomial of order 2 is given by: $P_2^0(x) = \frac{1}{2} (3x^2 - 1)$. The quadrupole field can therefore be written as:

$$V_{qua}(\mathbf{r}) = \frac{G}{r^3} (C - A) P_2^0(\cos \theta) . \qquad (20.66)$$

The term $(C - A)$ denotes the difference in the moments of inertia of the Earth around the rotation axis and around an axis through the equator, see figure 20.8. If the Earth were a perfect sphere, these moments of inertia would be identical and the quadrupole field would vanish. However, the rotation of the Earth causes the Earth to bulge at the equator. This departure from spherical symmetry is responsible for the Earth's quadrupole field.

If the Earth were spherical, the motion of satellites orbiting the Earth would satisfy Kepler's laws. The quadrupole term in the potential effects a measurable deviation of the trajectories of satellites from the orbits predicted by Kepler's laws. For example, if the potential is spherically symmetric, a satellite orbits in a fixed plane. The quadrupole field causes the plane in which the satellite orbits to precess slightly. Observations of the orbits of satellites can therefore be used to deduce the departure of the Earth's shape from spherical symmetry [47]. Using these techniques, it has been found that the difference $(C - A)$ in the moments of inertia has the numerical value [85]:

$$J_2 = \frac{(C - A)}{Ma^2} = 1.082626 \times 10^{-3} \, . \tag{20.67}$$

In this expression $a$ is the radius of the Earth, and the term $Ma^2$ is a measure of the average moment of inertia of the Earth. Expression (20.67) states therefore that the relative departure of the mass distribution of the Earth from spherical symmetry is of the order $10^{-3}$. This effect is small, but this number carries important information about the dynamics of our planet. In fact, the time derivative $\dot{J}_2$ of this quantity has been measured [104] as well! This quantity is of importance because the rotation rate of the Earth slowly decreases due to the braking effect of the tidal forces. Both the oceans and the solid Earth are subject to the tidal force that operates in the Earth–Moon system. There is a time lag in the response of both the ocean and the solid Earth and the tidal force that generates the tides. This time lag leads to a deceleration of the rotation of the Earth [85]; the Earth adjusts its shape to this deceleration. The measurement of $\dot{J}_2$ therefore provides important information about the response of the Earth to a time-dependent loading.

## 20.8 The fifth force

Gravity is the force in nature that was first understood by mankind through the discovery by Newton of the law of gravitational attraction. The reason that the gravitational force was understood first is that this force manifests itself in the macroscopic world in the motion of the Sun, Moon and planets. Later the electromagnetic force, and the strong and weak interactions were discovered. This means that presently four forces are thought to be operative in nature. Of these four forces, the electromagnetic force and the weak nuclear force can now be described by a single unified theory.

In the 1980s, geophysical measurements of gravity suggested that the gravitational field behaves in a different way over geophysical length scales (between meters and kilometers) than over astronomical length scales ($>10^4$ km). This has led to the speculation that this discrepancy is due to a fifth force in nature. This speculation and the observations that fuelled this idea are clearly described by Fishbach and Talmadge [32]. The central idea is that in Newton's theory of gravity the gravitational potential generated by a point mass $M$ is given by (20.38):

$$V_N(\mathbf{r}) = -\frac{GM}{r} . \tag{20.68}$$

The hypothesis of the fifth force presumes that a new potential should be added to this Newtonian potential: this fifth potential is given by

$$V_5(\mathbf{r}) = -\alpha\frac{GM}{r}e^{-r/\lambda} . \tag{20.69}$$

Note that this potential has almost the same form as the Newtonian potential $V_N(\mathbf{r})$, the main differences are that the fifth force decays exponentially with distance over a length $\lambda$ and that it is weaker than Newtonian gravity by a factor $\alpha$. This idea was prompted by measurements of gravity in mines, in the ice-cap of Greenland, on a 600 m high telecommunication tower and by a number of other experimental results that seemed to disagree with the gravitational force that follows from the Newtonian potential $V_N(\mathbf{r})$.

**Problem a:** Effectively, the fifth force leads to a change of the gravitational constant $G$ with distance. Compute the gravitational acceleration $\mathbf{g}(r)$ for the combined potential $V_N + V_5$ by taking the gradient and write the result as $-G(r)M\hat{\mathbf{r}}/r^2$ to show that the effective gravitational constant is given by:

$$G(r) = G\left[1 + \alpha\left(1 + \frac{r}{\lambda}\right)e^{-r/\lambda}\right] . \tag{20.70}$$

The fifth force thus effectively leads to a change of the gravitational constant over a characteristic distance $\lambda$. This effect is very small: in 1991 the value of $\alpha$ was estimated to be less than $10^{-3}$ for all estimates of $\lambda$ longer than 1 cm [32].

In doing geophysical measurements of gravity, one has to correct for perturbing effects such as the topography of the Earth's surface and density variations within the Earth's crust. It has now been shown that the uncertainties in these corrections are much larger than the observed discrepancy between the gravity measurements and Newtonian gravity [65]. This means that the issue of the fifth force seems be closed for the moment, and that the physical world appears again to be governed by only four fundamental forces.

# 21

---

# Cartesian tensors

In physics and mathematics, coordinate transformations play a very important role because many problems are much simpler when a suitable coordinate system is used. Furthermore, the requirement that physical laws do not change under certain transformations imposes constraints on the physical laws. An example of this is presented in section 21.11 where it is shown that the fact that the pressure in a fluid is isotropic follows from the requirement that some physical laws may not change under a rotation of the coordinate system. In this chapter how vectors and matrices change under coordinate transformations is derived. The derived transformation properties can be generalized to other mathematical objects which are called tensors. In this chapter, only transformations of rectangular coordinate systems are considered. Since these coordinate systems are called Cartesian coordinate systems, the associated tensors are called Cartesian tensors. The transformation properties of tensors in Cartesian and curvilinear coordinate systems are described in detail by Butkov [18] and Riley $et\ al.$ [72].

## 21.1 Coordinate transforms

In this section we consider the transformation of a coordinate system in two dimensions. In figure 21.1 an old coordinate system with coordinates $x^{old}$ and $y^{old}$ is shown. The unit vectors along the old coordinate axis are denoted by $\hat{\mathbf{e}}^{x,old}$ and $\hat{\mathbf{e}}^{y,old}$. In a coordinate transformation, these old unit vectors are transformed to new unit vectors $\hat{\mathbf{e}}^{x,new}$ and $\hat{\mathbf{e}}^{y,new}$, respectively. The matrix that maps the old unit vectors onto the new unit vectors is denoted by $\mathbf{C}^{-1}$, hence

$$\mathbf{C}^{-1}\hat{\mathbf{e}}^{x,old} = \hat{\mathbf{e}}^{x,new} , \quad \mathbf{C}^{-1}\hat{\mathbf{e}}^{y,old} = \hat{\mathbf{e}}^{y,new} . \tag{21.1}$$

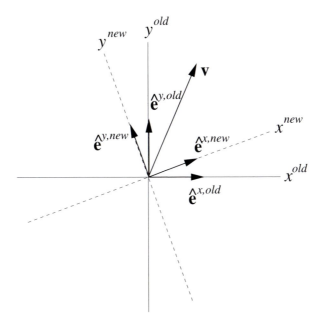

Fig. 21.1. Definition of vectors in the original and transformed coordinate systems.

The fact that we use the inverse of the matrix has no significance; the only reason we do so is that in the following sections we will mostly use the inverse of $\mathbf{C}^{-1}$ which is equal to $\mathbf{C}$. The elements of the matrix $\mathbf{C}^{-1}$ are labelled as follows

$$\mathbf{C}^{-1} = \begin{pmatrix} C_{11}^{-1} & C_{12}^{-1} \\ C_{21}^{-1} & C_{22}^{-1} \end{pmatrix}, \tag{21.2}$$

where it is understood that $C_{ij}^{-1}$ is not the inverse of the element $C_{ij}$, but is the $i, j$-element of the matrix $\mathbf{C}^{-1}$.

There is a close relation between the new basis vectors $\hat{\mathbf{e}}^{x,new}$ and $\hat{\mathbf{e}}^{y,new}$ and the matrix $\mathbf{C}^{-1}$. To see this we first use that in the old coordinate system the unit vector $\hat{\mathbf{e}}^{x,old}$ is the unit vector in the $x$-direction and that $\hat{\mathbf{e}}^{y,old}$ is the unit vector in the $y$-direction:

$$\hat{\mathbf{e}}^{x,old} = \begin{pmatrix} 1 \\ 0 \end{pmatrix}, \quad \hat{\mathbf{e}}^{y,old} = \begin{pmatrix} 0 \\ 1 \end{pmatrix}. \tag{21.3}$$

**Problem a:** Let the matrix $\mathbf{C}^{-1}$ act on these expressions and use expression (21.1) to show that the first column of $\mathbf{C}^{-1}$ is given by $\hat{\mathbf{e}}^{x,new}$ while the second column is equal to $\hat{\mathbf{e}}^{y,new}$:

$$\hat{\mathbf{e}}^{x,new} = \begin{pmatrix} C_{11}^{-1} \\ C_{21}^{-1} \end{pmatrix}, \quad \hat{\mathbf{e}}^{y,new} = \begin{pmatrix} C_{12}^{-1} \\ C_{22}^{-1} \end{pmatrix}. \tag{21.4}$$

**Problem b:** Use (21.2) to show that this implies that the matrix $\mathbf{C}^{-1}$ can also be written as

$$\mathbf{C}^{-1} = (\; \hat{\mathbf{e}}^{x,new} \quad \hat{\mathbf{e}}^{y,new} \;) . \tag{21.5}$$

In other words, the transformed basis vectors are the columns of the matrix that describes the coordinate transform.

Let us now consider a vector $\mathbf{v}$ as shown in figure 21.1. This vector is a physical quantity that denotes, for example, the position of an object relative to some specified origin. This vector is independent of any coordinate system since it is a physical quantity. However, when we describe this vector we use a system of basis vectors to give the components of this vector in directions defined by the basis vectors. For example, using the old coordinate vectors $\hat{\mathbf{e}}^{x,old}$ and $\hat{\mathbf{e}}^{y,old}$ the vector $\mathbf{v}$ can be decomposed into its components $v_x^{old}$ and $v_y^{old}$:

$$\mathbf{v} = v_x^{old} \hat{\mathbf{e}}^{x,old} + v_y^{old} \hat{\mathbf{e}}^{y,old} . \tag{21.6}$$

This same vector can also be decomposed into its components along the new coordinate system

$$\mathbf{v} = v_x^{new} \hat{\mathbf{e}}^{x,new} + v_y^{new} \hat{\mathbf{e}}^{y,new} . \tag{21.7}$$

The vector $\mathbf{v}$ is the same in each of these expressions. It is the goal of this section to derive the relation between the old components $v^{old}$ and the new components $v^{new}$.

**Problem c:** Use that (21.6) and (21.7) are expressions for the same vector and insert (21.3) and (21.4) for the unit vectors to show that the old and the new components are related by

$$v_x^{old} \begin{pmatrix} 1 \\ 0 \end{pmatrix} + v_y^{old} \begin{pmatrix} 0 \\ 1 \end{pmatrix} = v_x^{new} \begin{pmatrix} C_{11}^{-1} \\ C_{21}^{-1} \end{pmatrix} + v_y^{new} \begin{pmatrix} C_{12}^{-1} \\ C_{22}^{-1} \end{pmatrix} . \tag{21.8}$$

**Problem d:** Rewrite this result to derive that

$$\left. \begin{array}{l} C_{11}^{-1} v_x^{new} + C_{12}^{-1} v_y^{new} = v_x^{old} , \\ C_{21}^{-1} v_x^{new} + C_{22}^{-1} v_y^{new} = v_y^{old} . \end{array} \right\} \tag{21.9}$$

At this point it is useful to introduce the vectors $\mathbf{v}^{old}$ and $\mathbf{v}^{new}$ with the components of the vector in the old and new coordinate system respectively:

$$\mathbf{v}^{old} \equiv \begin{pmatrix} v_x^{old} \\ v_y^{old} \end{pmatrix} , \quad \mathbf{v}^{new} \equiv \begin{pmatrix} v_x^{new} \\ v_y^{new} \end{pmatrix} . \tag{21.10}$$

This harmless-looking definition belies a subtle complication. We now have three notations for the same vector. The notation $\mathbf{v}$ denotes the physical vector that is defined independently of any coordinate system, while the vectors $\mathbf{v}^{old}$ and $\mathbf{v}^{new}$ give the *representation* of this vector with respect to the old and the new coordinate systems, respectively. It is crucial to separate these three different vectors.

**Problem e:** Use the rules of matrix–vector multiplication to show that (21.9) implies that

$$\mathbf{C}^{-1}\mathbf{v}^{new} = \mathbf{v}^{old} , \qquad (21.11)$$

and derive from this that

$$\mathbf{v}^{new} = \mathbf{C}\mathbf{v}^{old} . \qquad (21.12)$$

This is the desired result because this expression prescribes how the old vector $\mathbf{v}^{old}$ is mapped in the coordinate transform onto the new vector $\mathbf{v}^{new}$. It is interesting to compare this expression with (21.1) which states that the old unit vectors are mapped onto the new basis vectors by the *inverse* matrix $\mathbf{C}^{-1}$. There is a good reason why the basis vectors transform with the inverse of the transformation matrix of the components. According to (21.6) the physical vector $\mathbf{v}$ is the product of the components of the vector and the basis vectors: $\mathbf{v} = v_x \hat{\mathbf{e}}^x + v_y \hat{\mathbf{e}}^y$. Since the physical vector $\mathbf{v}$ by definition is not changed by the coordinate transform, the *product* $v_x \hat{\mathbf{e}}^x + v_y \hat{\mathbf{e}}^y$ of the components and the basis vectors is invariant. This can only be the case when the transformation rule for the components is the inverse of the transformation rule of the basis vectors.

The treatment in this section was for a two-dimensional coordinate transform. Each step of the argument can be generalized to coordinate transforms in more than two dimensions.

## 21.2 Unitary matrices

In the previous section, it was tacitly assumed that the matrix $\mathbf{C}$ was defined in such a way that the new basis vectors $\hat{\mathbf{e}}^{x,new}$ and $\hat{\mathbf{e}}^{y,new}$ were of unit length. For a general coordinate transform $\mathbf{C}$ this is not true. However, we are only interested in orthonormal coordinate systems: these are coordinate systems in which the basis vectors have unit length and are mutually orthogonal.

**Problem a:** Let the unit vectors in a coordinate system be denoted by $\hat{\mathbf{e}}^i$. Show that for an orthonormal coordinate system

$$\left(\hat{\mathbf{e}}^i \cdot \hat{\mathbf{e}}^j\right) = \delta_{ij} , \qquad (21.13)$$

where $\delta_{ij}$ is the Kronecker delta defined as

$$\delta_{ij} = \begin{cases} 1 & \text{when } i = j \\ 0 & \text{when } i \neq j \end{cases} . \tag{10.15}$$

When the old coordinate system and the new coordinate system are orthonormal, both coordinate systems must satisfy (21.13). The condition is certainly satisfied when the inner product of two vectors does not change under the coordinate transformation. Such coordinate transformations are called *unitary transformations*. To be more specific, let $\mathbf{u}$ and $\mathbf{v}$ be two vectors in the old coordinate system that are mapped by the coordinate transform $\mathbf{C}$ to new vectors $\mathbf{u}' = \mathbf{Cu}$ and $\mathbf{v}' = \mathbf{Cv}$, respectively. The coordinate transform $\mathbf{C}$ is called unitary when $(\mathbf{u}' \cdot \mathbf{v}') = (\mathbf{u} \cdot \mathbf{v})$, or equivalently

$$(\mathbf{Cu} \cdot \mathbf{Cv}) = (\mathbf{u} \cdot \mathbf{v}) . \tag{21.14}$$

**Problem b:** Show that the fact that the length of a vector and the angle between vectors is preserved implies that the inner product of two vectors is preserved as well. Examples of unitary coordinate transformations are rotations, reflections in a plane or combinations of these. Draw a number of figures to convince yourself that these coordinate transformations do not change the length of the vector and the angle between two vectors so that these transformations are indeed unitary.

**Problem c:** The requirement (21.14) can be used to impose a constraint on $\mathbf{C}$. Write the matrix products and the inner product in (21.14) in terms of the components of the matrix $\mathbf{C}$ and the vectors $\mathbf{u}$ and $\mathbf{v}$ to derive that it is equivalent to

$$\sum_{i,j,k} C_{ij} u_j C_{ik} v_k = \sum_i u_i v_i . \tag{21.15}$$

Writing out a matrix expression in its components has an important advantage. In general one cannot interchange the order of two matrices, because in general the matrix product $\mathbf{AB}$ differs from the matrix product $\mathbf{BA}$.

**Problem d:** Verify this statement by taking

$$\mathbf{A} = \begin{pmatrix} 0 & 1 & 0 \\ -1 & 0 & 0 \\ 0 & 0 & 1 \end{pmatrix} \quad \text{and} \quad \mathbf{B} = \begin{pmatrix} 1 & 0 & 0 \\ 0 & 0 & 1 \\ 0 & -1 & 0 \end{pmatrix} \tag{21.16}$$

and showing that $\mathbf{AB} \neq \mathbf{BA}$.

**Problem e:** Geometrically the matrix **A** represents a rotation around the $z$-axis through 90 degrees while **B** represents a rotation around the $x$-axis through 90 degrees. Take this book and carry out these rotations in the two different orders to see that the final orientation of the book is indeed different in each case.

This means that in general one cannot interchange the order of matrices in a product. However, when a term $A_{ij}B_{jk}$ occurs in a summation, the terms $A_{ij}$ and $B_{jk}$ refer to individual matrix elements. These matrix elements are regular numbers that can be interchanged: $A_{ij}B_{jk} = B_{jk}A_{ij}$. This means that the decomposition of a matrix equation into its components allows us to move the individual components around in any way that we like.

In the following we will use the transpose $\mathbf{C}^T$ of a matrix $\mathbf{C}$ extensively. This is the matrix in which the rows and the columns are interchanged, so that

$$C_{ij}^T \equiv C_{ji} . \tag{21.17}$$

**Problem f:** Use the results in this section to show that (21.15) can be written as

$$\sum_{i,j,k} C_{ji}^T C_{ik} u_j v_k = \sum_i u_i v_i . \tag{21.18}$$

At this point it is important to realize that it is immaterial what we call a summation index. The right hand side of (21.18) can just as well be written as $\sum_j u_j v_j$ because this sum and $\sum_i u_i v_i$ both mean: 'take the corresponding elements of **u** and **v**, multiply them and sum over all components.'

**Problem g:** Show that (21.18) is equivalent to

$$\sum_{j,k} \left( \sum_i C_{ji}^T C_{ik} \right) u_j v_k = \sum_j u_j v_j . \tag{21.19}$$

**Problem h:** Use the definition (10.15) of the Kronecker delta to show that $\sum_j u_j v_j = \sum_{j,k} \delta_{jk} u_j v_k$.

**Problem i:** Use this result to show that **C** is unitary when it satisfies the following condition

$$\left( \sum_i C_{ji}^T C_{ik} \right) = \delta_{jk} . \tag{21.20}$$

At this point we use that $\delta_{jk}$ is just the $j,k$ element of the identity matrix $\mathbf{I}$ because the elements of this matrix are equal to 1 on the diagonal ($j = k$) and equal to zero off the diagonal ($j \neq k$). In matrix notation, (21.20) for a unitary matrix therefore implies that

$$\mathbf{C}^T \mathbf{C} = \mathbf{I} . \tag{21.21}$$

**Problem j:** Show that this implies that the inverse of a unitary matrix is equal to its transpose:

$$\mathbf{C}^{-1} = \mathbf{C}^T . \tag{21.22}$$

This result can save you a lot of work. Suppose you need to invert a matrix, and suppose that you suspect that the matrix is unitary, then according to (21.22) the inverse matrix is equal to the transpose. You can verify whether this is indeed the case by multiplying this 'guess' by the original matrix $\mathbf{C}$ to check whether the result is equal to the identity matrix; this amounts to applying (21.21).

**Problem k:** The matrix corresponding to a rotation over an angle $\varphi$ is given by

$$\mathbf{R} = \begin{pmatrix} \cos\varphi & \sin\varphi \\ -\sin\varphi & \cos\varphi \end{pmatrix} . \tag{21.23}$$

Compute the inverse of this matrix with a minimal amount of work. (This is also called the principle of maximal laziness.)

## 21.3 Shear or dilatation?

As an example of the importance of coordinate transforms we consider a geologist who observes a rock sample that is subjected to a shear as shown in figure 21.2. The height and the width of the rock are of unit length. In the shear motion the upper side of the rock is displaced rightward over a distance $d$ while the lower side of the rock is displaced leftward over a distance $-d$.

**Problem a:** Use figure 21.2 to show that the original unit vectors change during the deformation as

$$\begin{pmatrix} 1 \\ 0 \end{pmatrix} \rightarrow \begin{pmatrix} 1 \\ 0 \end{pmatrix} \quad \text{and} \quad \begin{pmatrix} 0 \\ 1 \end{pmatrix} \rightarrow \begin{pmatrix} d \\ 1 \end{pmatrix} . \tag{21.24}$$

**Problem b:** Let the deformation be described by a $2 \times 2$ matrix $\mathbf{D}$. Show that this matrix is given by

$$\mathbf{D} = \begin{pmatrix} 1 & d \\ 0 & 1 \end{pmatrix} . \tag{21.25}$$

Hint: check that this produces the deformation described by (21.24).

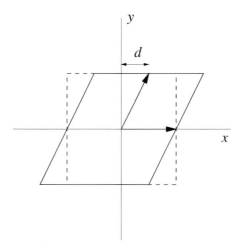

Fig. 21.2. The shear of a rock as seen by the geologist before (dashed lines) and after (solid lines) the shear.

The geologist is in general interested in the internal deformation of the rock. However, the transformation shown in figure 21.2 entails both a deformation within the rock sample as well as a rotation of the sample. We first need to unravel these two contributions. In the following we only considered very small shear ($d \ll 1$), the associated rotation is therefore also small.

**Problem c:** Use a Taylor expansion of (21.23) to show that the rotation matrix over an infinitesimal angle $\varphi$ is given by

$$\mathbf{R} = \mathbf{I} + \begin{pmatrix} 0 & \varphi \\ -\varphi & 0 \end{pmatrix}. \tag{21.26}$$

The shear described by (21.25) can be decomposed into this rotation plus an internal deformation. To make this explicit we decompose the matrix $\mathbf{D}$ into the matrix on the right hand side of (21.26) which accounts for the rotation and a new matrix $\mathbf{E}$:

$$\mathbf{D} = \begin{pmatrix} 0 & \varphi \\ -\varphi & 0 \end{pmatrix} + \mathbf{E}. \tag{21.27}$$

At this point we need to find the parameter $\varphi$ that characterizes the rotational component. It follows from (21.26) that the difference between the 12-component and the 21-component of the matrix is equal to $2\varphi$. According to (21.25) this difference is given by $D_{12} - D_{21} = d$; since this is equal to $2\varphi$ the rotation angle is given by $\varphi = d/2$.

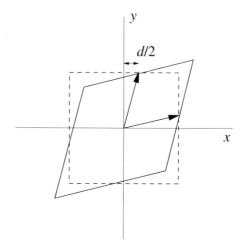

Fig. 21.3.   The same deformation as in the previous figure when the rotational component is removed. The undeformed sample is shown by the dashed lines, the deformed sample by the solid lines.

**Problem d:** Show that the deformation component of the transformation in figure 21.2 is given by

$$\mathbf{E} = \begin{pmatrix} 1 & d/2 \\ d/2 & 1 \end{pmatrix}. \tag{21.28}$$

Note that this matrix is symmetric $(E_{ij} = E_{ji})$, whereas the matrix $\mathbf{R}$ in (21.26) is antisymmetric $(R_{ij} = -R_{ji})$. The matrix $\mathbf{D}$ has thus been decomposed into a symmetric part and an antisymmetric part. This corresponds to a decomposition of the deformation into a rotational component plus a deformation. We have seen this before in chapter 6 where it was shown that the curl of a vector field in general has two contributions: rotation and shear. We return to this issue in section 21.9.

**Problem e:** We now focus on the internal deformation of the sample. Show that figure 21.3 displays the deformation $\mathbf{E}$ of the rock. Hint: determine first what happens to the unit vectors in the $x$- and $y$-directions when they are multiplied with the matrix $\mathbf{E}$.

Let us now consider a second geologist who is studying the same deforming rock but who is using a coordinate system that is rotated through 45 degrees with respect to the coordinate system used by the geologist who views the world as seen in figure 21.2. We ignore the effect of the rotation altogether because it does not lead to internal deformation of the rock. The $x'$-axis of the second geologist is aligned along the line $x = y$,

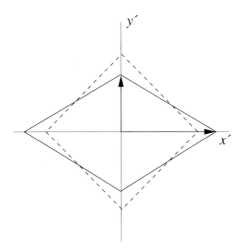

Fig. 21.4. The deformation of the previous figure seen by a geologist who uses a coordinate system that is rotated through 45 degrees.

whereas her $y'$-axis is aligned along the line $x = -y$. Hence the basis vectors of the second geologist are given by

$$\hat{\mathbf{e}}^{x'} = \frac{1}{\sqrt{2}} \begin{pmatrix} 1 \\ 1 \end{pmatrix}, \quad \hat{\mathbf{e}}^{y'} = \frac{1}{\sqrt{2}} \begin{pmatrix} 1 \\ -1 \end{pmatrix}. \tag{21.29}$$

**Problem f:** The simplest way to determine how she views the deformation is to determine what happens when $\mathbf{E}$ acts on the basis vectors of the second geologist. Show that this gives

$$\mathbf{E}\hat{\mathbf{e}}^{x'} = (1 + d/2)\,\hat{\mathbf{e}}^{x'}, \quad \mathbf{E}\hat{\mathbf{e}}^{y'} = (1 - d/2)\,\hat{\mathbf{e}}^{y'}. \tag{21.30}$$

**Problem g:** Use this result to show that the second geologist sees the deformation process as depicted in figure 21.4.

Note that the unit vectors of the second geologist are the eigenvectors of the matrix $\mathbf{E}$.

The second geologist would describe the deformation process seen in figure 21.4 as follows. The sample is extended in the $x'$-direction because the unit vector $\hat{\mathbf{e}}^{x'}$ is mapped onto itself with an amplification factor $(1 + d/2)$ that is greater than 1; in the $y'$-direction the sample is compressed because the unit vector $\hat{\mathbf{e}}^{y'}$ is mapped onto itself with an amplification factor $(1 - d/2)$ that is smaller than 1. This means that (apart from a rotation) the first geologist describes the transformation as pure shear whereas the second geologist describes the transformation as a combination of extension and compression. These two points of view must be equivalent. However, they can only be reconciled when we have techniques for switching from one coordinate system to another. This the goal of tensor calculus. Before we derive the transformation rules for matrices and more general tensors, we first introduce a notational convention that turns out to be very convenient.

## 21.4 The summation convention

In this section the summation convention is introduced; this convention is sometimes also referred to as the 'Einstein summation convention.' Let us consider the multiplication of the matrix $\mathbf{A}$ by a vector $\mathbf{x}$ that gives a new vector $\mathbf{y} = \mathbf{A}\mathbf{x}$.

**Problem a:** This is a vectorial notation; show that in component form this expression should be written as

$$\sum_j A_{ij} x_j = y_i \ . \tag{21.31}$$

You have seen in section 21.2 that in matrix and vector calculations one has to carry out a lot of summations. It is also shown in that section that working with the expressions in component form has advantages over working with the more abstract vector/matrix notation. We could save a lot of work by leaving out the summation signs in the equations. This is exactly what the summation convention does:

> *In the summation convention one leaves out the summation signs ($\sum$) with the understanding that one sums over any repeated index on either side of the equation.*

What does this mean for expression (21.31)? In this expression we sum over the index $j$. Note that this index appears twice on the left hand side. According to the summation convention we would write this expression as

$$A_{ij} x_j = y_i \ . \tag{21.32}$$

It takes a little time to get used to this notation. It should be kept in mind that under the summation convention the identity (21.32) does *not* imply that this expression holds for a single value of $j$. For example, (21.32) does not mean that $A_{i2} x_2 = y_i$ because a summation over the index $j$ is implied rather than that $j$ can take any fixed value $j = 1$ or $j = 2$ or $j = 3$.

As another example of the summation convention consider the inner product of two vectors. Normally one would write this as $(\mathbf{u} \cdot \mathbf{v}) = \sum_i u_i v_i$; using the summation convention one writes it as $(\mathbf{u} \cdot \mathbf{v}) = u_i v_i$. Note that the name of the summation index is irrelevant. This expression could therefore also be written as $(\mathbf{u} \cdot \mathbf{v}) = u_j v_j$. The matrix equation $\mathbf{C}^T \mathbf{C} = \mathbf{I}$ is written in component form as $\sum_k C_{ik}^T C_{kj} = \sum_k C_{ik} C_{kj} = \delta_{ij}$; in the summation convention this is written as $C_{ik}^T C_{kj} = C_{ik} C_{kj} = \delta_{ij}$.

**Problem b:** Write out the matrix product $\mathbf{AI}$ (with $\mathbf{I}$ is the identity matrix) in component form, and show that the result can be written as: $(\mathbf{AI})_{ij} = A_{ik} \delta_{kj}$.

In the last term a summation over the index $k$ is implied. The Kronecker delta is always zero, except when $k = j$. This means that the only nonzero contribution from the $k$-summation comes from the term $k = j$, so that $(\mathbf{AI})_{ij} = A_{ik} \delta_{kj} = A_{ij}$. This is simply the identity $\mathbf{AI} = \mathbf{A}$ in component form.

**Problem c:** Show that $v^2 = v_i v_i$ where $v$ is the length of the vector $\mathbf{v}$.

**Problem d:** The Laplacian is the divergence of the gradient, use this to write the Laplacian in the following form

$$\nabla^2 f = \partial_i \partial_i f , \qquad (21.33)$$

with $\partial_i \equiv \partial / \partial x_i$.

**Problem e:** Use the summation convention to rewrite expression (10.55) as

$$\frac{\partial(\rho v_i)}{\partial t} + \partial_j(\rho v_j v_i) = \mu \partial_k \partial_k v_i + F_i . \qquad (21.34)$$

**Problem f:** In some applications one carries out summations over several indices. As an example consider the double sum $\sum_{i,j} A_{ij} \delta_{ji}$. Carry out the summation and show that the result can be written as

$$A_{ij} \delta_{ji} = A_{ii} . \qquad (21.35)$$

Note that on the left hand side a summation over $i$ and $j$ is implied whereas on the right hand side a sum over $i$ is implied. The term on the right hand side is the sum of the diagonal elements of the matrix $\mathbf{A}$. This quantity plays an important role in a variety of applications and is called the *trace* of $\mathbf{A}$:

$$\mathrm{tr}\,\mathbf{A} \equiv \sum_i A_{ii} = A_{ii} . \qquad (21.36)$$

One caveat should be made with the summation convention. Suppose that one writes $M_{ii} = 0$. According to the summation convention this means that the trace of $\mathbf{M}$ is equal to zero. However, one could also mean to express that all the diagonal elements of $\mathbf{M}$ are equal to zero: $M_{11} = M_{22} = M_{33} = 0$. This is an example where an expression can be ambiguous. It is important to state explicitly whether one uses the summation convention. When one does, one can indicate that one deviates from this convention in an equation by explicitly stating that no summation is implied. For example, when one wants to express that all the diagonal elements of $\mathbf{M}$ are equal to zero one could write this in the summation convention as:

$$M_{ii} = 0 \qquad \text{(no summation over } i\text{)} . \tag{21.37}$$

**Problem g:** Use the properties of unitary matrices as derived in section 21.2 to show that for a unitary matrix $\mathbf{C}$:

$$C_{ij}C_{ik} = \delta_{kj} . \tag{21.38}$$

## 21.5 Matrices and coordinate transforms

In this section we determine the transformation properties of matrices under a coordinate transform $\mathbf{C}$ as introduced in section 21.1. In the remainder of this section we then restrict ourselves to unitary coordinate transformations. Let a matrix $\mathbf{D}$ map a vector $\mathbf{x}$ to a vector $\mathbf{y}$

$$\mathbf{y} = \mathbf{D}\mathbf{x} . \tag{21.39}$$

Consider this same operation in a new coordinate system and let us use a prime to denote the corresponding quantities is the new coordinate system. According to (21.12) the vector $\mathbf{x}$ in the new coordinate system is given by $\mathbf{x}' = \mathbf{C}\mathbf{x}$, and a similar expression holds for $\mathbf{y}$. Expression (21.39) is given in the new coordinate system by

$$\mathbf{y}' = \mathbf{D}'\mathbf{x}' . \tag{21.40}$$

In this section we determine the relation between the matrix $\mathbf{D}'$ in the new coordinate system and the old matrix $\mathbf{D}$.

**Problem a:** Use (21.12) to show that (21.40) can be written as

$$\mathbf{C}\mathbf{y} = \mathbf{D}'\mathbf{C}\mathbf{x} . \tag{21.41}$$

**Problem b:** Multiply this expression on the left by $\mathbf{C}^{-1}$, compare the result with (21.39) to derive that $\mathbf{D} = \mathbf{C}^{-1}\mathbf{D}'\mathbf{C}$. Multiply this last expression on the left and the right by suitable matrices to derive that

$$\mathbf{D}' = \mathbf{C}\mathbf{D}\mathbf{C}^{-1} . \tag{21.42}$$

This is the general transformation rule for matrices under a coordinate transform. It looks different from the transformation rule (21.12) for vectors because in (21.42) the inverse $\mathbf{C}^{-1}$ of the coordinate transformation appears as well. However, for unitary coordinate transforms one can eliminate this inverse and write (21.42) in a way that is similar to (21.12).

**Problem c:** Use that $\mathbf{C}$ is unitary and rewrite (21.42) in component form as

$$D'_{ij} = \sum_{k,l} C_{ik} C_{jl} D_{kl} \ . \tag{21.43}$$

(Remember that you cannot interchange the order of matrices in a multiplication, but that you can change the order of matrix *elements* that are multiplied.)

**Problem d:** Redo the calculation in the last problem using the summation convention at every step and rewrite the preceding result as

$$D'_{ij} = C_{ik} C_{jl} D_{kl} \ . \tag{21.44}$$

**Problem e:** To compare this result with the transformation rule (21.12) for vectors write $\mathbf{v}^{old} = \mathbf{v}$, $\mathbf{v}^{new} = \mathbf{v}'$ and use the summation convention to rewrite (21.12) using the summation convention as

$$v'_i = C_{ij} v_j \ . \tag{21.45}$$

Applying the matrix transformation (21.44) may appear to be complex to you. The easiest way to apply it in practice is to write it as

$$\mathbf{D}' = \mathbf{C} \mathbf{D} \mathbf{C}^T \ . \tag{21.46}$$

This expression is identical to (21.42) with the only exception that it is used that $\mathbf{C}$ is unitary: $\mathbf{C}^{-1} = \mathbf{C}^T$.

**Problem f:** According to (21.23) a rotation through 45 degrees of the coordinate system is described by the coordinate transform

$$\mathbf{C} = \frac{1}{\sqrt{2}} \begin{pmatrix} 1 & 1 \\ -1 & 1 \end{pmatrix} \ . \tag{21.47}$$

Check that this coordinate transformation transforms the matrix $\mathbf{E}$ in (21.28) to the matrix

$$\mathbf{E}' = \begin{pmatrix} 1 + d/2 & 0 \\ 0 & 1 - d/2 \end{pmatrix} \ . \tag{21.48}$$

Show also that this matrix describes the deformation shown in figure 21.4.

## 21.6 Definition of a tensor

Expressions (21.44) and (21.45) allow us to see a resemblance between the transformation rule (21.44) for a matrix and the rule (21.45) for a vector. In both equations the old vector or matrix is multiplied by the matrix $\mathbf{C}$; for the vector this happens once and for the matrix it happens twice. The *first* index of the quantity on the left hand side ($i$ in both cases) is the first index of the *first* matrix $\mathbf{C}$ on the right hand side as indicated by the arrows in the following expressions:

$$v'_i = C_{ij}v_j , \quad D'_{ij} = C_{ik}C_{jl}D_{kl} .$$

The *second* index of $\mathbf{D}'$ in (21.44) corresponds to the first index of the *second* term $\mathbf{C}$ as indicated by the arrows:

$$D'_{ij} = C_{ik}C_{jl}D_{kl} .$$

In both (21.44) and (21.45) the *second* index of each term $\mathbf{C}$ also occurs in the term $\mathbf{D}$ or $\mathbf{v}$ on which this matrix acts as shown by the arrows in the following expressions:

$$v'_i = C_{ij}v_j , \quad D'_{ij} = C_{ik}C_{jl}D_{kl} , \quad D'_{ij} = C_{ik}C_{jl}D_{kl} .$$

Note that according to the summation convention, a summation over these repeated indices is implied.

Some quantities that we use are not labelled by any index. Examples are temperature and pressure. Such a quantity is a pure number and it is called a *scalar*. A vector has a direction and is labelled with one index and a matrix has two indices. There is no reason not to define objects which are characterized with an arbitrary number of indices. In general a physical quantity can have any number of subscripts, and a quantity with $n$ subscripts can be denoted as $T_{i_1 i_2 \cdots i_n}$. We call such an object a tensor when it follows a transformation rule similar to (21.44) and (21.45) for a matrix and a vector, respectively.

**Definition:** The quantity $T_{i_1 i_2 \cdots i_n}$ is called a tensor of rank $n$ when it transforms under a unitary coordinate transform $\mathbf{C}$ with the following transformation rule:

$$T_{i_1 i_2 \cdots i_n} = C_{i_1 j_1} C_{i_2 j_2} \cdots C_{i_n j_n} T_{j_1 j_2 \cdots j_n} . \qquad (21.49)$$

Note that the summation convention is used so a summation over the indices $j_1, j_2, \ldots, j_n$ is implied.

**Problem a:** Convince yourself that the transformation rules (21.44) and
(21.45) for a matrix and a vector, respectively, are special cases of
this definition for the values $n = 2$ and $n = 1$, respectively.

This means that a vector that transforms according to (21.45) is a tensor of rank one and a matrix that transforms according to (21.44) is a
tensor of rank two. Note that a scalar does not change under coordinate
transformations. It has zero subscripts and one can say that when one
applies the coordinate transformation (21.49) the coordinate transform
**C** is applied zero times, for this reason one calls a scalar a tensor or rank
zero.

Let us consider some examples of tensors. The position vector **r** is a
tensor of rank one, as follows from the transformation rule (21.12) that
you derived in section 21.1. From this it follows that the velocity vector
**v** is a tensor of rank one provided that the coordinate transformation **C**
does not depend on time.

**Problem b:** Differentiate the transformation law $x'_i = C_{ij} x_j$ for the position vector with respect to time and use the result to show that
the velocity indeed transforms as a tensor of rank one.

**Problem c:** Show that acceleration is a tensor of rank one.

Now we can use Newton's law $\mathbf{F} = m\mathbf{a}$ and that the mass $m$ is a scalar.
Since the acceleration is a tensor of rank one, the force must also be a
tensor of rank one.

It is interesting to see what happens when one works in a rotating
coordinate system. In that case the transformation **C** which transforms
the fixed system to the rotating system depends on time.

**Problem d:** Go through the steps of the problems b and c and show
that in that case the acceleration transforms as

$$a'_i = C_{ij} a_j + 2\dot{C}_{ij} v_j + \ddot{C}_{ij} x_j , \qquad (21.50)$$

where the dot denotes a time derivative.

This means that for such a time-dependent coordinate transform the acceleration does *not* transform as a vector of rank one. Note that additional
terms appear that are proportional to the velocity and the position vector. These correspond to the Coriolis force and the centrifugal force as
treated in section 12.3.

Another example of a tensor of rank one is the gradient vector $\nabla_i =
\partial/\partial x_i$. The chain rule of differentiation states that

$$\nabla'_i = \frac{\partial}{\partial x'_i} = \frac{\partial x_j}{\partial x'_i} \frac{\partial}{\partial x_j} = \frac{\partial x_j}{\partial x'_i} \nabla_j , \qquad (21.51)$$

where to conform with the summation convention a summation over $j$ is implied.

**Problem e:** Use the transformation law of the position vector to derive that $\partial x_j / \partial x_i' = C_{ji}^{-1}$, and use the property that $\mathbf{C}$ is a unitary matrix to derive that

$$\nabla_i' = C_{ij} \nabla_j . \tag{21.52}$$

In other words, the gradient vector is a tensor of rank one.

As an example of a tensor of rank two we consider the identity matrix whose elements are the Kronecker delta $\delta_{ij}$. At first sight you might believe that since the elements of the Kronecker delta are simply equal to zero or one, it is a scalar. However, the identity matrix really transforms like a tensor of rank two.

**Problem f:** To see this, assume that the identity matrix follows the transformation rule (21.44) and use the property of the Kronecker delta to derive that

$$I_{ij}' = C_{ik} C_{jl} \delta_{kl} = C_{ik} C_{jk} . \tag{21.53}$$

**Problem h** Write the last term as $C_{kj}^T$ and use the property that $\mathbf{C}$ is unitary to derive that

$$I_{ij}' = \delta_{ij} . \tag{21.54}$$

In other words, the identity matrix transforms as a tensor of rank two and yet it takes the same form in every coordinate system!

## 21.7 Not every vector is a tensor

At this point you may think that any object with $n$-components is a tensor of rank one. However, keep in mind that not every vector transforms according to the transformation rule (21.49) for a tensor. Let us first consider an obvious example. One might be interested in the distribution of shoe sizes of students. In such a study one would measure the shoe size of each student in the group and could form a vector of the data that one obtains:

$$\mathbf{d} \equiv \begin{pmatrix} \text{shoe size of Marie} \\ \text{shoe size of Peter} \\ \text{shoe size of Klaas} \\ \vdots \end{pmatrix} . \tag{21.55}$$

This vector is not a tensor. There are two reasons for this. The coordinate transformation matrix $\mathbf{C}$ is by definition a square matrix with a dimension equal to the dimension of the coordinate system (usually 3 and for some

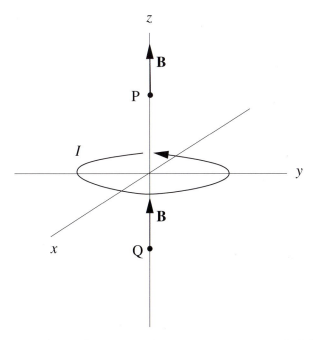

Fig. 21.5. A circular current that generates a magentic field **B**.

applications 4, see section 21.12). The vector **d** can have any dimension and the action of **C** in this vector is not defined. The second reason why **d** is not a tensor of rank one is that the shoe size of students does not depend on the coordinate system. In other words, all the elements of **d** are scalars, hence **d** is certainly not a tensor of rank one.

Another example of a vector that is not a tensor is the stress–displacement tensor that is used in the description of elastic wave propagation in elastic media [3]:

$$\mathbf{w} = \left( \begin{array}{c} u_y \\ \sigma_{yz} \end{array} \right) , \qquad (21.56)$$

where $u_y$ and $\sigma_{yz}$ are suitably chosen components of the displacement and the stress. This vector is not a tensor of rank one. It has a different dimension than **C**, hence the action of **C** on this vector is not defined. Furthermore the components of **w** are of different dimension, hence they can never be mixed in the transformation rule (21.49).

Let us now consider a more subtle example. At this point you might think that the magnetic field vector is a tensor of rank one. Consider the circular current in the $(x, y)$-plane shown in figure 21.5 that generates a magnetic field. Given the direction of this current the magnetic field on the $z$-axis is oriented in the direction of the positive $z$-axis. This holds for the points P and Q in figure 21.5.

**Problem a:** Let us perform the coordinate transformation in which the system is reflected in the $(x, y)$-plane. Make a drawing of the transformed current and make sure you understand that the current has not changed in the coordinate transform.

**Problem b:** Now that you know that the current has not changed, use figure 21.5 to draw the associated magnetic field vector at the points P and Q.

**Problem c:** Now add to your figure the image of the vectors **B** of figure 21.5 after they have been subjected to a reflection in the $(x, y)$-plane.

If you made the drawing correctly, you will have found that the magnetic field that you obtained after transforming the original magnetic field vector had the opposite direction to the magnetic field that was generated by the current that had been subjected to the coordinate transform.

The only conclusion is that the magnetic field does not behave like a tensor of rank one under coordinate transforms: it does not transform according to the transformation rule (21.45). In fact, the magnetic field behaves in a different way. It is an example of a *pseudo-tensor* indicating that it shares some of the properties of the tensors that we have explored, but that its transformation rules differ from the transformation rule (21.45) that we have been used to. More details about pseudo-tensors can be found in ref. [72]. The main thing to remember at this point is that not every vector is a tensor!

The next example shows that not every tensor is Cartesian. In section 3.2 we derived the transformation that maps the components $v_x$, $v_y$ and $v_z$ of a vector in Cartesian coordinates into its components $v_r$, $v_\theta$ and $v_\varphi$ in spherical coordinates:

$$\begin{pmatrix} v_r \\ v_\theta \\ v_\varphi \end{pmatrix} = \begin{pmatrix} \sin\theta\cos\varphi & \sin\theta\sin\varphi & \cos\theta \\ \cos\theta\cos\varphi & \cos\theta\sin\varphi & -\sin\theta \\ -\sin\varphi & \cos\varphi & 0 \end{pmatrix} \begin{pmatrix} v_x \\ v_y \\ v_z \end{pmatrix}. \qquad (3.19)$$

This transformation is of the same form as the general transformation rule (21.45) for a tensor of rank one. However, the matrix of the coordinate transform now depends on the position; this is a result of the fact that the spherical coordinate system is curvilinear. When we take a derivative of the vector, the dependence of the transformation matrix on the position gives additional terms. For this reason tensor calculus for tensors in curvilinear coordinate systems is significantly more complex than tensor calculus for Cartesian tensors.

## 21.8 The products of tensors

One can form products of vectors in different ways. You have probably seen the inner product and outer product of vectors. There are also other ways in which one can make products of tensors. The first product of tensors that we consider is the *contraction*. You have already seen examples of this: the inner product of two vectors is one:

$$(\mathbf{u} \cdot \mathbf{v}) = u_i v_i \ . \tag{21.57}$$

Another example is matrix-vector multiplication:

$$(\mathbf{M} \cdot \mathbf{v})_i = M_{ij} v_j \ . \tag{21.58}$$

Note that according to the summation convention a summation over $i$ is implied in (21.57) and a summation over $j$ in (21.58).

These operations can be extended. Let $U_{i_1 i_2 \cdots i_n}$ be a tensor of rank $n$ and $V_{j_1 j_2 \cdots j_m}$ a tensor of rank $m$. The contraction of $\mathbf{U}$ and $\mathbf{V}$ is defined by setting the last index of $\mathbf{U}$ equal to the first index of $\mathbf{V}$ and summing over this index:

$$(\mathbf{U} \cdot \mathbf{V})_{i_1 i_2 \cdots i_{n-1} k_2 \cdots k_m} = U_{i_1 \cdots i_{n-1} r} V_{r k_2 \cdots k_m} \ . \tag{21.59}$$

Note that we sum over the index $r$.

**Problem a:** Show that the rank of this tensor is $n + m - 2$.

**Problem b:** Show that the inner product of two vectors and matrix-vector multiplication are special cases of the contraction (21.59). Make sure you define $n$ and $m$ in each example. Verify that the rank of the result is indeed $n + m - 2$.

One can also apply a multiple contraction in which one sums over two or more indices of the tensors that one contracts. For example, the double contraction is given by

$$(\mathbf{U} : \mathbf{V})_{i_1 i_2 \cdots i_{n-2} k_3 \cdots k_m} = U_{i_1 \cdots i_{n-2} r s} V_{s r k_3 \cdots k_m} \ . \tag{21.60}$$

**Problem c:** Show that the resulting tensor is of rank $n + m - 4$.

A very important property is that *the contraction of two tensors is also a tensor*. The proof of this property is not difficult, it is mostly a tedious bookkeeping exercise, and the details can be found for example in refs. [18] and [72]. Here we will study some examples.

**Problem d:** Show that the double contraction of a matrix $\mathbf{A}$ with the identity matrix is equal to the trace of $\mathbf{A}$ as defined in (21.36):

$$(\mathbf{A} : \mathbf{I}) = A_{ii} = \operatorname{tr} \mathbf{A} \ . \tag{21.61}$$

The fact that the contraction of a tensor is also a tensor implies that the trace of a matrix is a tensor of rank zero: the trace of a matrix is a scalar. Therefore the trace of a matrix is invariant to unitary coordinate transformations.

**Problem e:** There is one coordinate transformation in which the new basis vectors are aligned with the eigenvectors of **A**. Show that in that particular coordinate system the trace is equal to the sum of the eigenvectors $\lambda_i$. Use this result to derive that in any coordinate system the trace of a matrix is equal to the sum of the eigenvalues:

$$\operatorname{tr} \mathbf{A} = \sum_i \lambda_i . \qquad (21.62)$$

A matrix that is often used in the *Hessian* of a function $f$ which is defined by the second partial derivatives of that function:

$$H_{ij} \equiv \frac{\partial^2 f}{\partial x_i \partial x_j} . \qquad (21.63)$$

**Problem e:** Show that **H** is a tensor of rank two by generalizing the derivation of problem e of section 21.6 to the transformation properties of the second derivative.

**Problem f:** Show that the trace of the Hessian is the Laplacian of $f$

$$\operatorname{tr} \mathbf{H} = \nabla^2 f . \qquad (21.64)$$

The trace is invariant for unitary coordinate transformations and the Laplacian is the trace of a tensor; therefore the Laplacian is invariant for unitary coordinate transformations as well.

The second type of product of tensors is the direct product. The direct product of two tensors is formed by multiplying the different elements of the two tensors without carrying out a summation over repeated indices. An example of this is the dyad of two vectors that you encountered for example in section 12.1 where the projection operator is written as the direct product of the unit vector $\hat{\mathbf{n}}$ with itself:

$$\mathbf{P} = \hat{\mathbf{n}} \hat{\mathbf{n}}^T . \qquad (12.5)$$

In component form this expression is given by

$$P_{ij} = n_i n_j . \qquad (21.65)$$

This idea can be generalized to form the direct product of a tensor **U** of rank $n$ and a tensor **V** of rank $m$. This direct product is usually denoted with the symbol $\otimes$:

$$(\mathbf{U} \otimes \mathbf{V})_{i_1 \cdots i_n j_1 \cdots j_m} = U_{i_1 \cdots i_n} V_{j_1 \cdots j_m} . \qquad (21.66)$$

Note that there are no repeated indices over which one sums.

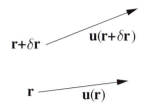

Fig. 21.6. The deformation of two nearby points **r** and **r**+δ**r**

**Problem g:** Show that the rank of this direct product is $n + m$.

**Problem h:** Show that the dyad (12.5) is a special example of the direct product (21.65).

**Problem i:** The direct product of two tensors is also a tensor. Show this property by applying the transformation rule (21.49) to the tensors on right hand side of (21.66).

**Problem j:** The gradient of a function is described in chapter 4. The gradient $\mathbf{G} = \nabla \otimes \mathbf{v}$ of a vector $\mathbf{v}$ is defined as

$$G_{ij} \equiv \frac{\partial v_j}{\partial x_i} \, . \tag{21.67}$$

Show that this is a tensor of rank two.

## 21.9 Deformation and rotation again

As an important example of the direct product we consider in this section the strain tensor which measures the state of deformation in a medium. Consider a medium that is subject to some deformation and let us focus on two nearby points **r** and **r** + δ**r** in the medium as shown in figure 21.6. During the deformation each point **r** is displaced by a vector **u**(**r**), which means that after the deformation the point **r** is located at **r** + **u**(**r**) and the point **r**+δ**r** is located at **r**+δ**r** + **u**(**r**+δ**r**). When the displacement of these neighboring points is the same (**u**(**r**+δ**r**) = **u**(**r**)) the relative positions of these points are not changed. Therefore, the deformation is associated with the variation of the displacement vector **u**(**r**) with position **r**. In order to describe the deformation we use the gradient tensor **D** = ∇⊗**u** of the displacement vector.

$$D_{ij} \equiv \frac{\partial u_j}{\partial x_i} \, . \tag{21.68}$$

**Problem a:** In a uniform translation (**u**(**r**) =*const.*) the medium is not deformed. Show that for this deformation the tensor **D** is indeed equal to zero.

The result from a previous problem reflects the fact that when the medium is not deformed the gradient tensor $\mathbf{D}$ is equal to zero. Unfortunately the reverse is not true; one may have a displacement that entails no deformation but the gradient tensor can be nonzero. We have seen in section 21.3 an example that the displacement can be decomposed into a rotation and a deformation. Let us assume for the moment that the displacement is due to a rigid rotation around a point $\mathbf{r}_0$ with rotation vector $\boldsymbol{\Omega}$. According to (12.24) the associated displacement is given by

$$\mathbf{u}^{rot}(\mathbf{r}) = \boldsymbol{\Omega} \times (\mathbf{r} - \mathbf{r}_0) \,. \tag{21.69}$$

**Problem b:** Show that in component form this displacement is given by

$$\mathbf{u}^{rot}(\mathbf{r}) = \begin{pmatrix} \Omega_y(z - z_0) - \Omega_z(y - y_0) \\ \Omega_z(x - x_0) - \Omega_x(z - z_0) \\ \Omega_x(y - y_0) - \Omega_y(x - x_0) \end{pmatrix} \,. \tag{21.70}$$

**Problem c:** Compute the partial derivatives of this vector to show that the associated gradient tensor is given by

$$\mathbf{D}^{rot} = \begin{pmatrix} 0 & \Omega_z & -\Omega_y \\ -\Omega_z & 0 & \Omega_x \\ \Omega_y & -\Omega_x & 0 \end{pmatrix} \,. \tag{21.71}$$

This is a remarkable result; although the displacement associated with the rotation depends on the position, the associated gradient tensor does not depend on the position. Note also that the gradient tensor does not depend on the point $\mathbf{r}_0$ around which the rotation takes place.

The gradient tensor $\mathbf{D}^{rot}$ is antisymmetric: each element is equal to the element on the other side of the main diagonal with an opposite sign: $D_{ij}^{rot} = -D_{ji}^{rot}$. Stated differently, the sum of each element and the element on the other side of the diagonal is equal to zero:

$$D_{ij}^{rot} + D_{ji}^{rot} = 0 \,. \tag{21.72}$$

In general, any tensor of rank two can be written as the sum of a symmetric tensor and an antisymmetric tensor by using the following identity:

$$D_{ij} = \underbrace{\frac{1}{2}(D_{ij} + D_{ji})}_{\text{deformation}} + \underbrace{\frac{1}{2}(D_{ij} - D_{ji})}_{\text{rigid rotation}} \,. \tag{21.73}$$

**Problem d:** Verify that this identity holds for any matrix $\mathbf{D}$. Show that the first term is symmetric and that the second term is antisymmetric.

According to (21.72) the rotational component of the displacement does not contribute to the first term of (21.73). For this reason, the first term of this expression is used to characterize the deformation of the medium, while the second term characterizes the rotational component of the displacement. The strain tensor that characterizes the deformation of the medium $\varepsilon$ is defined by the first term of (21.73).

**Problem e:** Show that according to this definition

$$\varepsilon_{ij} = \frac{1}{2} \left( \frac{\partial u_j}{\partial x_i} + \frac{\partial u_i}{\partial x_j} \right) . \tag{21.74}$$

**Problem f:** We know that **D** is a tensor because it is the direct product of the gradient vector and the displacement vector, but we have to show that $\varepsilon$ is a tensor. Do this by showing that when **D** is a tensor of rank two, then $\mathbf{D}^T$ also transforms as a tensor of rank 2. Then use that $\varepsilon = (1/2) \left( \mathbf{D} + \mathbf{D}^T \right)$ to show that $\varepsilon$ is a tensor of rank two as well.

The strain tensor plays a crucial role in continuum mechanics because it is a measure of the degree of deformation in a medium. As shown in the tutorial of Lister and Williams [50] the partitioning between rotation and shear plays a vital role in structural geology since it is crucial in the generation of faults and shear zones.

## 21.10 The stress tensor

In general, when a medium is deformed, reaction forces become operative that tend to counteract the deformation. However, what do we mean by 'reaction forces'? A force acts on something, but in a continuous medium there appears to be nothing to act on except a 'point' in the medium. Since a point has zero mass this would lead to an infinite acceleration. This paradox can be resolved by considering a hypothetical cube in the medium as shown in figure 21.7. The cube has six sides and we consider the force exerted by the rest of the medium on these six sides. Let us focus on the side on the right that is perpendicular to the $x$-direction. In other words the normal vector to this side is oriented in the $x$-direction.

The force on this surface depends on the size of the surface. A meaningful way to describe this is to use the *traction*, which is defined as the force per unit surface area. This traction $\mathbf{T}^{(x)}$ obviously has three components which are shown in figure 21.7; these are denoted by $T_x^{(x)}$, $T_y^{(x)}$ and $T_z^{(x)}$. The subscript refers to the component of the traction, the superscript refers to the fact that this is the traction on a surface perpendicular to

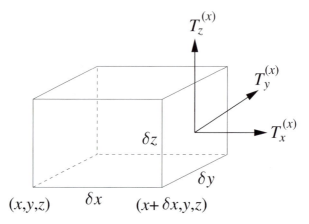

Fig. 21.7.   The traction acting on a surface perpendicular to the $x$-direction.

the $x$-direction. The component $T_x^{(x)}$ is normal to the surface and corresponds to a normal force, while the components $T_y^{(x)}$ and $T_z^{(x)}$ are parallel to the surface. The latter components are called the shear tractions because they cause shear motion of the medium.

When one considers the surface perpendicular to the $y$-direction one also finds a traction operating on that surface with components $T_x^{(y)}$, $T_y^{(y)}$ and $T_z^{(y)}$, and it is not clear at this point whether the traction $\mathbf{T}^{(y)}$ has any relation to the traction $\mathbf{T}^{(x)}$ that acts on a surface perpendicular to the $x$-direction. The tractions acting on the three surfaces can be grouped in a $3 \times 3$ matrix:

$$\boldsymbol{\sigma} \equiv \begin{pmatrix} \vdots & \vdots & \vdots \\ \mathbf{T}^{(x)} & \mathbf{T}^{(y)} & \mathbf{T}^{(z)} \\ \vdots & \vdots & \vdots \end{pmatrix} = \begin{pmatrix} T_x^{(x)} & T_x^{(y)} & T_x^{(z)} \\ T_y^{(x)} & T_y^{(y)} & T_y^{(z)} \\ T_z^{(x)} & T_z^{(y)} & T_z^{(z)} \end{pmatrix} . \tag{21.75}$$

The quantity $\boldsymbol{\sigma}$ is called the *stress tensor*.

The stress tensor has a remarkable property that is reminiscent of the property of the gradient which was introduced in section 4.1. The gradient is so useful because once one knows the three components of the gradient, one can compute the change of a function in *any* direction. The stress tensor gives the components of the traction to surfaces that are perpendicular to the coordinates axes. One can show that the traction on a surface that is perpendicular to an arbitrary unit vector $\hat{\mathbf{n}}$ is given by

$$\mathbf{T} = \boldsymbol{\sigma} \cdot \hat{\mathbf{n}} . \tag{21.76}$$

**Problem a:** The proof of this identity is actually very simple. Consult section 16.3 of Butkov [18] or your favorite book on continuum mechanics or mathematical physics (that is, aside from this book) for a proof.

We have called $\boldsymbol{\sigma}$ the stress tensor, but we have not shown that $\boldsymbol{\sigma}$ is a tensor. However, one can show that when a quantity is contracted with a tensor and the result is also a tensor, then this quantity must be a tensor as well. In (21.76) we know that $\mathbf{T}$ is a tensor because it is a force (normalized with surface area) and we have shown in section 21.6 that the force is a tensor of rank one. The normal vector $\hat{\mathbf{n}}$ is also a tensor of rank one because it transforms in the same way as the position vector. Hence we need to show from (21.76) and the fact that both $\mathbf{T}$ and $\hat{\mathbf{n}}$ are tensors of rank one that $\boldsymbol{\sigma}$ is a tensor as well.

In components (21.76) is written as

$$T_i = \sigma_{ij} n_j \,, \tag{21.77}$$

while the same expression in a transformed coordinate system is given by

$$T'_i = \sigma'_{ij} n'_j \,. \tag{21.78}$$

Before we determine the transformation property of the stress tensor $\boldsymbol{\sigma}$, we first express the original vector $\mathbf{v}$ in terms of the transformed vector $\mathbf{v}'$ using the transformation rule

$$v'_i = C_{ij} v_j \,. \tag{21.45}$$

**Problem b:** Multiply this expression on the left by $C_{ki}^{-1}$ and sum over $i$ to obtain $C_{ki}^{-1} v'_i = C_{ki}^{-1} C_{ij} v_j$. Use the fact that $\mathbf{C}$ is unitary on the left hand side and carry out the matrix multiplication on the right hand side to show that $C_{ik} v'_i = v_k$. Finally rename $i \to j$ and $k \to i$ to obtain

$$v_i = C_{ji} v'_j \,. \tag{21.79}$$

It is interesting to compare this expression for the original vector $\mathbf{v}$ in terms of the transformed vector $\mathbf{v}'$ with (21.45) which gives the transformed vector given the original vector. The only difference is the order of the subscripts in the coordinate transform $\mathbf{C}$. This is due to the fact that this matrix is unitary so that inversion amounts to interchanging the indices: $C_{ij}^{-1} = C_{ij}^T = C_{ji}$.

**Problem c:** Insert (21.79) into (21.77) in order to express the unprimed vectors $\mathbf{T}$ and $\hat{\mathbf{n}}$ in terms of their transformed vectors to obtain

$$C_{ki} T'_k = \sigma_{ij} C_{lj} n'_l \,. \tag{21.80}$$

**Problem d:** Write $C_{ki} = C_{ik}^T = C_{ik}^{-1}$ on the left hand side and multiply by $C_{mi}$ to obtain

$$T'_m = C_{mi}C_{lj}\sigma_{ij}n'_l . \tag{21.81}$$

**Problem e:** We want to compare this expression with (21.78) in order to find $\boldsymbol{\sigma}'$. However, the indices are different. As we noted earlier, the names of indices are irrelevant. Rename the indices in (21.81) so that it can directly be compared with (21.78) and use this to show that

$$\sigma'_{ij} = C_{ir}C_{js}\sigma_{rs} . \tag{21.82}$$

This is just the transformation rule (21.49) for a tensor of rank two. We have thus shown that the stress tensor is indeed a tensor of rank two. This is due to the fact that when an object is contracted with a tensor to give another tensor, this object must be a tensor as well. Using this property we can use a bootstrap procedure to find higher order tensors. An important example of this is the elasticity tensor which relates the stress to the strain:

$$\sigma_{ij} = c_{ijkl}\varepsilon_{kl} . \tag{21.83}$$

This expression generalizes the elastic force $F = -kx$ in a spring to continuous media; it is known as Hooke's law. The quantity $c_{ijkl}$ must be a tensor because we know that both the stress $\boldsymbol{\sigma}$ and the strain $\varepsilon$ are tensors. This means that $c_{ijkl}$ is a tensor of rank four.

## 21.11 Why pressure in a fluid is isotropic

Finally we have reached the point where we can use tensors to learn about physics. As a first example we consider the pressure in a fluid (or gas). The pressure is the normal force per unit surface area. It is an observational fact that in a gas or fluid this force does not depend on the *orientation* of this surface. (This is why the weatherman speaks about a pressure of 1020 mbar rather than saying that the pressure is 1015 mbar in the vertical direction and 1025 mbar in the horizontal direction.) In this section we discover why the pressure is independent of direction.

In order to understand this we return to the stress tensor (21.75) and we consider the traction acting on a surface perpendicular to the $x$-direction. As shown in figure 21.7, $T_x^{(x)}$ gives the traction normal to this surface while $T_y^{(x)}$ and $T_z^{(x)}$ give the shear traction that acts on this surface. This reasoning can be used for all the surfaces; the diagonal elements $T_x^{(x)}$, $T_y^{(y)}$ and $T_z^{(z)}$ of the stress tensor give the normal tractions while all the other elements give the shear tractions. In a fluid, there are no shear tractions because a fluid has zero shear strength. This means that in a fluid the stress tensor is diagonal:

$$\boldsymbol{\sigma} = -\begin{pmatrix} p_x & 0 & 0 \\ 0 & p_y & 0 \\ 0 & 0 & p_z \end{pmatrix}. \tag{21.84}$$

The diagonal elements $p_i$ denote the pressure in the three directions. The minus sign reflects the fact that a positive pressure corresponds to a force that is directed *inwards*. In this section we show that the pressure is isotropic, in other words that the diagonal elements are identical.

Let us see what happens to the stress tensor (21.84) when we rotate the coordinate system through 45 degrees around the $z$-axis. For a rotation in two dimensions the rotation is given by (21.23). Setting the rotation angle $\varphi$ to 45 degrees and extending the result to three dimensions gives the following matrix representation of this coordinate transformation:

$$\mathbf{C} = \begin{pmatrix} 1/\sqrt{2} & -1/\sqrt{2} & 0 \\ 1/\sqrt{2} & 1/\sqrt{2} & 0 \\ 0 & 0 & 1 \end{pmatrix}. \tag{21.85}$$

**Problem a:** Verify that this coordinate transformation is unitary.

**Problem b:** Use the transformation property of a tensor of rank two that the stress tensor (21.84) in the rotated coordinate system is given by

$$\boldsymbol{\sigma}' = -\begin{pmatrix} (p_x + p_y)/2 & (p_x - p_y)/2 & 0 \\ (p_x - p_y)/2 & (p_x + p_y)/2 & 0 \\ 0 & 0 & p_z \end{pmatrix}. \tag{21.86}$$

In a fluid, the stress tensor is diagonal in *any* coordinate system since the shear tractions vanish in any coordinate system. This means that the off-diagonal elements of $\boldsymbol{\sigma}'$ must be equal to zero, hence $p_x = p_y$.

**Problem c:** Find a suitable coordinate transform to show that $p_x = p_z$.

This means that all the diagonal elements are identical; this quantity is referred to as the pressure: $p_x = p_y = p_z = p$. In an acoustic medium such as a fluid or gas the stress tensor is therefore given by

$$\boldsymbol{\sigma} = -p\,\mathbf{I}, \tag{21.87}$$

so that pressure is indeed independent of direction.

Note that we have done something truly remarkable; we have derived a physical law ('the pressure is isotropic') from the invariance of a property ('the shear stress is zero in a fluid') under a coordinate transformation.

## 21.12 Special relativity

One of the most spectacular applications of tensor calculus is the theory of relativity which describes the physics of objects and fields at very high speeds. The theory of general relativity accounts for the fact that mass in the universe leads to a non-Cartesian structure of space-time [62]; by definition this cannot be treated with the Cartesian tensors used in this chapter. The theory of special relativity describes how different observers who both use Cartesian coordinate systems that move at great speeds with respect to each other describe the same physical phenomena. A clear physical description of the theory of special relativity is given by Taylor and Wheeler [87]. In this section we use the notation used by Muirhead [59] who uses a complex time variable.

Central to the theory of special relativity is the notion that space and time are intricately linked. The three position variables $x$, $y$ and $z$ as well as time $t$ are placed in a four-dimensional vector, called a four-vector:

$$\mathbf{x} = \begin{pmatrix} x \\ y \\ z \\ ict \end{pmatrix}. \tag{21.88}$$

In this expression $c$ is the speed of light and $i = \sqrt{-1}$. The fact that the last component is complex leads to the surprising result that the length of the four-vector can be negative because

$$|\mathbf{x}|^2 = (\mathbf{x} \cdot \mathbf{x}) = x^2 + y^2 + z^2 - c^2 t^2 \tag{21.89}$$

and there is no reason why the last term cannot dominate the other terms.

Suppose we have one observer who uses unprimed variables, and suppose that another observer moves in the $x$-direction with a relative velocity $v$ with respect to the first observer. The two coordinate systems of the observers are then related by a *Lorentz transformation* [59]:

$$\mathbf{L} = \begin{pmatrix} 1/\sqrt{1 - \dfrac{v^2}{c^2}} & 0 & 0 & i\dfrac{v}{c}/\sqrt{1 - \dfrac{v^2}{c^2}} \\ 0 & 1 & 0 & 0 \\ 0 & 0 & 1 & 0 \\ -i\dfrac{v}{c}/\sqrt{1 - \dfrac{v^2}{c^2}} & 0 & 0 & 1/\sqrt{1 - \dfrac{v^2}{c^2}} \end{pmatrix}. \tag{21.90}$$

**Problem a:** Show that the Lorentz transform is unitary by showing that $\mathbf{L}^T \mathbf{L} = \mathbf{I}$.

Note that the transpose $\mathbf{L}^T$ is used here and *not* the Hermitian conjugate $\mathbf{L}^\dagger$ that is defined as the transpose and the complex conjugate: $L_{ij}^\dagger \equiv L_{ji}^*$.

The theory of special relativity states that the four-vector **x** is a tensor of rank one. This implies that $|\mathbf{x}|^2$ is a scalar, which means that both observers will assign the same value to this property, so that

$$x'^2 + y'^2 + z'^2 - c^2 t'^2 = x^2 + y^2 + z^2 - c^2 t^2 . \tag{21.91}$$

**Problem b:** Use the fact that **x** is a tensor of rank one to show that after a Lorentz transformation the $x'$- and $t'$-coordinates are given by

$$\left.\begin{array}{l} x' = (x - vt) / \sqrt{1 - \dfrac{v^2}{c^2}} , \\[4mm] t' = \left(t - \dfrac{vx}{c^2}\right) / \sqrt{1 - \dfrac{v^2}{c^2}} . \end{array}\right\} \tag{21.92}$$

**Problem c:** Verify that this transformation (together with $y = y'$ and $z = z'$) indeed satisfies the identity (21.91).

The transformation (21.92) means that space and time are mixed in a Lorentz transformation. The distinction between space and time largely disappears in theory of relativity! The term $(1 - v^2/c^2)^{-1/2}$ on the right hand side leads to the clocks of the two observers running at different speeds, contracting rods and other phenomena that are counterintuitive but which have been confirmed experimentally [87]. Note that the contraction terms $(1 - v^2/c^2)^{-1/2}$ are crucial in problem c in establishing that (21.91) is satisfied.

The theory of special relativity has many surprises. We have already seen in section 21.7 that a magnetic field does not behave as a tensor of rank one under coordinate transformations in three dimensions. In fact, when we also consider coordinate transformations between moving coordinate systems the electric field and the magnetic fields are mixed! According to the theory of special relativity [59] the magnetic field **B** and the electric field **E** transform as the following tensor of rank two:

$$\mathbf{F} = \begin{pmatrix} 0 & B_z & -B_y & -iE_x/c \\ -B_z & 0 & B_x & -iE_y/c \\ B_y & -B_x & 0 & -iE_z/c \\ iE_x/c & iE_y/c & iE_z/c & 0 \end{pmatrix} . \tag{21.93}$$

The momentum of a particle is also given by a four-vector; this momentum-energy vector [59] is given by

$$\mathbf{p} = \begin{pmatrix} p_x \\ p_y \\ p_z \\ iE/c \end{pmatrix} , \tag{21.94}$$

where $E$ is the energy of the particle. This four-vector transforms as a tensor of rank one under a Lorentz transformation [59].

**Problem d:** Use the property that $\mathbf{p}$ is a tensor of rank one to show that $p^2 - E^2/c^2$ is invariant, where $p^2 = p_x^2 + p_y^2 + p_z^2$.

**Problem e:** In general the energy of a body is given by the energy of that body at rest plus a contribution due to the motion of the body. This rest energy is denoted by $E_0$. Show that for the particle at rest the norm of the four-vector $\mathbf{p}$ is given by $|\mathbf{p}|^2 = -E_0^2/c^2$ and use the result of problem d to show that

$$E = \sqrt{E_0^2 + p^2 c^2} \ . \tag{21.95}$$

This expression holds for any value of the momentum. For the moment we consider a particle that moves much slower than the speed of light ($p = mv \ll mc$). We shall show that for such a particle the rest energy $E_0$ is much larger than the energy $pc$ that is due to the motion.

**Problem f:** Make a Taylor series expansion of (21.95) in the parameter $pc/E_0$ to show that for slow speeds

$$E = E_0 + \frac{c^2 p^2}{2E_0} - \frac{c^4 p^4}{8E_0^3} + \cdots . \tag{21.96}$$

The first term on the right hand side corresponds to the rest energy which we do not yet know. The second term is quadratic in the momentum. For small velocities the laws of classical mechanics hold and the kinetic energy is given by $\frac{1}{2}m_0 v^2$. Using the classical relation $p = m_0 v$, the kinetic energy is in classical mechanics given by $p^2/2m_0$. In classical mechanics the terms of order $(cp/E_0)^4$ are ignored. The kinetic energy $p^2/2m_0$ in classical mechanics is therefore equal to the second term on the right hand side of (21.96).

**Problem g:** Derive from this statement that

$$E_0 = m_0 c^2 . \tag{21.97}$$

You have just derived the famous relation that relates the rest energy of a particle to its rest mass. Note that the only ingredients that you have used were the fact that the four-vector $\mathbf{p}$ transforms like a tensor of rank one under coordinate transforms and some elementary results from classical mechanics! The implications of (21.97) are profound. It reflects the fact that matter can be seen as a condensed form of energy. In nuclear reactions this is used for the benefit (or demise) of mankind because the mass of the end-result of some nuclear reactions is smaller than the mass of the ingredients for that reaction. This mass difference is released in the form of radiation and the kinetic energy (heat) of the resulting particles.

**Problem h:** In the derivation of (21.97) we assumed that the rest energy in the classical limit is much larger than the kinetic energy. Show that the ratio of the second term on the right hand side of (21.96) to the rest mass is given by

$$\frac{c^2 p^2}{2E_0} / E_0 = \frac{1}{2} (v/c)^2 \ . \tag{21.98}$$

**Problem i:** The velocity with which a rocket can overcome the attraction of the Earth is called the *escape velocity* [44]; it has the numerical value of 11 184 km/s. Compute the ratio $v/c$ for a rocket that leaves the Earth at the escape velocity and compute the ratio of the third term to the second term on the right hand side of (21.96).

The third term on the right hand side (21.96) is a relativistic correction term that gives the leading order correction to the classical kinetic energy. The calculation you have just carried out shows that relativistic effects are small for fast-moving rockets.

At this point you might think that the theory only has implications for high-velocity bodies in cosmological problems. However, this is not true; the electrons in atoms move so fast that relativistic effects leave an imprint on microscopic bodies as well. For example, the relativistic correction term $c^4 p^4 / 8E_0^3$ in (21.96) leads in quantum theory to a measurable shift in the frequency of light emitted by excited hydrogen atoms that is called *fine-structure splitting* [76].

# 22

---

# Perturbation theory

From this book and most other books on mathematical physics you may have obtained the impression that most equations in the physical sciences can be solved. This is actually not true; most textbooks (including this book) give an unrepresentative state of affairs by only showing the problems that *can* be solved in closed form. It is an interesting paradox that as our theories of the physical world become more accurate, the resulting equations become more difficult to solve. In classical mechanics the problem of two particles that interact with a central force can be solved in closed form, but the three-body problem in which three particles interact has no analytical solution. In quantum mechanics, the one-body problem of a particle that moves in a potential can be solved only for a very limited number of situations: for the free particle, the particle in a box, the harmonic oscillator, and the hydrogen atom. In this sense the one-body problem in quantum mechanics has no general solution. This shows that as a theory becomes more accurate, the resulting complexity of the equations makes it often more difficult to actually find solutions.

One way to proceed is to determine numerical solutions of the equations. Computers are a powerful tool and can be extremely useful in solving physical problems. Another approach is to find approximate solutions to the equations. In chapter 11, scale analysis was used to drop from the equations terms that appear to be irrelevant. In this chapter, a systematic method is introduced to account for terms in the equations that are small but that make the equations difficult to solve. The idea is that a complex problem is compared to a simpler problem that can be solved in closed form, and to consider these small terms as a perturbation to the original equation. The theory of this chapter then makes it possible to determine how the solution is perturbed by the perturbation in the original equation; this technique is called *perturbation theory*. A classic reference on perturbation theory has been written by Nayfeh [60]. The book by

388

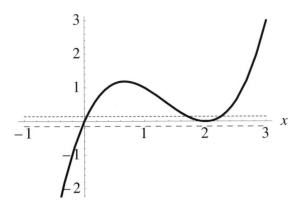

Fig. 22.1. The polynomial $x^3 - 4x^2 + 4x$ (thick solid line) and the lines $\varepsilon = 0.15$ (dotted line) and $\varepsilon = -0.15$ (dashed line)

Bender and Orszag [11] gives a very useful and illustrative overview of a wide variety of perturbation methods.

The central idea of perturbation theory is introduced for an algebraic equation in section 22.1. Sections 22.2, 22.3 and 22.5 contain important applications of perturbation theory to differential equations. As shown in section 22.4, perturbation theory has a limited domain of applicability, and this may depend on the way the perturbation problem is formulated. Finally, it is shown in section 22.7 that not every perturbation problem is well behaved; this leads to singular perturbation theory.

## 22.1 Regular perturbation theory

As an introduction to perturbation theory let us consider the following equation

$$x^3 - 4x^2 + 4x = 0.01 . \tag{22.1}$$

Let us for the moment assume that we don't know how to find the roots of a third order polynomial, so we cannot solve this equation. The problem is the small term 0.01 on the right hand side. If this term were equal to zero, the resulting equation can be solved; $x^3 - 4x^2 + 4x = 0$ is equivalent to $x(x^2 - 4x + 4) = x(x - 2)^2 = 0$, which has the solutions $x = 0$ and $x = 2$. In figure 22.1 the polynomial of (22.1) is shown by the thick solid line; it is indeed equal to zero for $x = 0$ and $x = 2$.

The problem that we face is that the right hand side of (22.1) is *not* equal to zero. In perturbation theory one studies the perturbation of the solution under a perturbation of the original equation. In order to do this, we replace the original equation (22.1) by the more general equation

$$x^3 - 4x^2 + 4x = \varepsilon. \tag{22.2}$$

When $\varepsilon = 0.01$ this equation is identical to the original problem, while for $\varepsilon = 0$ it reduces to the unperturbed problem that we can solve in closed form. It may appear that we have made the problem more complex because we still need to solve the same equation as our original equation, but it now contains a new variable $\varepsilon$ as well! However, this is also the strength of this approach.

The solution of (22.2) is a function of $\varepsilon$ so that

$$x = x(\varepsilon). \tag{22.3}$$

In section 2.1 the Taylor series was used to approximate a function $f(x)$ by a power series in the variable $x$:

$$f(x) = f(0) + x \frac{df}{dx}(x = 0) + \frac{x^2}{2!} \frac{d^2 f}{dx^2}(x = 0) + \cdots. \tag{2.11}$$

When the solution $x$ of (22.2) depends in a regular way on $\varepsilon$, this solution can also be written as a similar power series by making the substitutions $x \to \varepsilon$ and $f \to x$ in (2.11):

$$x(\varepsilon) = x(0) + \varepsilon \frac{dx}{d\varepsilon}(\varepsilon = 0) + \frac{\varepsilon^2}{2!} \frac{d^2 x}{d\varepsilon^2}(\varepsilon = 0) + \cdots. \tag{22.4}$$

This expression is not very useful because we need the derivative $dx/d\varepsilon$ and higher derivatives $d^n x/d\varepsilon^n$ as well; in order to compute these derivatives we need to find the solution $x(\varepsilon)$ first, but this is just what we are trying to do. There is, however, another way to determine the series (22.4). Let us write the solution $x$ as a power series in $\varepsilon$

$$x = x_0 + \varepsilon x_1 + \varepsilon^2 x_2 + \cdots. \tag{22.5}$$

The coefficients $x_n$ are not known at this point, but once we know them the solution $x$ can be found by inserting the numerical value $\varepsilon = 0.01$. In practice one truncates the series (22.5); it is this truncation that makes perturbation theory an approximation.

When the series (22.5) is inserted into (22.2) one needs to compute $x^2$ and $x^3$ when $x$ is given by (22.5). Let us first consider the $x^2$-term. The square of a sum of terms is given by

$$\begin{aligned} (a + b + c + \cdots)^2 &= a^2 + b^2 + c^2 + \cdots \\ &\quad + 2ab + 2ac + 2bc + \cdots. \end{aligned} \tag{22.6}$$

Let us apply this to the series (22.5) and retain only the terms up to order $\varepsilon^2$, this gives

$$\left(x_0 + \varepsilon x_1 + \varepsilon^2 x_2 + \cdots\right)^2 = x_0^2 + \varepsilon^2 x_1^2 + \varepsilon^4 x_2^2 + \cdots$$
$$+2\varepsilon x_0 x_2 + 2\varepsilon^2 x_0 x_2 + 2\varepsilon^3 x_1 x_2 + \cdots . \quad (22.7)$$

If we are only are interested in retaining the terms up to order $\varepsilon^2$, the terms $\varepsilon^4 x_2^2$ and $2\varepsilon^3 x_1 x_2$ in this expression can be ignored. Collecting terms of equal powers of $\varepsilon$ then gives

$$\left(x_0 + \varepsilon x_1 + \varepsilon^2 x_2 + \cdots\right)^2 = x_0^2 + 2\varepsilon x_0 x_1 + \varepsilon^2\left(x_1^2 + 2x_0 x_2\right) + O(\varepsilon^3). \quad (22.8)$$

A similar expansion in powers of $\varepsilon$ can be used for the term $x^3$. This expansion is based on the identity

$$(a + b + c + \cdots)^3 = a^3 + b^3 + c^3 + \cdots$$
$$+3a^2 b + 3ab^2 + 3a^2 c + 3ac^2 + 3b^2 c + 3bc^2 + \cdots . \quad (22.9)$$

**Problem a:** Apply this identity to the series (22.5), collect together all the terms with equal powers of $\varepsilon$ and show that up to order $\varepsilon^2$ the result is given by

$$\left(x_0 + \varepsilon x_1 + \varepsilon^2 x_2 + \cdots\right)^3 = x_0^3 + 3\varepsilon x_0^2 x_1 + 3\varepsilon^2\left(x_0 x_1^2 + x_0^2 x_2\right) + O(\varepsilon^3). \quad (22.10)$$

**Problem b:** At this point we can express all the terms in (22.2) in a power series of $\varepsilon$. Insert (22.5), (22.8) and (22.10) into the original equation (22.2) and collect together terms of equal powers of $\varepsilon$ to derive that

$$x_0^3 - 4x_0^2 + 4x_0$$
$$+\varepsilon\left(3x_0^2 x_1 - 8x_0 x_1 + 4x_1 - 1\right)$$
$$+\varepsilon^2\left(3x_0 x_1^2 + x_0^2 x_2 - 4x_1^2 - 8x_0 x_2 + 4x_2\right) + \cdots = 0 . \quad (22.11)$$

In this and subsequent expressions the dots denote terms of order $O(\varepsilon^3)$. The term $-1$ in the term that multiplies $\varepsilon$ comes from the right hand side of (22.2).

At this point we use that $\varepsilon$ does not have a fixed value, but that it can take any value within certain bounds. This means that expression (22.11) must be satisfied for a range of values of $\varepsilon$. This can only be the case when the coefficients that multiply the different powers $\varepsilon^n$ are equal to zero. This means that (22.11) is equivalent to the following system of equations which consists of the terms that multiply the terms $\varepsilon^0$, $\varepsilon^1$ and $\varepsilon^2$ respectively:

$$\left.\begin{array}{ll} O(1)\text{-terms:} & x_0^3 - 4x_0^2 + 4x_0 = 0 , \\ O(\varepsilon)\text{-terms:} & 3x_0^2 x_1 - 8x_0 x_1 + 4x_1 - 1 = 0 , \\ O(\varepsilon^2)\text{-terms:} & 3x_0 x_1^2 + x_0^2 x_2 - 4x_1^2 - 8x_0 x_2 + 4x_2 = 0 . \end{array}\right\} \quad (22.12)$$

You may wonder whether we have not made the problem more complex. We started with a single equation for a single variable $x$, and now we have a system of coupled equations for many variables. However, we could not solve (22.2) for the single variable $x$, while it is not difficult to solve (22.12).

**Problem c:** Show that (22.12) can be rewritten in the following form:

$$\left.\begin{array}{l} x_0^3 - 4x_0^2 + 4x_0 = 0 , \\ \left(3x_0^2 - 8x_0 + 4\right) x_1 = 1 , \\ \left(x_0^2 - 8x_0 + 4\right) x_2 = (4 - 3x_0) x_1^2 . \end{array}\right\} \quad (22.13)$$

The first equation is simply the unperturbed problem, this has the solutions $x_0 = 0$ and $x_0 = 2$. For reasons that will become clear in section 22.7 we focus here only on the solution $x_0 = 0$. Given $x_0$, the parameter $x_1$ follows from the second equation because this is a linear equation in $x_1$. The last equation is a linear equation in the unknown $x_2$ which can easily be solved once $x_0$ and $x_1$ are known.

**Problem d:** Solve (22.13) in this way to show that the solution near $x = 0$ is given by

$$x_0 = 0 , \qquad x_1 = \frac{1}{4} , \qquad x_2 = \frac{1}{16} . \qquad (22.14)$$

Now we are close to the final solution of our problem. The coefficients of the previous expression can be inserted into the perturbation series (22.5) so that the solution as a function of $\varepsilon$ is given by

$$x = 0 + \frac{1}{4}\varepsilon + \frac{1}{16}\varepsilon^2 + O(\varepsilon^3) . \qquad (22.15)$$

At this point we can revert to the original equation (22.1) by inserting the numerical value $\varepsilon = 0.01$, which gives:

$$x = \frac{1}{4} \times 10^{-2} + \frac{1}{16} \times 10^{-4} + O(10^{-6}) = 0.002506 . \qquad (22.16)$$

It should be noted that this is an approximate solution because the terms of order $\varepsilon^3$ and higher have been ignored. This is indicated by the term

$O(10^{-6})$ in (22.16). Assuming that the error made by truncating the perturbation series is of the same order as the first term that is truncated, the error in the solution (22.16) is of the order $10^{-6}$. For this reason the number on the right hand side of (22.16) is given to six decimals; the last decimal is of the same order as the truncation error.

If this result is not sufficiently accurate for the application that one has in mind, then one can easily extend the analysis to higher powers $\varepsilon^n$ in order to reduce the truncation error of the truncated perturbation series. Although the algebra resulting from doing this can be tedious, there is no reason why this analysis cannot be extended to higher orders. A truly formal analysis of perturbation problems can be very difficult. For example, the perturbation series (22.5) only converges for sufficiently small values of $\varepsilon$. It is often not clear whether the employed value of $\varepsilon$ (in this case $\varepsilon = 0.01$) is sufficiently small to ensure convergence. Even when a perturbation series does not converge for a given value of $\varepsilon$, one can often obtain a useful approximation to the solution by truncating the perturbation series at a suitably chosen order [11]. In this case one speaks of an *asympotic series*.

When one has obtained an approximate solution of a perturbation problem, one can sometimes substitute it back into the original equation to verify whether this solution indeed satisfies the equation with an acceptable accuracy. For example, inserting the numerical value $x = 0.002506$ in (22.1) gives

$$x^3 - 4x^2 + 4x = 0.0099989 = 0.01 - 0.0000011 \ . \tag{22.17}$$

This means that the approximate solution satisfies (22.1) with a *relative* error that is given by $0.0000011/0.01 = 10^{-4}$. This is a very accurate result given the fact that only three terms were retained in the perturbation analysis of this section.

## 22.2 The Born approximation

In many scattering problems one wants to account for the scattering of waves by the heterogeneities in the medium. Usually these problems are so complex that they cannot be solved in closed form. Suppose one has a background medium in which scatterers are embedded. When the background medium is sufficiently simple, one can solve the wave propagation problem for this background medium. For example, in section 18.3 we computed the Green's function for the Helmholtz equation in a homogeneous medium.

In this section we consider the Helmholtz equation with a *variable* velocity $c(\mathbf{r})$ as an example of the application of perturbation theory to scattering problems. This means we consider the wave field $p(\mathbf{r}, \omega)$ in the frequency domain that satisfies the the following equation:

$$\nabla^2 p(\mathbf{r}, \omega) + \frac{\omega^2}{c^2(\mathbf{r})} p(\mathbf{r}, \omega) = S(\mathbf{r}, \omega) . \tag{22.18}$$

In this expression $S(\mathbf{r}, \omega)$ denotes the source that generates the wave field. In order to facilitate a systematic perturbation analysis we decompose $1/c^2(\mathbf{r})$ into a term $1/c_0^2$ that accounts for a homogeneous reference model and a perturbation:

$$\frac{1}{c^2(\mathbf{r})} = \frac{1}{c_0^2} [1 + \varepsilon n(\mathbf{r})] . \tag{22.19}$$

In this expression $\varepsilon$ is a small parameter which measures the strength of the heterogeneity. The function $n(\mathbf{r})$ gives the spatial distribution of the heterogeneity. Combining the previous expressions it follows that the wave field satisfies the following expression:

$$\nabla^2 p(\mathbf{r}, \omega) + \frac{\omega^2}{c_0^2} [1 + \varepsilon n(\mathbf{r})] \, p(\mathbf{r}, \omega) = S(\mathbf{r}, \omega) . \tag{22.20}$$

The solution $p(\mathbf{r}, \omega)$ of this expression is a function of the scattering strength $\varepsilon$; for sufficiently small values of $\varepsilon$ it can be written as a power series in $\varepsilon$:

$$p = p_0 + \varepsilon p_1 + \varepsilon^2 p_2 + \cdots . \tag{22.21}$$

**Problem a:** Insert the perturbation series (22.21) into (22.20), collect together the terms that multiply equal powers of $\varepsilon$ and show that the terms that multiply the different powers of $\varepsilon$ give the following equations:

$$\left.\begin{aligned}
O(1) : \qquad & \nabla^2 p_0(\mathbf{r}, \omega) + \frac{\omega^2}{c_0^2} p_0(\mathbf{r}, \omega) = S(\mathbf{r}, \omega) , \\[2mm]
O(\varepsilon) : \qquad & \nabla^2 p_1(\mathbf{r}, \omega) + \frac{\omega^2}{c_0^2} p_1(\mathbf{r}, \omega) = -\frac{\omega^2}{c_0^2} n(\mathbf{r}) p_0(\mathbf{r}, \omega) , \\[2mm]
O(\varepsilon^2) : \qquad & \nabla^2 p_2(\mathbf{r}, \omega) + \frac{\omega^2}{c_0^2} p_2(\mathbf{r}, \omega) = -\frac{\omega^2}{c_0^2} n(\mathbf{r}) p_1(\mathbf{r}, \omega) , \\
& \qquad\qquad\qquad\vdots
\end{aligned}\right\} \tag{22.22}$$

The first expression gives the Helmholtz equation for a homogeneous medium. The source of the unperturbed wave $p_0$ is given by $S(\mathbf{r}, \omega)$ which is also the source of the perturbed problem. The source of the first order perturbation $p_1$ is given by the right hand side of the second equation, hence the source of $p_1$ is given by $- (\omega^2/c_0^2) \, n(\mathbf{r}) p_0(\mathbf{r}, \omega)$. This means that the source of $p_1$ is proportional to the inhomogeneity $n(\mathbf{r})$ of the medium. Physically this corresponds to the fact that the heterogeneity is the source of the scattered waves. The source of $p_1$ is also proportional to the unperturbed wave field $p_0$. The reason for this is that the generation of the scattered waves depends on the local perturbation of the medium as well as the strength of the wave field at the location of the scatterers.

Each of the equations in (22.22) is of the form $\nabla^2 p(\mathbf{r}, \omega) + \omega^2/c_0^2 p(\mathbf{r}, \omega) = F(\mathbf{r}, \omega)$. According to the theory of section 17.4 the solution to this expression is given by

$$p(\mathbf{r}, \omega) = \int G_0(\mathbf{r}, \mathbf{r}'; \omega) F(\mathbf{r}', \omega) dV' , \qquad (22.23)$$

where the unperturbed Green's function $G_0(\mathbf{r}, \mathbf{r}'; \omega)$ is the response in a homogeneous medium at location $\mathbf{r}$ due to a point source at location $\mathbf{r}'$:

$$\nabla^2 G_0(\mathbf{r}, \mathbf{r}'; \omega) + \frac{\omega^2}{c_0^2} G_0(\mathbf{r}, \mathbf{r}'; \omega) = \delta(\mathbf{r} - \mathbf{r}') . \qquad (22.24)$$

The specific form of the unperturbed Green's function in one, two and three dimensions is given in (18.43). From this point on it is not shown explicitly that the solution and the Green's function depend on the angular frequency $\omega$, but it should be kept in mind that all the results in this section depend on frequency.

**Problem c:** Use these results to show that the solution of (22.22) is given by

$$\left.\begin{array}{l} p_0(\mathbf{r}) = \int G_0(\mathbf{r}, \mathbf{r}') S(\mathbf{r}') dV' , \\[2mm] p_1(\mathbf{r}) = -\dfrac{\omega^2}{c_0^2} \int G_0(\mathbf{r}, \mathbf{r}') n(\mathbf{r}') p_0(\mathbf{r}') dV' , \\[2mm] p_2(\mathbf{r}) = -\dfrac{\omega^2}{c_0^2} \int G_0(\mathbf{r}, \mathbf{r}') n(\mathbf{r}') p_1(\mathbf{r}') dV' , \\[2mm] \vdots \end{array}\right\} \qquad (22.25)$$

**Problem d:** Insert the expression for the unperturbed wave $p_0$ into the second equation (22.25) to show that the first order perturbation is given by

$$p_1(\mathbf{r}) = -\frac{\omega^2}{c_0^2} \iint G_0(\mathbf{r}, \mathbf{r}_1) n(\mathbf{r}_1) G_0(\mathbf{r}_1, \mathbf{r}_0) S(\mathbf{r}_0) dV_1 dV_0 . \qquad (22.26)$$

Note that the integration variable $\mathbf{r}'$ has been relabelled as $\mathbf{r}_0$ and $\mathbf{r}_1$ respectively.

| total | unperturbed | single | double |
| wave | wave | scattered | scattered |
| | | wave | wave |

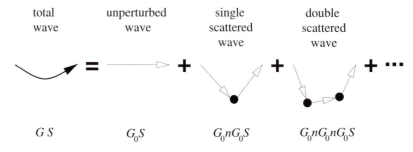

$$G\,S \qquad\qquad G_0 S \qquad\qquad G_0 n G_0 S \qquad\qquad G_0 n G_0 n G_0 S$$

Fig. 22.2.  Decomposition of the total wave field (thick solid line) in the unperturbed wave, the single scattered wave, the double scattered wave and higher order scattering events. The total Green's function $G$ is shown by a thick line, the unperturbed Green's function $G_0$ by thin lines, and each scattering event by the heterogeneity $n$ is indicated by a solid dot.

**Problem e:** Insert this result in the last line of (22.25) to derive that the second order perturbation is given by

$$p_2(\mathbf{r}) = \frac{\omega^4}{c_0^4} \iiint G_0(\mathbf{r}, \mathbf{r}_2) n(\mathbf{r}_2) G_0(\mathbf{r}_2, \mathbf{r}_1) n(\mathbf{r}_1) G_0(\mathbf{r}_1, \mathbf{r}_0) S(\mathbf{r}_0) dV_2 dV_1 dV_0 .$$

$$(22.27)$$

Inserting this result and (22.25) in the perturbation series (22.21) finally gives the following perturbation series for the scattered waves:

$$p(\mathbf{r}, \omega) = \int G_0(\mathbf{r}, \mathbf{r}_0) S(\mathbf{r}_0) dV_0$$

$$- \frac{\omega^2}{c_0^2} \iint G_0(\mathbf{r}, \mathbf{r}_1) n(\mathbf{r}_1) G_0(\mathbf{r}_1, \mathbf{r}_0) S(\mathbf{r}_0) dV_1 dV_0$$

$$+ \frac{\omega^4}{c_0^4} \iiint G_0(\mathbf{r}, \mathbf{r}_2) n(\mathbf{r}_2) G_0(\mathbf{r}_2, \mathbf{r}_1) n(\mathbf{r}_1) G_0(\mathbf{r}_1, \mathbf{r}_0) S(\mathbf{r}_0) dV_2 dV_1 dV_0$$

$$+ \cdots . \qquad\qquad (22.28)$$

This expansion is shown graphically in figure 22.2. Reading each of these lines from right to left one can follow the 'life-history' of the waves that are scattered in the medium. The top line of the right hand side gives the unperturbed wave. This wave is excited by the source at location $\mathbf{r}_0$; this is described by the source term $S(\mathbf{r}_0)$. The wave then propagates through the unperturbed medium to the point $\mathbf{r}$; this is accounted for by the term $G_0(\mathbf{r}, \mathbf{r}_0)$. Graphically this is shown by the first diagram after the equality sign in figure 22.2. In this figure the thin arrows denote the unperturbed Green's function $G_0$. The second line in (22.28) physically describes a wave that is generated at the source $S(\mathbf{r}_0)$; this wave then propagates through the unperturbed medium with the Green's function $G_0(\mathbf{r}_1, \mathbf{r}_0)$ to a scatterer at location $\mathbf{r}_1$. The wave is then scattered; this is accounted for by the terms $-(\omega^2/c_0^2)\, n(\mathbf{r}_1)$. In figure 22.2 this scattering interaction is indicated by a solid dot. The wave then travels through the unperturbed medium to the point $\mathbf{r}$; this is accounted for by the term $G_0(\mathbf{r}, \mathbf{r}_1)$.

**Problem f:** Describe in a similar way the 'life-history' of the double scattered wave that is given by the last line of (22.28) and convince yourself that this corresponds to the right-most diagram in figure 22.2.

The analysis in this section can be continued to any order. The resulting series is called the *Neumann series*. It gives a decomposition of the total wave field into single scattered waves, double scattered waves, triple scattered waves, etc. In practice it is often difficult to compute the waves that are scattered more than once. In the *Born approximation* one simply truncates the perturbation series after the second term. Using the notation of (22.25) this means that in the Born approximation the wave field is given by

$$p_B(\mathbf{r}) = p_0(\mathbf{r}) - \frac{\omega^2}{c_0^2} \int G_0(\mathbf{r}, \mathbf{r}')n(\mathbf{r}')p_0(\mathbf{r}')dV' \ . \tag{22.29}$$

This expression is extremely useful for a large variety of applications. For sufficiently weak perturbations $\varepsilon n(\mathbf{r})$ it allows the analytical computation of the (single) scattered waves. In the Born approximation the scattered waves are given by the last term in (22.29). This last term gives a linear(ized) relation between the scattered waves and the perturbation of the medium. In many applications one measures the scattered waves and one wants to retrieve the perturbation of the medium. The Born approximation provides a linear relation between the scattered waves and the perturbation of the medium. Methods from linear algebra can then be used to infer the perturbation $n(\mathbf{r})$ of the medium from measurements of the scattered waves. The Born approximation provides the basis for most of the techniques used in reflection seismology for the detection of hydrocarbons in the Earth, see for example refs. [20] and [103]. The Born approximation also forms the basis of the imaging techniques used with radar [41] and a variety of other applications. In fact, it has been argued that imaging with multiple-scattered waves is not feasible in practice [20]; a discussion of this controversial issue can be found in ref. [77].

In this section the Born approximation for the Helmholtz equation was derived. However, this derivation can readily be generalized to other scattering problems. The only required ingredient is that, when one divides the medium into an unperturbed medium and a perturbation, one can compute the Green's function for the unperturbed medium. The Born approximation is used in quantum mechanics [55], electromagnetic wave scattering [42], scattering of elastic body waves [102] and elastic surface waves [81].

There is a famous application of the Born approximation. According to (22.29) the scattered waves are multiplied by $\omega^2$ compared to the unperturbed waves. This is also the case for the scattering of electromagnetic waves [42]. This term $\omega^2$ explains why the sky is blue. The scattered waves are proportional to $\omega^2$ compared to the unperturbed waves. This means that light with a high frequency is scattered more strongly than light with a lower frequency. In other words, blue light is scattered more strongly than red light. The blue light that comes from the Sun is scattered more effectively out of the light beam from the Sun to an observer than the red light. When this blue light is scattered again by small particles in the atmosphere it travels to an observer as blue light that comes from the sky from a different location than the Sun. We perceive this as 'the blue sky.' This argument should, however, be used with caution because the Green's function in (22.29) also depends on frequency.

## 22.3 Linear travel time tomography

An important tool for determining the interior structure of the Earth and other bodies is travel time tomography. In this technique one measures the travel time of waves between a large number of sources and receivers. When the coverage with rays is sufficiently dense, one can determine the velocity of seismic waves in the Earth from the recorded travel times. Detailed descriptions of seismic tomography can be found in refs. [40] and [61]. The travel time along a ray is given by the integral

$$\tau = \int \frac{1}{c(\mathbf{r})} ds . \tag{22.30}$$

Since the integral is proportional to $1/c(\mathbf{r})$, it is convenient to use the *slowness* $u(\mathbf{r}) = 1/c(\mathbf{r})$ rather than the velocity. Using this quantity the travel time is given by

$$\tau = \int_{\mathbf{r}[u]} u(\mathbf{r}) ds . \tag{22.31}$$

The last expression suggests a linear relation between the measured travel time $\tau$ and the unknown slowness $u(\mathbf{r})$. Such a linear relation is ideal for solving the inverse problem of the determination of the slowness because one can resort to techniques from linear algebra. However, integral (22.31) is taken over the ray that joins the source and the receiver.

The rays are curves of stationary travel time and themselves depend on the slowness. This dependence effectively makes the relation between the slowness and the travel time nonlinear. In this section we perturb both the slowness and the travel time to derive a linearized relation between the travel time *perturbation* and the slowness *perturbation*. The travel time along rays follows from geometric ray theory as shown in section 11.4. It is shown in that section that the travel time $\tau(\mathbf{r})$ from a given source to location $\mathbf{r}$ is given by the eikonal equation (11.25) which can be written as

$$|\nabla\tau(\mathbf{r})|^2 = u^2(\mathbf{r}). \tag{22.32}$$

Let us now assume that we have a reasonable guess $u_0(\mathbf{r})$ for the slowness, and that we seek a small perturbation of the slowness; this perturbation is denoted as $\varepsilon u_1(\mathbf{r})$. The slowness can then be written as

$$u(\mathbf{r}) = u_0(\mathbf{r}) + \varepsilon u_1(\mathbf{r}). \tag{22.33}$$

Again, the parameter $\varepsilon$ only serves to systematically set up the perturbation treatment. When the slowness is perturbed, the travel time changes as well and it can be expanded in a perturbation series of the parameter $\varepsilon$:

$$\tau = \tau_0 + \varepsilon\tau_1 + \varepsilon^2\tau_2 + \cdots. \tag{22.34}$$

In this section we seek the relation between the first order travel time perturbation $\tau_1$ and the slowness perturbation $u_1$.

**Problem a:** Insert (22.33) and (22.34) into the eikonal equation (22.32), use that $|\nabla\tau|^2 = (\nabla\tau \cdot \nabla\tau)$ and collect together the terms in $O(1)$ and $O(\varepsilon)$ to show that the first and zeroth order travel time perturbation are given by

$$|\nabla\tau_0(\mathbf{r})|^2 = u_0^2(\mathbf{r}), \tag{22.35}$$
$$(\nabla\tau_0 \cdot \nabla\tau_1) = u_0 u_1. \tag{22.36}$$

The first equation is nothing but the eikonal equation for the unperturbed problem. This expression states that the length of the vector $\nabla\tau_0$ is equal to $u_0$. This means that $\nabla\tau_0$ can be written as

$$\nabla\tau_0 = u_0\hat{\mathbf{n}}_0. \tag{22.37}$$

In this expression the unit vector $\hat{\mathbf{n}}_0$ gives the direction of $\nabla\tau_0$ as shown in figure 22.3. It was shown in section 4.1 that the gradient $\nabla\tau_0$ is perpendicular to the surfaces of constant travel time, see figure 22.3. It is also shown in that figure that the rays are the curves that are everywhere perpendicular to the surfaces of constant travel time.

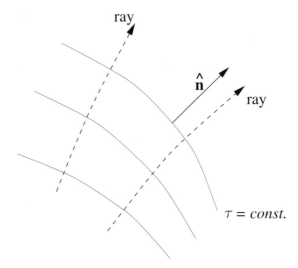

Fig. 22.3.    Wavefronts as the surfaces of constant travel time $\tau$ (solid lines), and
the rays that are the curves perpendicular to the travel time surfaces (dashed
lines). The unit vector **n** is perpendicular to the wavefronts.

**Problem b:** Use (4.20) to show that the derivative of $\tau_0$ along the $\mathbf{r}_0$ of
the reference medium is given by

$$\frac{d\tau_0}{ds_0} = u_0 , \tag{22.38}$$

where $s_0$ denotes the arclength along the ray in the reference medium.

This last expression can be integrated to give

$$\tau_0 = \int_{\mathbf{r}_0[u_0]} u_0(\mathbf{r}) ds_0 . \tag{22.39}$$

This expression is identical to (22.31) with the exception that all quantities are for the reference medium $u_0$ and its associated rays $\mathbf{r}_0[u_0]$.

**Problem c:** In order to derive the first order travel time perturbation,
insert (22.37) into (22.36) to derive that

$$(\hat{\mathbf{n}}_0 \cdot \nabla \tau_1) = u_1 . \tag{22.40}$$

Note that the unit vector $\hat{\mathbf{n}}_0$ is directed along the rays in the reference medium $u_0$ and that it is therefore independent of the slowness
perturbation $u_1$.

**Problem d:** Integrate the last expression to give

$$\tau_1 = \int_{\mathbf{r}_0[u_0]} u_1(\mathbf{r}) ds_0 . \tag{22.41}$$

In this expression the integration is along the rays $\mathbf{r}_0[u_0]$ in the reference medium. Since these rays are assumed to be known, (22.41) constitutes a linearized relation between the travel time perturbation $\tau_1$ and the slowness perturbation $u_1$. Techniques from linear algebra can then be used to determine the unknown slowness perturbation from the measured travel time perturbations $\tau_1$. In many textbooks (e.g. ref. [61]) this result is derived from Fermat's theorem. However, the treatment in this section (which was proposed by Aldridge [4]) is conceptually much simpler. In fact, the treatment in this section can easily be extended to compute the travel time perturbation to *any* order [83].

## 22.4 Limits on perturbation theory

Perturbation theory is a powerful tool; in principle it provides a systematic way to derive the perturbation to any desired order. In this section we will discover that for a given order of truncation of the perturbation series the accuracy of the obtained result may depend strongly on the value of certain parameters of the problem that one is considering. This is illustrated with a simple problem which we can also solve analytically. We consider the differential equation

$$\ddot{x} + \omega_0^2 \left(1 + \varepsilon\right) x = 0 , \tag{22.42}$$

with the initial conditions

$$x(0) = 1 , \qquad \dot{x}(0) = 0 . \tag{22.43}$$

This equation describes a harmonic oscillator in which the frequency is perturbed.

**Problem a:** Show that the exact solution of this problem is given by

$$x(t) = \cos\left(\omega_0 \sqrt{1 + \varepsilon}\, t\right) . \tag{22.44}$$

**Problem b:** The solution $x(t)$ is a function of the perturbation parameter $\varepsilon$, it can therefore be written as a perturbation series in this parameter:

$$x(t) = x_0(t) + \varepsilon x_1(t) + \varepsilon^2 x_2(t) + \cdots . \tag{22.45}$$

Insert this series into (22.42) and collect together the terms of equal powers in $\varepsilon$ to show that the terms $x_n(t)$ satisfy the following equations:

$$\left.\begin{array}{rcl} \ddot{x}_0 + \omega_0^2 x_0 & = & 0 , \\ \ddot{x}_1 + \omega_0^2 x_1 & = & -\omega_0^2 x_0 , \\ \ddot{x}_2 + \omega_0^2 x_2 & = & -\omega_0^2 x_1 , \\ & \vdots & \end{array}\right\} \tag{22.46}$$

**Problem c:** In order to solve these equations one must also consider the initial conditions of $x_n(t)$. Obtain these condition by inserting the perturbation series (22.45) into the initial conditions (22.43) and derive that

$$\left. \begin{array}{ll} x_0(0) = 1 , & \dot{x}_0(0) = 0 , \\ x_n(0) = 0 , & \dot{x}_n(0) = 0 \qquad \text{for } n \geq 1 . \end{array} \right\} \qquad (22.47)$$

**Problem d:** Solve the differential equation for $x_0$ for the boundary condition of (22.47) and show that the solution is given by

$$x_0(t) = \cos(\omega_0 t) . \qquad (22.48)$$

Note that this expression is identical to the exact solution (22.44) when one switches off the perturbation by setting $\varepsilon = 0$. Inserting this solution into the second line of (22.46) one finds that the first order perturbation satisfies the following differential equation

$$\ddot{x}_1 + \omega_0^2 x_1 = -\omega_0^2 \cos(\omega_0 t) . \qquad (22.49)$$

This equation describes a harmonic oscillator with eigenfrequency $\omega_0$ which is driven by a force on the right hand side. This driving force also oscillates with frequency $\omega_0$. This means that the oscillator $x_1$ is driven at its resonance frequency, which in general leads to a motion that grows with time.

**Problem e:** In order to solve (22.49) write the perturbation $x_1(t)$ as

$$x_1(t) = f(t) \cos(\omega_0 t) + g(t) \sin(\omega_0 t) . \qquad (22.50)$$

Insert this expression into (22.49) and collect together the terms that multiply $\cos(\omega_0 t)$ and $\sin(\omega_0 t)$ to show that the unknown functions $f(t)$ and $g(t)$ obey the following differential equations:

$$\left. \begin{array}{l} \ddot{f} + 2\omega_0 \dot{g} = -\omega_0^2 , \\ \ddot{g} - 2\omega_0 \dot{f} = 0 . \end{array} \right\} \qquad (22.51)$$

**Problem f:** Show that these equations are satisfied by the following solution:

$$\dot{f} = 0 , \qquad \dot{g} = -\frac{\omega_0}{2} , \qquad (22.52)$$

and integrate these equations to derive a particular solution that is given by

$$f = 0 , \qquad g(t) = -\frac{1}{2}\omega_0 t . \qquad (22.53)$$

This solution can be used in (22.50). The general solution is given by this particular solution plus the general solution of the homogeneous equation:

$$x_1(t) = -\frac{1}{2}\omega_0 t \, \sin(\omega_0 t) + A \cos(\omega_0 t) + B \sin(\omega_0 t) \,. \qquad (22.54)$$

In this expression $A$ and $B$ are integration constants. It follows from (22.47) that the initial conditions for $x_1(t)$ are given by $x_1(0) = 0$, $\dot{x}(0) = 0$.

**Problem g:** Derive from these initial conditions that the integration constants are given by $A = B = 0$, so that the solution is given by

$$x_1(t) = -\frac{1}{2}\omega_0 t \, \sin(\omega_0 t) \,. \qquad (22.55)$$

It was noted earlier that (22.49) describes an oscillator that is driven at its resonance frequency. Solution (22.55) grows linearly with time. This growth with time is called the *secular growth*. It is an artifact of the perturbation technique employed because the original problem (22.42) does not contain a resonant driving force at all. Inserting the first order perturbation (22.55) and the unperturbed solution (22.48) into the perturbation series (22.45) finally gives

$$x(t) = \cos(\omega_0 t) - \frac{\varepsilon}{2}\omega_0 t \, \sin(\omega_0 t) + O(\varepsilon^2) \,. \qquad (22.56)$$

Let us first verify that this expression is indeed the first order expansion of the exact solution (22.44). Using the series (2.12), (2.13) and (2.16), the exact solution (22.44) is to first order in $\varepsilon$ given by

$$
\begin{aligned}
x(t) &= \cos\left(\omega_0\sqrt{1+\varepsilon}\,t\right) \\
&= \cos\left\{\omega_0\left[1+\frac{\varepsilon}{2}+O(\varepsilon^2)\right]t\right\} \\
&= \cos(\omega_0 t)\cos\left(\frac{\varepsilon}{2}\omega_0 t\right) - \sin(\omega_0 t)\sin\left(\frac{\varepsilon}{2}\omega_0 t\right) + O(\varepsilon^2) \\
&= \cos(\omega_0 t) - \frac{\varepsilon}{2}\omega_0 t \sin(\omega_0 t) + O(\varepsilon^2) \qquad (22.57)
\end{aligned}
$$

This result is indeed identical to the first order expansion (22.56) that was obtained from perturbation theory. However, the fact that this result is correct does not imply that this result is also useful. Perturbation theory is based on the premise that the truncated perturbation series gives a good approximation to the true solution. The truncation of this series only makes sense when the subsequent terms in the perturbation expansion rapidly become smaller. However, the first order term $(\varepsilon\omega_0 t/2) \sin(\omega_0 t)$

in (22.46) is as large as the zeroth order term $\cos{(\omega_0 t)}$ when $\varepsilon\omega_0 t/2 \sim 1$. This means that the first order perturbation series (22.56) ceases to be a good approximation to the true solution when

$$t \sim \frac{1}{\varepsilon\omega_0} \, . \tag{22.58}$$

The upshot of this example is that even though a truncated perturbation series may be correct, it may only be useful for a restricted range of parameters. In this example the first order perturbation series is only a good approximation when $t \ll 1/\varepsilon\omega_0$. For the problem in this section it would be more appropriate to make a perturbation series of the phase and the amplitude of the oscillator. Systematic techniques such as *multiple-scale analysis* [11] have been developed for this purpose. In the following section we carry out the transformation $p = e^S$ to derive the perturbations of the amplitude and phase of a wave that propagates through an inhomogeneous medium.

## 22.5 The WKB approximation

In this section we analyze the propagation of a wave through a one-dimensional acoustic medium. In the frequency domain the pressure satisfies the following differential equation:

$$\rho\frac{d}{dx}\left(\frac{1}{\rho}\frac{dp}{dx}\right) + \frac{\omega^2}{c^2}p = 0 \, . \tag{22.59}$$

In this expression both the density $\rho$ and the velocity $c$ vary with the position $x$. This equation can only be solved in closed form for a special form of the functions $\rho(x)$ and $c(x)$. It should be noted that the treatment in this section is also applicable to the Schrödinger equation

$$\frac{d^2\psi}{dx^2} + \frac{2m}{\hbar^2}\left[E - V(x)\right]\psi = 0 \, , \tag{22.60}$$

by making the substitutions

$$\rho(x) \to 1 \, , \qquad p \to \psi \, , \qquad 1/c(x) \to \sqrt{2m\left(E - V(x)\right)}/\hbar \, . \tag{22.61}$$

At this point there is no small perturbation parameter yet. We restrict our attention to media in which the length scale of the variation in $\rho$ and $c$ is much larger than the wavelength $\lambda$ of the wave. Physically this type of medium does not contain strong inhomogeneities on the scale of a wavelength, so that the waves are not reflected strongly by the heterogeneity. Let the length scale of the heterogeneity be denoted by $L$. When this length scale is much larger than the wavelength $\lambda = 2\pi\omega/c$, the following parameter is small:

$$\varepsilon = \frac{c}{\omega L} \ll 1. \tag{22.62}$$

As noted in the previous section, when the perturbation affects mostly the phase of a wave, it is advantageous to perturb the phase (and amplitude) of the wave rather than the solution $p$ itself. This can be achieved by making the transformation

$$p = e^S. \tag{22.63}$$

When $S$ is complex this transformation is without any loss of generality.

**Problem a:** Insert this relation into (22.59) and derive that $S(x)$ satisfies the following differential equation:

$$\frac{d^2S}{dx^2} - \frac{1}{\rho}\frac{d\rho}{dx}\frac{dS}{dx} + \left(\frac{dS}{dx}\right)^2 + \frac{\omega^2}{c^2} = 0. \tag{22.64}$$

At this point we have only made the problem more complex because this equation is nonlinear in the unknown function $S(x)$ whereas the original equation (22.59) is linear in the pressure $p(x)$. However, we have not yet applied the perturbation technique. Before we do this, let us first reflect on the transformation (22.63). If the medium were homogeneous, the solution would be given by $p = Ae^{ikx}$, with the wavenumber given by $k = \omega/c$. This special solution corresponds to $S = \ln(A) + ikx$, so that $dS/dx = ik = i\omega/c$. For an inhomogeneous medium one may expect the derivative of the phase to be close to this value, therefore we make the following substitution:

$$\frac{dS}{dx} = \frac{i\omega}{c(x)}F(x). \tag{22.65}$$

**Problem b:** Show that this transformation transforms (22.64) into the following differential equation for $F$:

$$\frac{dF}{dx} - \frac{1}{\rho c}\frac{d(\rho c)}{dx}F + \frac{i\omega}{c}F^2 = \frac{i\omega}{c}. \tag{22.66}$$

Now we use perturbation analysis by using that the parameter $\varepsilon$ defined in (22.62) is much smaller than unity. This can be achieved by transforming the distance $x$ to a dimensionless distance $\xi$ defined by

$$\xi \equiv x/L, \tag{22.67}$$

where $L$ is the characteristic length scale of the velocity and density variations.

**Problem c:** Under this transformation the derivative $d/dx$ changes to $d/dx = d/d(\xi L) = (1/L)\, d/d\xi$. Use this to show that $F(\xi)$ satisfies the following differential equation:

$$\frac{c}{\omega L}\frac{dF}{d\xi} - \frac{c}{\omega L}\frac{1}{\rho c}\frac{d\,(\rho c)}{d\xi}F + iF^2 = i. \qquad (22.68)$$

This equation contains the small dimensionless parameter $c/\omega L$ defined in (22.62), so that this equation is equivalent to

$$\varepsilon\frac{dF}{d\xi} - \varepsilon\frac{1}{\rho c}\frac{d\,(\rho c)}{d\xi}F + iF^2 = i. \qquad (22.69)$$

Now we have an equation that looks similar to those in the perturbation problems we have seen in this chapter. We solve this equation by inserting the following perturbation series for $F(\xi)$:

$$F(\xi) = F_0(\xi) + \varepsilon F_1(\xi) + \cdots. \qquad (22.70)$$

**Problem d:** Insert this perturbation series into the differential equation (22.69) to derive that $F_0(\xi)$ and $F_1(\xi)$ satisfy the following equations:

$$\left.\begin{array}{c} F_0^2 = 1, \\[2mm] 2iF_0F_1 = -\dfrac{dF_0}{d\xi} + \dfrac{1}{\rho c}\dfrac{d\,(\rho c)}{d\xi}F_0. \end{array}\right\} \qquad (22.71)$$

The first of these equations has the solutions $F_0 = \pm 1$. According to (22.65) this corresponds to the phase derivative $dS/dx = \pm i\omega/c$. The plus sign denotes a right-going wave, and the minus sign a left-going wave. We focus here on a right-going wave so that $F_0 = +1$.

**Problem e:** Insert this solution into the second line of (22.71) and show that $F_1(x')$ is given by

$$F_1(\xi) = -\frac{i}{2}\frac{1}{\rho c}\frac{d\,(\rho c)}{d\xi}. \qquad (22.72)$$

This means that the first order perturbation series for $F(\xi)$ is given by

$$F(\xi) = 1 - \frac{i\varepsilon}{2}\frac{1}{\rho c}\frac{d\,(\rho c)}{d\xi} + O(\varepsilon^2). \qquad (22.73)$$

**Problem f:** Now that we have obtained this solution as a function of the transformed distance $\xi$, transform back to the original distance $x$ by using (22.67). Use (22.62) to show that the solution (22.73) is equivalent to

$$F(x) = 1 - \frac{i}{2}\frac{1}{\rho\omega}\frac{d\,(\rho c)}{dx} + \cdots. \qquad (22.74)$$

**Problem g:** Use (22.65) to convert this into an equation for $dS/dx$. Integrate this equation to show that the solution $S(x)$ is given by

$$S(x) = i \int_{-\infty}^{x} \frac{\omega}{c(x')} dx' + \frac{1}{2} \ln \left[ \rho(x)c(x) \right] + B \,. \tag{22.75}$$

**Problem h:** Use the transformation (22.63) to show that this solution corresponds to the following pressure field

$$p(x) = A\sqrt{\rho(x)c(x)} \, \exp \left( i \int_{-\infty}^{x} \frac{\omega}{c(x')} dx' \right) , \tag{22.76}$$

with the new constant defined by $A = e^B$.

This solution states that the wave propagates to the right with an amplitude that is proportional to $\sqrt{\rho(x)c(x)}$. The local wavenumber $k(x)$ of the wave is given by the derivative of the phase of the wave; it is therefore given by

$$k(x) = \frac{d}{dx} \int_{-\infty}^{x} \frac{\omega}{c(x')} dx' = \frac{\omega}{c(x)} \,. \tag{22.77}$$

This means that the local wavenumber at a location $x$ is given by the wavenumber $\omega/c(x)$ that the medium would have if it were homogeneous with the properties of the medium at that location $x$. The solution (22.76) is known as the WKB solution (named after Wentzel, Kramers and Brillouin). Seismologists prefer to call this solution the WKBJ approximation because of the contribution of Lord Jeffreys [43].

In most textbooks (e.g. ref. [55]) this solution is derived for the Schrödinger equation rather than the acoustic wave equation; with the transformation (22.61) the derivations are equivalent.

**Problem i:** Use the correspondence (22.61) to show that in quantum mechanics the WKB solution is given by

$$\psi(x) = \frac{A}{[E - V(x)]^{1/4}} \, \exp \left\{ i \int_{-\infty}^{x} \frac{\sqrt{2m\left[ E - V(x') \right]}}{\hbar} dx' \right\} \,. \tag{22.78}$$

**Problem j:** Show that this approximation is infinite at the *turning points* of the motion. These are the points where the total energy of the particle is equal to the potential energy. This means that the WKB solution breaks down at the turning points.

A clear account of the WKB approximation with a large number of applications is given by Bender and Orszag [11].

## 22.6 The need for consistency

In perturbation theory one derives an approximate solution to a problem. In many applications this approximation is then used in subsequent calculations. In doing so, one must keep in mind that the solution obtained from perturbation theory is not the true solution, and that it is pointless to carry out the subsequent calculations with an accuracy which is higher than the accuracy of the solution obtained from perturbation theory.

As an example we consider in this section the WKB solution (22.76) for the pressure field $p(x)$ and compute the particle velocity $v(x)$ that is associated with this pressure field. We assume that the motion is sufficiently small that the equation of motion can be linearized so that Newton's law gives $\rho \partial v/\partial t = F$. As shown in section 4.2 the pressure force is given by $F = -\partial p/\partial x$, so that in the time domain Newton's law is given by $\rho \partial v/\partial t = -\partial p/\partial x$.

**Problem a:** Show that with the Fourier convention (14.42) the corresponding equation is given in the frequency domain by

$$i\omega\rho v = \partial p/\partial x. \tag{22.79}$$

**Problem b:** Apply this result to the WKB solution (22.76) and show that the velocity is given by

$$i\omega\rho v = \frac{1}{2}\frac{1}{\rho c}\frac{d\,(\rho c)}{dx}p + \frac{i\omega}{c}p. \tag{22.80}$$

**Problem c:** Use the estimate of the derivative in section 11.2 to show that the first term on the right hand side is of the order $p/L$, where $L$ is the characteristic length scale over which the density and the velocity vary.

The second term on the right hand side of (22.80) is of the order $\omega p/c$. This means that the ratio of the first term to the second term is given by $(p/L)/(\omega p/c) = c/(\omega L) = \varepsilon$, where the parameter $\varepsilon$ is defined in (22.62). In the previous section we assumed that the medium varies so smoothly that $\varepsilon \ll 1$. This means that under the assumptions which underlie the WKB approximation the first term on the right hand side of (22.80) can be ignored with respect to the second term.

**Problem d:** Show that in this approximation the velocity is given by

$$v = \frac{p}{\rho c}. \tag{22.81}$$

Note that ignoring the first term is consistent with the terms that we have ignored in the WKB approximation.

This last expression has an interesting interpretation. The quantity $\rho c$ is called the *acoustic impedance*. This term is reminiscent of the theory of electromagnetism. For a resistor, Ohm's law $I = V/R$ relates the current $I$ that is generated by a voltage $V$. For a general linear electric system the current and the voltage are related by

$$I = \frac{V}{Z} , \qquad (22.82)$$

where $Z$ is a generalization of the resistance that is called the *impedance*. The impedance gives the strength of the current for a given potential. Similarly, the acoustic impedance $\rho c$ in (22.81) gives the particle velocity for a given pressure.

Combining (22.76) with (22.81) shows that the velocity, and hence the particle motion, is proportional to $1/\sqrt{\rho c}$. This means that the particle motion increases when the acoustic impedance decreases. This has important implications for earthquake hazards. For soft soils, both the density $\rho$ and the wave velocity $c$ are small. This means that the acoustic impedance is much smaller in soft soils than in hard rock. This in turn means that the ground motion during earthquakes is much more severe in soft soils than in hard rock. (The motion in the Earth is governed by the elastic wave equation rather than the acoustic wave equation. However, one can show [3] that for elastic waves also the displacement is inversely proportional to $1/\sqrt{\rho c}$, where $c$ is the propagation velocity of the elastic wave under consideration.)

The fact that the ground motion is inversly proportional to the square-root of the impedance is one of the factors that made the 1985 earthquake along the west coast of Mexico cause so much damage in Mexico City. This city is constructed on very soft sediments which have filled the swamp onto which the city is built. The very small value of the associated elastic impedance was one the causes of the extensive damage in Mexico City after this earthquake.

## 22.7 Singular perturbation theory

In section 22.1 we analyzed the behavior of the root of the equation $x^3 - 4x^2 + 4x = \varepsilon$ that was located near $x = 0$. As shown in that section, the unperturbed problem also has a root $x = 2$. The roots $x = 0$ and $x = 2$ can be seen graphically in figure 22.1 because for these values of $x$ the polynomial shown by the thick solid line is equal to zero. In figure 22.1 the value $\varepsilon = +0.15$ is shown by a dotted line while the value $\varepsilon = -0.15$ is indicated by the dashed line. There is a profound difference between the two roots when the parameter $\varepsilon$ is nonzero. The root near $x = 0$ depends in a continuous way on $\varepsilon$, and (22.2) has for the root near $x = 0$ a solution regardless of whether $\varepsilon$ is positive or negative. This situation is completely different for the root near $x = 2$. When $\varepsilon$ is positive (the dotted line), the polynomial has *two* intersections with the dotted line, whereas when $\varepsilon$ is negative the polynomial does *not* intersect the dashed line at all. This means that depending on whether $\varepsilon$ is positive or negative, the solution has two or zero solutions, respectively. This behavior cannot be described by a regular perturbation series of the form (22.5) because this expansion assigns *one* solution to each value of the perturbation parameter $\varepsilon$.

Let us first diagnose where the treatment of section 22.1 breaks down when we apply it to the root near $x = 2$.

**Problem a:** Insert the unperturbed solution $x_0 = 2$ into the second line of (22.13) and show that the resulting equation for $x_1$ is

$$0 \cdot x_1 = 1 . \tag{22.83}$$

This equation obviously has no finite solution. This is related to the fact that the tangent of the polynomial at $x = 2$ is horizontal. First order perturbation theory effectively replaces the polynomial by the straight line that is tangent to the polynomial. When this tangent line is horizontal, it can never have a value that is nonzero.

This means that the regular perturbation series (22.5) is not the appropriate way to study the behavior of root near $x = 2$. In order to find out how this root behaves, let us set

$$x = 2 + y . \tag{22.84}$$

**Problem b:** Show that under the substitution (22.84) the original problem (22.2) transforms to

$$y^3 + 2y^2 = \varepsilon . \tag{22.85}$$

We will not yet carry out a systematic perturbation analysis, but we will first determine the dependence of the solution $y$ on the parameter $\varepsilon$. For small values of $\varepsilon$, the parameter $y$ is also small. This means that the term $y^3$ can be ignored with respect to the term $y^2$. Under this assumption (22.85) is approximately equal to $2y^2 \approx \varepsilon$ so that $y \approx \sqrt{\varepsilon/2}$. This means that the solution does not depend on integer powers of $\varepsilon$ as in the perturbation series (22.5), but that it does depend on the square-root of $\varepsilon$. The square-root of $\varepsilon$ is shown in figure 22.4. Note that for $\varepsilon = 0$ the tangent of this curve is vertical and that for $\varepsilon < 0$ the function $\sqrt{\varepsilon}$ is not defined for

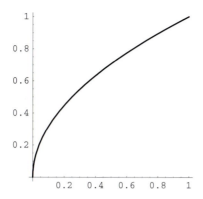

Fig. 22.4.   Graph of the function $\sqrt{\varepsilon}$.

real values of $\varepsilon$.* This reflects the fact that the roots near $x = 2$ depend in a very different way on $\varepsilon$ than the root near $x = 0$.

We know now that a regular perturbation series (22.5) is not the correct tool to use to analyze the root near $x = 2$. However, we do not yet know what type of perturbation series we should use for the root near $x = 2$; we only know that the perturbation depends to leading order on $\sqrt{\varepsilon}$. To wit, let us make the following substitution:

$$x = 2 + \sqrt{\varepsilon}\, z \,. \tag{22.86}$$

**Problem c:** Insert this solution into (22.2) and show that $z$ satisfies the following equation:

$$\sqrt{\varepsilon} z^3 + 2z^2 = 1 \,. \tag{22.87}$$

Now we have a new perturbation problem with a small parameter. However, this small parameter is not the original perturbation parameter $\varepsilon$, but it is the square-root $\sqrt{\varepsilon}$. The perturbation problem in this section is a *singular perturbation problem*. In a singular perturbation problem the solution is not a well-behaved function of the perturbation parameter. This has the result that the corresponding perturbation series cannot be expressed in powers $\varepsilon^n$, where $n$ is a positive real integer. Instead, negative or fractional powers of $\varepsilon$ are present in the perturbation series of a singular perturbation problem.

**Problem d:** Since the small parameter in (22.87) is $\sqrt{\varepsilon}$, it makes sense to seek an expansion of $z$ in this parameter:

$$z = z_0 + \varepsilon^{1/2} z_1 + \varepsilon z_2 + \cdots \,. \tag{22.88}$$

---

* When one allows a complex solution $x(\varepsilon)$ of the equation, there are always two roots near $x = 2$. However, these complex solutions also display a fundamental change in their behavior when $\varepsilon = 0$ which is characterized by a bifurcation.

Collect together the coefficients of equal powers of $\varepsilon$ when this series is inserted into (22.87) and show that this leads to the following equations for the coefficients $z_0$ and $z_1$:

$$\left.\begin{array}{ll} O(1)\text{-terms:} & 2z_0^2 - 1 = 0 , \\ O(\varepsilon^{1/2})\text{-terms:} & z_0^3 + 4z_0 z_1 = 0 . \end{array}\right\} \qquad (22.89)$$

**Problem e:** The first equation of (22.89) obviously has the solution $z_0 = \pm 1/\sqrt{2}$. Show that for both the plus and the minus signs $z_1 = -1/2$. Use these results to derive that the roots near $x = 2$ are given by

$$x = 2 \pm \frac{1}{\sqrt{2}}\sqrt{\varepsilon} - \frac{1}{2}\varepsilon + O(\varepsilon^{3/2}) . \qquad (22.90)$$

It is illustrative to compute the numerical values of these roots for the original problem (22.1), where $\varepsilon = 0.01$; this gives for the two roots:

$$x = 1.924 \qquad \text{and} \qquad x = 2.065 . \qquad (22.91)$$

In these numbers only three decimals are shown. The reason is that the error in the truncated perturbation series is of the order of the first truncated term, hence the error is of the order $(0.01)^{3/2} = 0.001$. When these solutions are compared with the perturbation solution (22.16) for the root near $x = 0$, it is striking that the singular perturbation series for the root near $x = 2$ converges much less rapidly than the regular perturbation series (22.16) for the root near $x = 0$. This is a consequence of the fact that the solution near $x = 2$ is a perturbation series in $\sqrt{\varepsilon}(= 0.1)$ rather than $\varepsilon(= 0.01)$. When the roots (22.91) are inserted into the polynomial (22.1) the following solutions are obtained for the two roots:

$$\left.\begin{array}{ll} x = 1.924 : & x^3 - 4x^2 + 4x = 0.0111 = 0.01 + 0.0011 , \\ x = 2.065 : & x^3 - 4x^2 + 4x = 0.0087 = 0.01 - 0.0012 . \end{array}\right\} \qquad (22.92)$$

Note that these results are much less accurate than the corresponding result (22.17) for the root near $x = 0$. Again this is a consequence of the singular behavior of the roots near $x = 2$.

The singular behavior of the roots of the polynomial (22.1) near $x = 2$ corresponds to the fact that the solution changes in a discontinuous way when the perturbation parameter $\varepsilon$ goes to zero. It follows from figure 22.1 that for the perturbation problem in this section the problem has one root near $x = 2$ when $\varepsilon = 0$, it has no roots when $\varepsilon < 0$ and there are two roots when $\varepsilon > 0$. Such a discontinuous change in the character of the solution also occurs in fluid mechanics in which the equation of motion is given by

$$\frac{\partial(\rho \mathbf{v})}{\partial t} + \nabla \cdot (\rho \mathbf{v} \mathbf{v}) = \mu \nabla^2 \mathbf{v} + \mathbf{F} . \qquad (10.55)$$

In this expression the viscosity of the fluid gives a contribution $\mu \nabla^2 \mathbf{v}$, where $\mu$ is the viscosity. This viscous term contains the highest spatial derivatives of the velocity that are present in the equation. When the viscosity $\mu$ goes to zero, the equation for fluid flow becomes a first order differential equation rather than a second order differential equation. This changes the number of boundary conditions that are needed for the solution, and hence it drastically affects the mathematical structure of the solution. This has the effect that boundary-layer problems are, in general, singular perturbation problems [92].

# 23

## Epilogue, on power and knowledge

*We all continue to feel a frustration because of our inability to foresee the soul's ultimate fate. Although we do not speak about it, we all know that the objectives of our science are, from a general point of view, much more modest than the objectives of, say, the Greek sciences were; that our science is more successful in giving us power than in giving us knowledge of truly human interest.* [E.P. Wigner, 1972, The place of consciousness in modern physics, in *Consciousness and reality*, Eds. C. Muses and A. M. Young, Outerbridge and Lizard, New York, pp. 132–141].

In this book we have explored many methods of mathematics as used in the physical sciences. Mathematics plays a very important role in the physical sciences because it is the only language we have for expressing quantitative relations in the world around us. In fact, mathematics not only allows us to express phenomena in a quantitative way, it also has a remarkable predictive power in the sense that it allows us to deduce the consequences of natural laws in terms of measurable quantities. In fact, we don't quite understand *why* mathematics gives such an accurate description of the world around us [100].

It is truly stunning how accurate some of the predictions in (mathematical) physics have been. The orbits of the planetary bodies can now be computed with an extreme accuracy. Morrison and Stephenson [58] compared the path a solar eclipse at 181 BC with historic descriptions made in a city in eastern China which was located in the path of the solar eclipse. According to the computations, the path of the solar eclipse passed 50 degrees west of site of this historic observation. This eclipse took place about 2000 years ago; this means that Earth has rotated through about $2.8 \times 10^8$ degrees since the eclipse. The *relative* error in the path of the eclipse over the Earth is thus only $1.7 \times 10^{-7}$. In fact, this discrepancy of 50 degrees can be explained well by the observed deceleration of the Earth [58] due to the braking effect Earth's tides.

The light emitted by hydrogen atoms has discrete spectral lines which are due to the fact that electrons behave as standing waves. Every electron is coupled to the field of electromagnetic radiation (light). There is a small chance that an electron will emit and re-absorb a virtual photon [76]. (Photons are the light-quanta.) This leads to the so-called *Lamb shift* of the spectral lines of light radiated by excited hydrogen atoms. For hydrogen atoms the shift for the transition between the 2s and $2p_{1/2}$ state is 1060 MHz. This corresponds to a reciprocal wavelength of $0.035$ cm$^{-1}$, the reciprocal wavelength that corresponds to the ionization energy is $2700$ cm$^{-1}$. Compared to the ionization energy of the ground state this corresponds therefore to a *relative* frequency shift of $1.3 \times 10^{-6}$. This is in very good agreement with observations. It is interesting to note that this prediction of the Lamb shift is based on second order perturbation theory [76]. This means that an approximate theory provides a stunningly accurate prediction of the Lamb shift.

A third example of the extreme accuracy of mathematics in the physical sciences is the perihelion precession of Mercury [62]. According to the laws of Newton, a planet will orbit in a fixed ellipsoidal orbit around the Sun. The general theory of relativity predicts that this ellipse slowly changes its position; the point of the ellipse closest to the Sun (the perihelion) slowly precesses around the Sun. According to the theory of general relativity this precession is given by 42.98 arcsec/century, whereas the observed precession rate is $43.1 \pm 0.1$ arcsec/century. Note that this precession rate is extremely small, but that it is well predicted from theory.

Mathematics not only provides us with valuable and stunningly accurate insights in the world around us, it is also an indispensable tool in making technical innovations. The design and implementation of rockets, aircraft, chemical plants, water treatment systems, modern electronics, information technology and many other innovations would have been impossible without mathematics. Mathematics and the physical sciences have created many new opportunities for mankind. For this reason one can state that mathematics and the physical sciences have greatly increased our power to modify the world in which we live; see also the quote of Nobel prize laureate Wigner [101] at the beginning of this section.

The problem with releasing power is that it can be used for good and for bad purposes. To make matters worse, there is no objective standard for 'good' and 'bad'. Science is objective in the sense that a certain theory is either consistent with observations, or it is not. However, scientific knowledge does not come with the moral standard that tells us *how* the power that we release in our scientific efforts should be used. It is essential that each of us develops such a standard, so that the fruits of our knowledge can be used for the benefit of mankind and the world we inhabit.

# References

[1] Abramowitz, M. and I.A. Stegun, 1965, *Handbook of mathematical functions*, Dover Publications, New York.

[2] Aharonov, Y., and D. Bohm, 1959, Significance of electromagentic potentials in the quantum theory, *Phys. Rev.*, **115**, 485–491.

[3] Aki, K. and P.G. Richards, 1980, *Quantitative seismology*, volume 1, Freeman and Company, San Francsisco.

[4] Aldridge, D.F., 1994, Linearization of the eikonal equation, *Geophysics*, **59**, 1631–1632.

[5] Arfken, G.B., 1995, *Mathematical methods for physicists*, Academic Press, San Diego.

[6] Backus, M.M., 1959, Water reverberations – their nature and elimination, *Geophysics*, **24**, 233–261.

[7] Barish, B.C. and R. Weiss, 1999, LIGO and the detection of gravitational waves, *Phys. Today*, **52(10)**, 44–55.

[8] Barton, G., 1989, *Elements of Green's functions and propagation, potentials, diffusion and waves*, Oxford Scientific Publications, Oxford.

[9] Beissner, K., 1998, The acoustic radiation force in lossless fluids in Eularian and Lagrangian coordinates, *J. Acoust. Soc. Am.*, **103**, 2321–2332.

[10] Bellman, R. and R. Kalaba, 1960, Invariant embedding and mathematical physics I. Particle processes, *J. Math. Phys.*, **1**, 280–308.

[11] Bender, C.M. and S.A. Orszag, 1978, *Advanced mathematical methods for scientists and engineers*, McGraw-Hill, New York.

[12] Berry, M.V. and S. Klein, 1997, Transparent mirrors: rays, waves and localization, *Eur. J. Phys.*, **18**, 222–228.

[13] Berry, M.V. and C. Upstill, 1980, Catastrophe optics: Morphologies of caustics and their diffraction patterns, *Prog. Optics*, **18**, 257–346.

[14] Blakeley, R.J., 1995, *Potential theory in gravity and magnetics*, Cambridge Univ. Press, Cambridge.

416

[15] Boas, M.L., 1983, *Mathematical methods in the physical sciences*, 2nd edition, Wiley, New York.

[16] Brack, M. and R.K. Bhaduri, 1997, *Semiclassical physics*, Addison-Wesley, Reading MA.

[17] Broglie, L. de, 1952, *La théorie des particules de spin 1/2*, Gauthier-Villars, Paris.

[18] Butkov, E., 1968, *Mathematical physics*, Addison Wesley, Reading MA.

[19] Claerbout, J.F., 1976, *Fundamentals of geophysical data processing*, McGraw-Hill, New York.

[20] Claerbout, J.F., 1985, *Imaging the Earth's interior*, Blackwell, Oxford.

[21] Chopelas, A., 1996, Thermal expansivity of lower mantle phases MgO and $MgSiO_3$ perovskite at high pressure derived from vibrational spectroscopy, *Phys. Earth. Plan. Int.*, **98**, 3–15.

[22] Coveney, P., and R. Highfield, 1991, *The arrow of time*, Harper Collins, London.

[23] Dahlen, F.A., 1979, The spectra of unresolved split normal mode multiples, *Geophys. J.R. Astron. Soc.*, **58**, 1–33.

[24] Dahlen, F.A. and I.H. Henson, 1985, Asymptotic normal modes of a laterally heterogeneous Earth, *J. Geophys. Res.*, **90**, 12653–12681.

[25] Dahlen, F.A., and J. Tromp, *Theoretical global seismology*, Princeton Univ. Press, Princeton, 1998.

[26] DeSanto, J.A., *Scalar wave theory; Green's functions and applications*, Springer Verlag, Berlin, 1992.

[27] Dziewonski, A.M., and J.H. Woodhouse, 1983, Studies of the seismic source using normal-mode theory, in *Earthquakes: Observations, theory and intepretation*, edited by H. Kanamori and E. Boschi, North Holland, Amsterdam, pp. 45–137.

[28] Earnshaw, S., 1842, On the nature of molecular forces which regulate the constitution of the luminiferous ether, *Trans. Camb. Phil. Soc.*, **7**, 97–112.

[29] Edmonds, A.R., 1974, *Angular momentum in quantum mechanics*, 3rd edition, Princeton Univ. Press, Princeton.

[30] Feynman, R.P., 1975, *The character of physical law*, MIT Press, Cambridge (MA).

[31] Feynman, R.P. and A.R. Hibbs, 1965, *Quantum mechanics and path integrals*, McGraw-Hill, New York.

[32] Fishbach, E. and C. Talmadge, 1992, Six years of the fifth force, *Nature*, **356**, 207–214.

[33] Fletcher, C., 1996, *The complete walker III*, Alfred A. Knopf, New York.

[34] Gradshteyn, I.S. and I.M. Rhyzik, 1965, *Tables of integrals, series and products*, Academic Press, New York.

[35] Gubbins, D. and R. Snieder, 1991, Dispersion of P waves in subducted lithosphere: Evidence for an eclogite layer, *J. Geophys. Res.*, **96**, 6321–6333.

[36] Guglielmi, A.V. and O.A. Pokhotelov, 1996, *Geoelectromagnetic waves*, Inst. of Physics Publ., Bristol.

[37] Halbwachs, F., 1960, *Théorie relativiste des fluides a spin*, Gauthier-Villars, Paris.

[38] Hildebrand, A.R., M. Pilkington, M. Conners, C. Ortiz-Aleman and R.E. Chavez, 1995, Size and structure of the Chicxulub crater revealed by horizontal gravity and cenotes, *Nature*, **376**, 415–417.

[39] Holton, J.R., 1992, *An introduction to dynamic meteorology*, Academic Press, San Diego.

[40] Iyer, H.M. and K. Hirahara (Eds.), 1993, *Seismic tomography, theory and practice*, Prentice Hall, London.

[41] Ishimaru, A., 1997, *Wave propagation and scattering in random media*, Oxford Univ. Press, Oxford.

[42] Jackson, J.D., 1975, *Classical electrodynamics*, Wiley, New York.

[43] Jeffreys, H., 1924, On certain approximate solutions of linear differential equations of second order, *Phil. London Math. Soc.*, **23**, 428–436.

[44] Kermode, A.C., 1972, *Mechanics of flight*, Longman Singapore.

[45] Kline, S.J., 1965, *Similitude and approximation theory*, McGraw-Hill, New York.

[46] Kravtsov, Yu.A., 1988, Ray and caustics as physical objects, *Prog. Optics*, **26**, 228–348.

[47] Lambeck, K., 1988, *Geophysical geodesy*, Oxford Univ. Press, Oxford.

[48] Lauterborn, W., T. Kurz and M. Wiesenfeldt, 1993, *Coherent optics*, Springer Verlag, Berlin.

[49] Lin, C.C., L.A. Segel and G.H. Handelman, 1974, *Mathematics applied to deterministic problems in the natural sciences*, Macmillan, New York.

[50] Lister, G.S., and P.F. Williams, 1983, The partitioning of deformation in flowing rock masses, *Tectonophysics*, **92**, 1–33.

[51] Love, S.G., and D.E. Brownlee, 1993, A direct measurement of the terrestrial mass accretion rate of cosmic dust, *Science*, **262**, 550–553.

[52] Madelung, E., 1926, Quantentheorie in hydrodynamischer form, *Z. Phys.*, **40**, 322.

[53] Marchaj, C.A., 1993, *Aerohydrodynamics of sailing*, 2nd edition, Adlard Coles Nautical, London.

[54] Marsden, J.E. and A.J. Tromba, 1988, *Vector calculus*, Freeman and Company, New York.

[55] Merzbacher, E., 1970, *Quantum mechanics*, Wiley, New York.

[56] Moler, C. and C. van Loan, 1978, Nineteen dubious ways to compute the exponential of a matrix, *SIAM Review*, **20**, 801–836.

[57] Morley, T., 1985, A simple proof that the world is three dimensional, *SIAM Review*, **27**, 69–71.

[58] Morrison, L. and R. Stephenson, 1988, The sands of time and the Earth's rotation, *Astronomy and Geophysics*, **39(5)**, 8–13.

[59] Muirhead, H., 1973, *The special theory of relativity*, Macmillan, London.

[60] Nayfeh, A.H., 1981, *Introduction to perturbation techniques*, Wiley, New York.

[61] Nolet G. (Ed.), 1987, *Seismic tomography, with applications in global seismology and exploration geophysics*, Kluwer, Dordrecht.

[62] Ohanian, H.C. and R. Ruffini, 1994, *Gravitation and spacetime*, Norton, New York.

[63] Olson, P., 1989, Mantle convection and plumes, in *The encyclopedia of solid earth geophysics*, Ed. D.E. James, Van Nostrand Reinholt, New York.

[64] Oort, A.H. and J.P. Peixoto, 1992, *Physics of climate*, Springer-Verlag, New York.

[65] Parker, R.L. and M.A. Zumberge, 1989, An analysis of geophysical experiments to test Newton's law of gravity, *Nature*, **342**, 29–31.

[66] Parsons, B. and J.G. Sclater, 1977, An analysis of the variation of the ocean floor bathymetry and heat flow with age, *J. Geophys. Res.*, **32**, 803–827.

[67] Pedlosky, J., 1979, *Geophysical fluid dynamics*, Springer Verlag, Berlin.

[68] Popper, K., 1956, The arrow of time, *Nature*, **177**, 538.

[69] Press, W.H., B.P. Flannery, S.A. Teukolsky and W.T. Vetterling, 1986, *Numerical recipes*, Cambridge Univ. Press, Cambridge.

[70] Price, H., 1996, *Time's arrow and Archimedes' point, New directions for the physics of time*, Oxford Univ. Press, New York.

[71] Rayleigh, Lord, 1917, On the reflection of light from a regularly stratified medium, *Proc. Roy. Soc. Lon.*, **A93**, 565–577.

[72] Riley, K.F., M.P. Hobson, and S.J. Bence, 1998, *Mathematical methods for physics and engineering*, Cambridge Univ. Press, Cambridge.

[73] Rossing, T.D., 1990,, *The science of sound*, Addison Wesley, Reading (MA).

[74] Rummel, R., 1986, Satellite gradiometry, in *Mathematical techniques for high-resolution mapping of the gravitational field*, Lecture notes in the Earth Sciences, vol. 7, Ed. H. Suenkel, Springer Verlag, Berlin.

[75] Robinson, E.A. and S. Treitel, 1980, *Geophysical signal analysis*, Prentice Hall, Englewood Cliffs (NJ).

[76] Sakurai, J.J., 1978, *Advanced quantum mechanics*, Addison Wesley, Reading (MA).

[77] Scales, J.A. and R. Snieder, 1997, Humility and nonlinearity, *Geophysics*, **62**, 1355–1358.

[78] Schneider, W.A., 1978, Integral formulation for migration in two and three dimensions, *Geophysics*, **43**, 49–76.

[79] Silverman, M.P., 1993, *And yet it moves; strange systems and subtle questions in physics*. Cambridge Univ. Press, Cambridge.

[80] Snieder, R.K., 1985, The origin of the 100,000 year cycle in a simple ice age model, *J. Geophys. Res.*, **90**, 5661–5664.

[81] Snieder, R., 1986, 3D Linearized scattering of surface waves and a formalism for surface wave holography, *Geophys. J.R. astr. Soc.*, **84**, 581–605.

[82] Snieder, R., 1996, Surface wave inversions on a regional scale, in *Seismic modelling of Earth structure*, Eds. E. Boschi, G. Ekstrom and A. Morelli, Editrice Compositori, Bologna, pp. 149–181.

[83] Snieder, R. and D.F. Aldridge, 1995, Perturbation theory for travel times, *J. Acoust. Soc. Am.*, **98**, 1565–1569.

[84] Snieder, R. and G. Nolet, 1987, Linearized scattering of surface waves on a spherical Earth, *J. Geophys.*, **61**, 55–63.

[85] Stacey, F.D., 1992, *Physics of the Earth*, 3rd edition, Brookfield Press, Brisbane.

[86] Tabor, M., 1989, *Chaos and integrability in nonlinear dynamics*, John Wiley, New York.

[87] Taylor, E.F., and J.A. Wheeler, 1966, *Spacetime physics*, Freeman, San Francisco.

[88] Thompson, P.A., 1972, *Compressible-fluid dynamics*, McGraw-Hill, New York.

[89] Tritton, D.J., 1982, *Physical fluid dynamics*, Van Nostrand Reinhold, Wokingham (UK).

[90] Tromp, J. and R. Snieder, 1989, The reflection and transmission of plane P- and S-waves by a continuously stratified band: a new approach using invariant embedding, *Geophys. J.*, **96**, 447–456.

[91] Turcotte, D.L. and G. Schubert, 1982, *Geodynamics*, Wiley, New York.

[92] van Dyke, M., 1978, *Perturbation methods in fluid mechanics*, Parabolic Press, Stanford CA.

[93] Virieux, J., 1996, Seismic ray tracing, in *Seismic modelling of Earth structure*, Eds. E. Boschi, G. Ekstrom and A. Morelli, Editrice Compositori, Bologna, pp. 221–304.

[94] Vogel, S., Exposing life's limits with dimensionless numbers, *Physics Today*, **51(11)**, 22–27.

[95] Watson, T.H., 1982, A real frequency, wave-number analysis of leaking modes, *Bull. Seismol. Soc. Am.*, **62**, 369–394.

[96] Webster, G.M. (Ed.), 1981, *Deconvolution*, Geophysics reprint series, vol. 1, Society of Exploration Geophysicists, Tulsa.

[97] Weisskopf, V.F., 1939, On the self-energy and the electric field of the electron, *Phys. Rev.*, **56**, 72–85.

[98] Whitaker, S., 1968, *Introduction to fluid mechanics*, Prentice-Hall, Englewood Cliffs.

[99] Whitham, G.B., 1974, *Linear and nonlinear waves*, Wiley, New York.

[100] Wigner, E.P., 1960, The unreasonable effectiveness of mathematics in the natural sciences, *Comm. Pure Appl. Math.*, **13**, 222–236.

[101] Wigner, E.P, 1972, The place of consciousness in modern physics, in *Consciousness and reality*, Eds. C. Muses and A.M. Young, Outerbridge and Lizard, New York, pp. 132–141.

[102] Wu, R.S. and K. Aki, 1985, Scattering characteristics of elastic waves by an elastic heterogeneity, *Geophysics*, **50**, 582–595.

[103] Yilmaz, O., 1987, *Seismic data processing, Investigations in geophysics*, vol. 2, Society of Exploration Geophysicists, Tulsa.

[104] Yoder, C.F., J.G. Williams, J.O. Dickey, B.E. Schutz, R.J. Eanes and B.D. Tapley, 1983, Secular variation of the Earth's gravitational harmonic $J_2$ coefficient from LAGEOS and nontidal acceleration of the Earth rotation, *Nature*, **303**, 757–762.

[105] Ziolkowski, A., 1991, Why don't we measure seismic signatures? *Geophysics*, **56**, 190–201.

[106] Zumberge, M.A., J.R. Ridgway and J.A. Hildebrand, 1997, A towed marine gravity meter for near-bottom surveys, *Geophysics*, **62**, 1386–1393.

# Index